the PHILOSOPHERS' STONE

the PHILOSOPHERS' STONE

ALCHEMY and the SECRET RESEARCH
for
EXOTIC MATTER

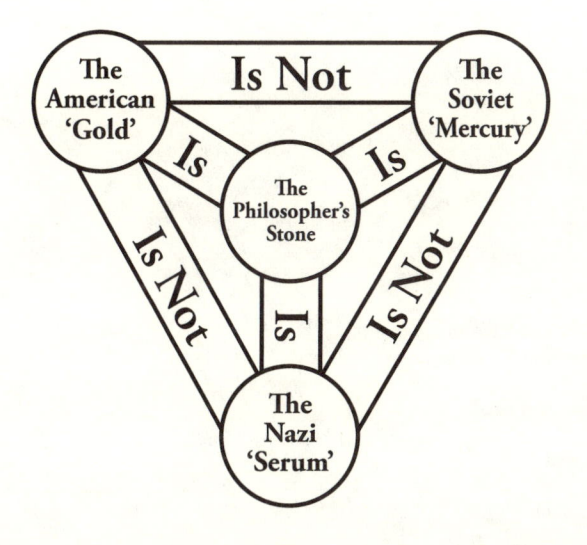

JOSEPH P. FARRELL

FERAL HOUSE

For T.S.F., and J.W.S.
And all the other "extended Inklings" out there.

*The Philosopher's Stone: Alchemy and
the Secret Research for Exotic Matter*

© 2009 by Joseph P. Farrell

All rights reserved.

A Feral House book.

ISBN: 978-1-932595-40-6

Feral House

1240 W. Sims Way Suite 124

Port Townsend WA 98368

www.FeralHouse.com

Book design by Jacob Covey.

*"AS ONE OF THE GREAT ALCHEMISTS
fittingly observed, man's quest for gold is often his
undoing, for he mistakes the alchemical processes,
believing them to be purely material. He does not realize
that the Philosopher's Gold, the Philosopher's Stone,
and the Philosopher's Medicine exist in each of the four
worlds and that the consummation of the experiment
cannot be realized until it is successfully carried on in
four worlds simultaneously according to one formula."*

Manly P. Hall,
The Secret Teachings of All Ages,
Reader's Edition, p. 508.

ACKNOWLEDGEMENTS

This book owes much of its contents to the unflagging research efforts of a friend of mine who has asked that her name not be mentioned, who tracked down many of the physics papers I knew were out there, but which, due to my isolated circumstances, would have been rather difficult for me to obtain. Presented with a list of obscure journals and academic papers, Ms. "LD" went to work and dug them all up within a rather adventurous day in the libraries of a university on the west coast.

I also owe a very large debt of gratitude to Mr. Richard C. Hoagland, of the enterprisemission.com, for many stimulating discussions via numerous emails over these and related topics, and for his extremely insightful observations on the Nazi "IRR Xerum 525" and the Farm Hall transcripts that are presented in part four.

And a large "thank you" also to George Ann Hughes, Jack W. Smith, and all the others who have helped and encouraged along the way. Many thanks are also due to all those who have been regular readers and donators at my website, www.gizadeathstar.com, for supporting my continued research and writing. Work of this sort does not occur in a vacuum, and can be – as I know all too well – very time-consuming, and consequently, would not see the light of day without the generous support of so many good people willing to assist financially in this effort with their donations. Many have the opinion that authors, because they are published, are, if not rich, at least well off. But the reality of publishing is that, with but rare exceptions, authors generally do not see much return for their investment of time and research. Accordingly, a word of thanks must surely also go to Mr. Adam Parfrey, publisher of this book, whose kindness and willingness to take a chance have brought it to fruition: thank you Adam!

Finally, and especially, a great debt of gratitude to my dear friend Tracy Fisher, to whom this work is dedicated, for your consistent encouragement and prayers.

God bless you all with every good thing and with His compassion.

Joseph P. Farrell
Spearfish, South Dakota
2008

PREFACE

Every now and then, one has to pause and consider the synchronicities in one's life. As a boy, I, like many other American boys, loved to read comic books. But unlike the majority of my comic-book-reading colleagues, I did not spend my time with Superman, Batman, the X-men, Tarzan, or comics of the superhero genre. Rather, I spent it with — and the reader is permitted a laugh, for I too regard it with some amusement — Disney's Scrooge McDuck, the world's richest duck, with a "money bin" full of three cubic acres of money, making him more-than-fabulously wealthy.

Scrooge and his nephew, Donald, and grandnephews Huey, Dewey, and Louie, would set off on fantastic adventures all over the world in search of this or that famous object or artifact. More to the point, the adventures which stimulated my imagination and thought the most were always those penned and composed by the cartoonist Carl Barks, a master storyteller. Barks was certainly familiar with the odd niche and avenue within human history, archaeology, and esotericism. I remember well specific comics where the Duck family would go off in search of the Seven Cities of Cibola, or where they would find lost Inca tribes (of ducks, naturally), living high in the Andes among ruins that, later, I would identify as having a more than coincidental resemblance to Machu Picchu.

In one of these Barks-inspired adventures, after Scrooge has assiduously searched through numerous self-evidently obscure and ancient texts, the intrepid ducks went in search of, and actually found, the ancient alchemical Philosophers' Stone, an object reputedly capable of transmuting base metals into gold. In other words, Barks merely used the Duck family characters as a template for his adventures, which in turn sometimes functioned merely as a means to educate and communicate legends and myths a young boy might not otherwise encounter in the bland and atrophied curriculum of modern American public school "education." The concept of the Philosophers' Stone thus lodged in my mind at a very early age, and I purposed to learn more about it and, perhaps, to write my own comic book about it some day.

Well, this book is obviously not a comic book, but it is the fruit of my interest and researches on the subject of the Philosophers' Stone and alchemy in general. And there are "synchronicities" in it that readers of my previous books will readily recognize. The whole theme of exotic matter or states of matter was first enunciated way back in my first book on alternative science and history, *The Giza Death Star,* followed up in *The Giza Death Star Destroyed,* and pursued in, of all places, *The SS Brotherhood of the Bell: NASA's Nazis, JFK, and M<small>AJIC</small>-12.* The idea of a transmutative physical medium, similarly, was enunciated in the first book and a whole chapter appendix was devoted to it in *The Giza Death Star Destroyed.* The whole idea of the non-locality of the physical medium and its strange hyper-dimensional properties, and the peculiar interplay of such hyper-dimensional physics with human history, has been a consistent leitmotif interwoven on many detailed levels throughout all my books.

One of the themes of my book *Secrets of the Unified Field: The Philadelphia Experiment, the Nazi Bell, and the Discarded Theory,* is a major theme again in this one: torsion. Therefore, a word is necessary on how I use this term "torsion" in the current work. Torsion, in its normal mathematical and physics meaning, has a very specific description and yields certain very discrete, usually very minute, values, whether one is referring to the Einstein-Cartan or the Ricci torsion tensors. I am not using the word in this specific manner, though I certainly do not exclude these meanings, when appropriate, from my usage here. Rather, I use the word to signify the *concept* that torsion *represents*: the folding and pleating of space that occurs in a spiraling rotating system of space. The analogy I used to explain the concept in *Secrets of the Unified Field* was that of wringing an empty aluminum soda can like one was wringing a dishrag. The analogy is applied here as well.

All this being said, this book concentrates on three exotic "materials" explored in my previous books, namely the monatomic American "Gold" of David Hudson, the ballotechnic Soviet "Mercury" of the "red mercury" nuclear scare of the 1990s, and the torsion and high-spin-state Nazi "Serum" of the Nazi Bell device. While the previous books did not go into great depth with their exploration of each of these substances, this one does so, with a view to exhibiting their common underlying features, and the torsion-based physics from which they are derived and upon which they are based.

Finally a word about my whole approach in this book is warranted, for in some respects it is different than in the other books. Many "paranormal" phenomena or "esoteric arts" can be scientifically rationalized by appeals to the various standard models of physics. So why appeal to alchemy and less well-known, and certainly "non-standard" theories in physics to explain it? And why write about these three substances — the American "Gold," the Soviet "Mercury," and the Nazi "Serum" — in particular, when so much more could be said about Dark Matter, Dark Energy, and all the wonderful scenery and special effects on the entertaining screen of the standard models of physics?

To ask these questions is to imply the answers, for so much *has* been written about the scientific rationalizations and connections with other "esoteric subjects," to do so again would only add to the volume of burgeoning and oftentimes mediocre literature about them. Besides, the esoteric "science" of alchemy made certain very specific claims and assertions, as do the claims and assertions surrounding the legends of those three very alchemical modern substances that one will *never* find mentioned in that burgeoning literature and their tired rehearsals of the same well-known fads of popularized standard model physics. These substances are best rationalized by more "off the books" and less well-known ideas, ideas equally deserving of popular treatment. As a result of this approach, one implication of this study is that there would indeed seem to be adumbrations of confirmations of ancient esoteric views of the physical medium, and that shadowy groups in the contemporary world are intent upon recovering that lost science, implementing the whole range of technologies that it implies, and monopolizing it for themselves.

In recounting this story, a little-known physics theory is explored, one which, unlike the fads of modern theoretical physics, has every appearance of being able to explain the odd behavior of these materials, and to make its own specific *testable* predictions. And interestingly enough, that theory, as we shall see, leads us once more back to Germany, and to what the Nazis may actually have been up to. It is, as will also be seen, a hyper-dimensional unified field theory, complete with sub-spaces, and spin orientation characteristics, but one which, unlike certain other splendiferous mathematical scholasticisms within contemporary theoretical physics, has enough *testable predictive power* that it can be, and has been, actually tested and to some extent verified.

And that, interestingly enough, will lead us in turn, once again, to NASA, and NASA's Nazis....

Joseph P. Farrell
Spearfish, South Dakota
2008

TABLE OF CONTENTS

Acknowledgements	7
Author's Preface	9

PART ONE
De Materia Prima Philosophorum

1. The Alchemical Wedding: Mediaeval Esotericism and Modern Physics	25
A. The Origins and Heyday of Alchemy	27
1. Back to the "Esoteric Gap": The Real Significance of Alchemy for Esoteric History	27
2. The Origins and Heyday of Alchemy	29
B. Exoteric and Esoteric Alchemy	31
1. The Ambiguous Nature of the Technical Language of Alchemy	33
C. The Goal and Quest of Alchemy: the Transmutative Medium of the *Materia Prima*	36
1. The Problem of Alchemy's Survival: The Triune Stone and the Augustinian Trinity	43
2. Alchemical References to the Augustinized Trinity	48
a. Philippus Theophrastus Areolus Bombastus von Hohenheim, a.k.a. Paracelsus	51
(1) On the Paleoancient Very High Civilization	52
(2) On the Relation between Astronomy and Alchemy	54
(3) Crystals and Metals: Sapphire and Mercury	55
(4) And the Augustinized Trinity and Alchemy	57
b. The Ultimate *Reductio* and the Clincher	60
2. The Indestructible Stone: The Persistence of Alchemy and the Powers of the Philosophers' Stone	63
A. Stories of Alchemical Success	63
1. The Swedish General	64
2. A Provincial Frenchman	64
3. The Hapsburg Emperors Ferdinand III and Leopold I	65

B. The Alchemical Reading of Particular Texts	67
1. Ovid's *Metamorphosis*	67
2. The Nine Philosophers, The *Enneads*, The Council of the Gods, and the Egyptian *Neters*	68
3. The Alchemical Reading of the Episode of the Golden Calf in the Old Testament	69
C. Stone, Powder, Elixir, and the Sequence of Colors	71
D. The Claimed Properties of the Philosophers' Stone	76
1. Longevity and Healing	76
2. Occupying No Space	77
3. Ability to Affect Action at a Distance	78
4. Indestructibility: A Link to the Sumerian Tablets of Destinies?	78
E. Conclusions	80

PART TWO
The American "Gold"

3. Scorched Gold in the Arizona Desert: David Hudson and the Beginnings of Monatomic Gold	85
A. Acid Baths for the Soil and an Explosive Discovery	86
1. Sunlight and a Pencil	87
2. Reductions, Beads, and Anomalous Shattering	87
3. The Cornell Labs Episode	88
B. Spectroscopic Analysis: Enter the Russians	89
1. The Platinum Group Metals	90
2. Anomalies in Neutron Activation Analysis	91
a. White to Red to Rhodium	92
b. Platinum in the Arizona Desert	94
C. General Electric Sees the Explosions: New Energy Technology?	95
D. The Stunning Mass Anomaly	96
4. Transmutations: Torsion Superdeformities and the New Nuclear Physics	99
A. Getting Serious with Hal Puthoff and the 44% Mass Loss Anomaly	99
B. Superconductivity	100
1. Consciousness and the Superconductive Behavior of DNA	101
2. Gamma Ray Bursts	103
3. Hudson Discovers the Alchemical Connection	104
C. Superdeformities and the New Nuclear Physics	105
1. Hudson's Physics Papers Sources and a Methodology	108
2. Formal Definition of Superdeformity	109
3. Properties of Superdeformity	110

a. Spontaneous Fission and New Models of Fission in Superdeformed States	110
b. Superconductivity	112
4. The All-Important Principle of Superdeformity: Rotation and Angular Momentum	113
5. Mercury	114
5. CONCLUSIONS TO PART TWO	117

PART THREE
The Soviet "Mercury"

6. "RED" MERCURY: HOAX, CODE NAME, INTELLIGENCE OPERATION, OR GENUINE ARTICLE?	123
A. Various Explanations of the Red Mercury Scare	124
1. The "Simple Hoax" Explanation	124
2. The Anti-Terrorism Counterintelligence Hoax	125
3. The BBC's Problematical Statements of 2006	126
B. Its Alleged Uses	129
1. "Stealth Paint"	129
2. "Ballotechnic Explosive"	130
3. A "Stimulated Gamma Ray Emitter"	136
4. The Strange Contradiction of DuPont's Mercury Antimony Oxide	137
5. The Platinum Metals Group Again: Hudson and the Russian Involvement Reconsidered	138
C. Its Strangely Alchemical Recipe and the Strange Behavior it Conjured	140
1. The Alleged Classified Russian Report on Red Mercury	140
2. The Press Campaign and Subsequent Russian Debunking	141
3. The Strangely Contradictory Statements of British Physicist Frank Barnaby	142
4. Various Recipes and Stable Features	144
D. Conclusions and Connections	147
7. TIME IS NOT A SCALAR: NIKOLAI KOZYREV'S CAUSAL MECHANICS	151
A. A Brief *Curriculum Vitae* of Dr. Nikolai Kozyrev	152
B. An Alchemical Aside: Paracelsus on the Incorporation of Celestial Geometries into Alchemical Apparatus	156
C. Tensors, Time, and Torsion	156
1. Various Views	157
2. The Strength of Torsion in the Einstein-Cartan Torsion Tensor	157
3. The Ricci Torsion Tensor	157

4. Dynamic Torsion, Vortex Mechanics, and Time	158
D. Dr. Nikolai Kozyrev on the Nature of Time: The Physical Mechanics of Cause and Effect	158
1. Time is Not a Scalar	160
2. A Theological Aside	161
3. Time and Kozyrev's Formally Explicit Definitions of Cause and Effect	162
4. Quasi-Quantized Time-Space Spin Orientations	167
5. The Alchemical Implication	168
8. Of Gyroscopes, Sponges, and Hydrogen Bombs: Kozyrev's Experiments and Wilcock's Analogies	171
A. Of Sponges: David Wilcock's Analogy	172
1. Submerged Sponges, Sub-Spaces and Hyper-Spaces: An Alchemical Parallel	173
2. Spinning, Vibrating, Heating, Cooling: Another Alchemical Parallel	173
B. Of Gyroscopes and Other Things: The Experimental World of Dr. Kozyrev	174
1. The Simple Gyroscope Experiments and Their Breathtaking Results	174
2. Time Imparts a Spin Orientation to the Total System	176
3. The "Intensity of Time" and Seasonal Variations of Experimental Results	177
4. The "Cause-Effect Dipole," Torsion, "pre-Action" and Superluminal Information Transfer	179
5. Kozyrev's "Torsion Telescope"	181
a. Kozyrev's Progeny and the Reverse Engineering of Time: The Alleged Russian "Scalar Spheres"	182
b. Torsion, System Memory, and Psychometry	183
c. Peculiar Parallels: Quantized Results	185
(1) Quantized Results and the Gravitor Experiments of Thomas Townsend Brown	186
(2) Quantized Results and the Alchemists	187
(3) Torsion, System Memory, Crystal Growth and Defects: A Philosophers' Stone in the Egyptian Desert	188
d. The Charge of Pseudo-Science: Curious Echoes of the "Red Mercury" Story	189
C. Back to "Red" Mercury: Of Hydrogen Bombs, Far-From-Equilibrium Systems, and Torsion	190
9. Conclusions to Part Three	195

PART FOUR
The Nazi "Serum"

10. **Final Farm Hall Farce: Wirtz, Diebner, and the Mysterious Photochemical Process** — 205
    - A. The Farce at Farm Hall: Wirtz and Diebner on the Photochemical Process of Isotope Enrichment — 206
        1. The "Artificial Rubies" Passage from the Farm Hall Transcripts — 207
        2. The Farm Hall Transcripts' Indications of a German Photochemical Process of Isotope Separation and Enrichment — 209
    - B. Richard C. Hoagland's Analysis and Hypothesis — 213
        1. Hoagland's Analysis and Hypothesis: The First Email — 213
        2. What Hartek Saw at I.G. Farben and What Mr. Hoagland Did Not Then Know: Back to the Auschwitz "Buna Factory" — 218
        3. The Modern Version of Tunable Lasers — 225
    - C. Other Mysteries that Tend to Corroborate Hoagland's Proposal — 231
        1. Vast Quantities of Enriched Uranium in Nazi Germany Implied by the Evidence — 231
        2. The Alleged Test of a Small Critical Mass High-Yield Nuclear Device at Ohrdruf in March, 1945 — 234
        3. The Thorium and Radium Mysteries — 237
        4. The Laser Mystery — 239
    - D. A Tentative Conclusion — 245

11. **The Greater German Alchemical Reich: The Goldmaker, the Gold, Gerlach, and "Himmler's Rasputin"** — 247
    - A. The Strange Case and Alchemical Beliefs of "Himmler's Rasputin": SS Brigadier General Karl Maria Wiligut — 248
        1. Himmler's Rasputin: — 250
            a. Brief Biography — 250
            b. And the SS *Ahnenerbedienst* — 253
                (1) The *Ahnenerbe*, "Atlantis," and Esoteric Geopolitics — 254
                (2) The *Ahnenerbe* and the Scientific Decoding of Esoteric Lore — 255
            c. Wiligut and the Bloodline: Otto Rahn — 258
        2. Wiligut's Alchemical Views: The "Primeval Twist" — 259
        3. Wiligut's Version of the Augustinized Trinity and the Tripartite Stone — 260
        4. Wiligut's Opposing Spirals: The "Swastika Tensor" Revisited — 263
    - B. The Tausend Affair and Prof. Dr. Gerlach's Article — 267
        1. Prof. Dr. Walther Gerlach Takes Note of an Alchemical Paper: Excerpt from *The SS Brotherhood of the Bell* — 267
        2. Alchemical Papers by One Japanese and Two German Physicists — 270

3. …The Franz Tausend Affair	273
C. Conclusions	275
12. THE NAZI SERUM: "IRR XERUM 525," THE NAZI BELL, AND THE RECIPE	277
A. The Black Projects Reich within the Reich: The *Kammlerstab* Think Tank	278
1. Agoston and Dr. Wilhelm Voss	279
2. The *Reichprotektorat* of Bohemia and Moravia	280
3. The *Forschungen, Entwicklungen, und Patente*	281
4. The SS *Entwicklungstelle IV*	282
B. The Nazi Bell, its Operation, Effects, and Scientific Rationalization: A Review	283
1. Kaluza-Klein, Torsion, and the Layout of the Counter-Rotating Drums	287
a. Wiligut	287
b. Rudolf Hess and Nazi Hermeticism: The Egyptian, Alchemical and Unified Field Theory Connections	288
2. Two Types of Electrical Potentials and Ultra-High Speed Mechanical Rotation	291
a. A Curious Parallel: the Varo Annotated Edition	294
b. Another Curious Parallel: Lyne and Tesla	294
3. Cryogenic Cooling and Superconductors	295
a. Professor Nagaoka's Observation	295
b. A Brief Note on the Bell and Nazi Centrifuge Isotope Separation Technologies	296
4. The Operative Principle in Evidence Thus Far: Maximizing Torsion Sheer Effects in the Medium	296
C. The Alchemical Serum: The Provisional Role of IRR Xerum 525 in Light of the Previous Analysis	297
1. The Thorium and Radium Mysteries	297
a. One Explanation: Thorium Isomers	298
b. Another Highly Speculative Explanation: Super-dense Matter	299
2. Hoagland's Measurement Model: Torsion Effects, Radioactive Decay Rates and Shielding, and a Proposal	303
3. A Prototypical Technology and Three Potential Applications	304
D. The Mystery of Two More Scientists	305
1. The Strange Case of Dr. Hubertus Strughold	306
2. The Even Stranger Case of Pascual Jordan	307
a. His Mysterious Absence from Wartime Nazi Secret Weapons Research	308
b. A Connection to Hilgenberg and Gerlach?	309
E. Conclusions and Speculations	310

13. NASA Shows an Interest (With a Little Help From the Nazis)	313
A. Heim Theory	314
1. Getting Acquainted: The Book	314
2. Getting Acquainted: The Documentary	317
3. The Basics of Heim Theory	319
a. Motivations of the Theory	319
b. The Geometrization and Quantization of Space-Time Itself	321
c. Spin-Orientation Potential of the Lattice Structure of Space-Time Itself	322
d. The European Space Agency's and the U.S. Air Force's Anti-Gravity Experiment, Versus Heim Theory	324
B. The Mystery of Heim's World War II Research	326
1. The Accident	326
2. Heim and Heisenberg on a "Clean" Hydrogen Bomb: Ballotechnic Explosive Work?	327
3. His Curiously Alchemical Reference: Quintessence, and His Publication of his Theory with a New Age Publisher	328
4. Pascual Jordan and Wernher Von Braun	328
14. Epilogue is Prologue: The Holy Grail of Physics and Alchemy	331
A. Otto Rahn, the Grail, and the Languedoc	333
1. The Grail, the Stone, the Fleece, and the Cosmic War	334
2. The First Rumor: The Vril Society and the Time Project	335
3. The Second Rumor: Of Superconductors	335
B. World War II: An Alchemical War to Control the Technology of Time?	337
Bibliography	341

the PHILOSOPHERS' STONE

Part One
DE MATERIA PRIMA PHILOSOPHORUM

∴

"Ah, Charles the German, where is your treasure? Where are your philosophers? Where are your doctors? Where are your decocters of woods, who at least purge and relax? Is your heaven reversed? Have yours stars wandered out of their course, and are they straying in another orbit, away from the line of limitation...?"

Philippus Theophrastus Areolus Bombastus von Hohenheim,
a.k.a. Paracelsus, *The Treasure of Treasures for Alchemists*,
From *Paracelsus and his Aurora, & Treasure of the Philosophers*,
As also the Water-Stone of the Wise Men; Describing the matter of,
and manner how to attain the universal Tincture, Faithfully
Englished and Published by J.J. Oxon,
London, Giles Calvert, 1659.

The Alchemical Wedding
Mediaeval Esotericism and Modern Physics

•••

"The general possibility of an 'alchemy' is no longer in need of proof..."
Nobel Physics Laureate Prof. Dr. Walther Gerlach[1]

What do NASA, the Soviet Academy of Sciences, and the Nazi *Allgemeine SS* all have in common?

Answer: alchemy.

What do string theory, loop quantum gravity, or even — for the cognoscenti — Heim theory all have in common?

Answer: alchemy.

What do the Moon, Mars, Mercury, and the Sun have in common?

Answer: alchemy.

And what do the three exotic substances that I have explored in my previous books — the American "gold," the Soviet "mercury," and the Nazi "serum" — all have in common?

Answer: alchemy.

It may seem a strange or even outlandish answer and assertion....

Alchemy.

---

1   Prof. Dr. Walther Gerlach, „Die Verwandlung von Quecksilber in Gold" ("The Transmutation of Mercury into Gold"), *Frankfurter Zeitung*, Freitag, 18, Juli 1924 (Friday, July 18, 1924), cited in my *SS Brotherhood of the Bell*, p. 273. The full English translation is given on pp. 272–273, with the German original of the newspaper article reproduced on p. 274.

Only speak this potent word and it conjures images — a magical act in and of itself in our supposedly non-magical, scientific age — images of charlatans and frauds bent over primitive flasks and beakers and furnaces and crucibles, inhaling the noxious and toxic fumes of sulfur and mercury, trying by a variety of obscure and murky processes only darkly understood by their practitioners to turn base metals into gold, and going quite mad in the process, its adepts dying insanely young. It is the last gasp of a dying, unscientific, mediaeval age that is desperate for fresh air in an atmosphere choked with vials of vile philters and smoggy, greasy arts and vapors.

Alchemy.

Only speak the word and one will make a "real" scientist distinctly uncomfortable, perhaps even somewhat belligerent.[2]

For alchemy — in its quest for the Great Elixir, the Philosophers' Stone, capable of transmuting base metals into "gold" — with its bewildering array of coded symbols, astrological lore, geometric diagrams and charts, is really a quest for *exotic matter* or exotic *states* of matter. In this, it is curiously much like modern physics with its *own* bewildering array of obscure hierophants scratching the coded symbols of higher mathematics on blackboards, working in their own laboratories of arcane equipment, poring over their own charts of computer-generated models and geometries of atoms, paying enormous sums for obscure volumes of wisdom and recipes of equations, all in aid of its own quest to confect exotic matter, a dark *materia prima,* able to manipulate the fabric of space-time itself.

And like alchemy, it seeks the patronage of the wealthy and caters to the powerful, all the while speaking its own coded language, trying to keep its secrets to itself and away from the great masses of the people.

Little, if anything at all, has really changed.

And in this resemblance, of course, there lies a tale, for the truth of the matter is, perhaps, that the resemblance is more than coincidental, for it concerns more than the similarities of their accidentals, but lies chiefly in the substance and methods of the quest itself. Why, indeed, should modern science and ancient alchemy be after essentially the same thing? Whence, beyond the desires of the human heart and mind itself, does this quest originate? And more importantly, where has it taken us thus far? The answer to these questions is what this book, in part, is about.

More particularly, this book is also about three of those modern alchemi-

---

2 Even such a well-known alchemist and the man that many regard as also laying the foundations for modern science, Sir Francis Bacon, Lord Verullam, complains of the practice of alchemy being "full of Error and Imposture," *On the Making of Gold,* from Century IV of *Sylva Sylvarum, or a Naturall Historie in ten Centures,* London, 1627. This work was incorporated in Bacon's great unfinished work, the *Great Instauration or Instauratio Magna.* Q.v. www.levity.com/alchemy/bacongld.html, p. 1.

cal "exotic materials": the American "Gold," the Soviet "Mercury," and the Nazi "Serum." I have written briefly about each of these substances in my previous books,[3] but in each case, only in connection to the other subjects under discussion. Here however the exploration is couched in terms of these exotic, "alchemical" materials themselves. As will be seen in this discussion, new details will emerge, details that convey the strong impression that a very different and exotic form of physics exists and is being pursued, off the books, at great expense and in uttermost secrecy, in the occulted laboratories of the world's major powers.

But why call this modern quest for such materials "alchemical" at all? Why invoke the deliberate comparison between modern physics and mediaeval esotericism? To answer this question will require us to have a basic familiarity with what alchemy really was and what it was really after, and not what the popular imagination *says* it was and *says* it was after. And along the way, we shall even uncover a peculiar theological connection.

A. The Origins and Heyday of Alchemy

1. Back to the Esoteric "Gap": The Real Significance of Alchemy for Esoteric History

To put it succinctly, the very *existence* of alchemy is testament to the fact that, from the death of the last great Neoplatonist magician, Iamblichus, to the rise of the Renaissance with its own strong esoteric preoccupations, there was more or less a continuous underground current of esoteric thought deliberately trying to "turn the stream" and to recover ancient lost science. Alchemy was thought to be both the means of recovering that lost science and also the embodiment of it.

But what does this mean?

In the third book of my *Giza Death Star* trilogy, *The Giza Death Star Deployed*, I wrote that there is a historical "gap" of some eight centuries' duration between the death of the last great Neoplatonic magician and practitioner of the "ancient arts," Iamblichus, in 302 A.D. in Rome, and the rise, centuries later, of the first great occidental esoteric society, the Templars ca. 1118 A.D. The question was, what happened to the esoteric and occult tradition during that period? How does one "fill it in"?[4] "Various theories," I noted,

---

3 For the American "gold," see my *Giza Death Star Destroyed* (Kempton, Illinois: Adventures Unlimited Press, 2005), pp. 151–162; for the Soviet "mercury," see my *The SS Brotherhood of the Bell* (Kempton, Illinois, 2006), pp. 278–296; for the Nazi Serum, see Nick Cook, *The Hunt for Zero Point* (London: Century, 2001), pp. 191–192; and Igor Witkowski, *The Truth About the Wunderwaffe*, (Farnborogh, England: 2003), pp. 232, 247, 249,2 54; and my *SS Brotherhood of the Bell*, pp. 278–296.

4 Joseph P. Farrell, *The Giza Death Star Destroyed*, pp. 84–85.

May be and have been advanced to fill in this gap, with the usual contenders for carrying on the tradition being Jewish Qabbalists in Moorish Spain, the Cathars in the Languedoc in southern France, to the more ingenious explanations of some that the Templars, by coming into contact with Byzantine humanists, with their access to the rich archives of Constantinople, or with Arab-Muslim esoteric societies such as the Assassins, which had preserved apart from the prying eyes of Muslim "orthodoxy" the ancient mysteries.[5]

But this still does not satisfactorily explain the "gap," for the gap exists not on the Spanish or Byzantine or Muslim peripheries of the Western European heartland, but in that heartland itself, otherwise, the sudden reemergence of esotericism in the royal courts of Renaissance Europe makes little sense. So once again the important question must be asked: Who, or what entities, preserved this tradition during the period of the "gap"?

To have been able to preserve such a tradition for so long would have meant that any entities doing so had to have had rather powerful and *stable* institutional protectors and mentors, and there are really only two places that offer themselves for consideration in this respect, and one of them is the papal court in the Lateran Palace itself,

> For one conceivable place with access to repositories of knowledge not accessible to the general public would have been the papal court....

But there was a caveat:

> I am not suggesting that the papal court *itself* was this entity, rather, that within the papal court there may have existed, for some centuries following the collapse of the Western Roman Empire, a hidden group of cardinal-deacons and other clergy, *constituted of old Roman senatorial families*, whose allegiance to Christianity may have been only superficial. Control of the growing papal archives, some of which included old Roman imperial archives, would have given access to the type of knowledge that guided the early Templar excavations at the Temple mount. And only the Church would have been in a position of sufficient wealth to reward the Templars for their (discoveries in Palestine) — or to pay them enough to keep quiet about it.[6]

It is this link to the old Roman senatorial families — who directed much of the inner machinations of Papal statecraft during the early Middle Ages — that in turn leads to the second candidate for powerful and stable patronage of the

---
5   Ibid., p. 85.
6   Farrell, *The Giza Death Star Deployed,* pp. 85–86, emphasis added.

esoteric tradition during the period of the "esoteric gap," and that entity is the older royal and noble houses of Europe, some with ties, perhaps, to the Jewish diaspora after the fall of Jerusalem in 70 A.D.[7] And so I concluded that "the best place to look to fill in the 'gap' between the ancient mystery schools and the first modern esoteric schools and secret societies is either in Rome itself, or in Europe's ancient royal houses."[8]

It is alchemy itself, then, that constitutes the precise vehicle by which the practitioners of the esoteric arts carried forward this tradition during the "esoteric gap." They not only did so in a more or less continuous and uninterrupted stream, but, in confirmation of the thesis advanced in my *Giza Death Star Destroyed*, they did so not only under the loose patronage of the Church, but more importantly, *under the direct patronage of some of the most well-known royal houses of Europe,* as will be seen a little later on in this chapter.

2. The Origins and Heyday of Alchemy

Most scholars of alchemy are agreed that its origins lie in ancient Egypt. The English scholar E.J. Holmyard, however, expressed a more cautious outlook toward the view that would connect the science too exclusively with Egypt itself:

> The word alchemy is derived from the Arabic name of the art, *alkimia*, in which 'al' is the definite article. On the origin of 'kimia' there are differences of opinion. Some hold that it is derived from *kmt* or *chem*, the ancient Egyptians' name for their country; this means 'the black land,' and is a reference to the black alluvial soil bordering the Nile as opposed to the tawny-coloured desert sands. *In the early days of alchemy it was much practised in Egypt,* and if this derivation is accepted the name would mean 'the Egyptian art.' Against this etymology is the fact that in ancient texts *kmt* or *chem* is never associated with alchemy, and it is perhaps more likely that *kimia* comes from the Greek *chyma*, meaning to fuse or cast a metal.[9]

Thus, while the etymology of the word is suggestively obscure, there is little doubt that the actual *practice* of the art is connected to Egypt in some form.

This becomes more apparent as one searches for the earliest mentions of the art. And in this search for the earliest mention, another strange twist is added to the story, for a new contender for the origin of the practice enters the scene: China.

---

7 Ibid., p. 87.
8 Farrell, *The Giza Death Star Destroyed*, p. 87.
9 E.J. Holmyard, *Alchemy* (Minneola, New York: Dover Publications, 1990), p. 19, emphasis added.

There is some doubt concerning the earliest mention of alchemy, for a reference to it occurs in a Chinese edict of 44 B.C., while a book on alchemical matters was written in Egypt by Bolos Democritos at a date that cannot be more precisely fixed than about 200 B.C. However, whether the honour should go to China, or whether Egypt established a slight lead, there is no uncertainty about the fact that the main line of development of alchemy began in Hellenistic Egypt, and particularly Alexandria and other towns of the Nile delta.[10]

That both Egypt and China should record some of the earliest mentions of alchemy highlights once again the possibility that the art may be *very* ancient indeed, and stem from some hitherto unknown contact between the two civilizations, or alternatively, may be the declined legacy in each case of an even older common civilization, of extreme antiquity, from which Egypt and China derived it.[11]

In any case, by the time of the famous Greek alchemist Zosimos of Panopolis, or Akhmim, in Egypt in 300 A.D. "alchemical speculation (had run) riot."[12]

We now find in it a bewildering confusion of Egyptian magic, Greek philosophy, Gnosticism, Neo-Platonism, Babylonian astrology, Christian theology, and pagan mythology, together with the enigmatical and allusive language that makes the interpretation of alchemical literature so difficult and so uncertain.[13]

By the time of the Byzantine alchemist Stephanos of Alexandria, active during the reign of the Emperor Heraclius I (610–641),[14] alchemy had considerably toned down the Gnostic and Neoplatonic elements, but the situation in general remained more or less the same as far as the confusion and ambiguity present in alchemical texts and their "technological" vocabulary were concerned.

The mention of Heraclius I in this regard is significant of another thing, a point which Holmyard neglects to mention. Heraclius, intent upon the final destruction of the Persian Empire, and the restoration and recovery of Egypt to the Roman Empire, had embarked on a military and theological campaign

---

10 Ibid., p. 25.
11 This civilization I intentionally named "paleoancient" in my *Giza Death Star* trilogy (q.v. *The Giza Death Star*, p. 2) in order to indicate its extreme antiquity. As argued throughout those books, and more recently in my book *The Cosmic War: Interplanetary Warfare, Modern Physics, and Ancient Texts,* the "sciences" of these civilizations appear to be the considerably declined legacy of a much older, much more sophisticated and exact physics in the modern sense. In this respect, alchemy would be but another, and the most obvious, example of this phenomenon.
12 E.J. Holmyard, *Alchemy*, p. 27.
13 Ibid.
14 Ibid., p. 29.

designed to return Egypt's Monophysite Christians and reconcile them to the Ecumenical Councils of the Imperial Church. Theologically this effort was spearheaded by his erudite patriarch, Sergius of Constantinople. This effort failed largely because, under Sergius' tutelage, a new heresy broke out known as Monotheletism,[15] but for our purposes the campaign indicates that Stephanòs the alchemist, whose activities were surely known to the Emperor, was tolerated. And this implies, of course, a kind of tacit royal tolerance that in practical terms fell just short of patronage. This connection of alchemy to royal tolerance or outright patronage is a pattern that we shall see recur over and over again.

By the time of alchemy's heyday, "from about A.D. 800 to the middle of the seventeenth century,"[16] its practitioners included everyone from

> kings, popes, and emperors to minor clergy, parish clerks, smiths, dyers, and tinkers. Even such accomplished men as Roger Bacon, St Thomas Aquinas, Sir Thomas Browne, John Evelyn, and Sir Isaac Newton were deeply interested in it, and Charles II had an alchemical laboratory built under the royal bedchamber with access by a private staircase. Other alchemical monarchs were Herakleios I of Byzantium, James IV of Scotland, and the Emperor Rudolf II.[17]

In other words, alchemy had the patronage not only of some popes, but more importantly, of the powerful royal houses of Hapsburg and Stuart, whose own connections to Masonry and other esoteric societies and doctrines is a matter of some record.[18]

B. Exoteric and Esoteric Alchemy

What exactly *is* alchemy then?

Most people are aware of the fact that alchemy is the "quest to make the Philosophers' Stone," and most know that this in turn is a stone which purportedly has the power to "transmute base metals into pure gold," either through touching them with it, or via some other operation involving it. But here popular knowledge usually stops and fantasy, or ignorance, begins, for all is not as simple as the popular imagination would make it out to be, for the quest for the Philosophers' Stone really involves the whole system of alchemical belief regarding the properties of matter, and their derivation from the first act of creation itself.

---

15 Monotheletism is the belief that Christ had only one will, a hybridized "divine-human" will. This teaching was combated in large part by the Patriarch of Jerusalem and by the famous Byzantine saint, Maximus the Confessor.
16 E.J. Holmyard, *Alchemy*, p. 15.
17 Ibid.
18 Q.v. for example, Sir Laurence Gardner's most recent work, *The Shadow of Solomon*.

The first thing to be noticed about alchemy is its persistent "dual" nature at almost every level, from the ambiguity of its technical terminology to its overall framework involving both exoteric and esoteric pursuits. In the latter respect, Holmyard observes that

> Alchemy is of a twofold nature, an outward or exoteric and a hidden or esoteric. Exoteric alchemy is concerned with attempts to prepare a substance, the philosophers' stone, or simply the Stone, endowed with the power of transmuting the base metals lead, tin, copper, iron, and mercury into the precious metals gold and silver.... The belief that it could be obtained only by divine grace and favour led to the development of esoteric or mystical alchemy, and this gradually developed into a devotional system where the mundane transmutation of metals became merely symbolic of the transformation of sinful man into a perfect being through prayer and submission to the will of God. The two kinds of alchemy were often inextricably mixed; however, in some of the mystical treatises it is clear that the authors are not concerned with material substances but are employing the language of exoteric alchemy for the sole purpose of expressing theological, philosophical, or mystical beliefs and aspirations.[19]

This dual exoteric-esoteric aspect of alchemy, however, is so intricately intertwined that its exoteric aspect "cannot properly be appreciated if the other aspect is not always borne in mind."[20]

The "dual" aspect of alchemy is replicated in its exoteric practice as well. The mediaeval alchemist Petrus Bonus, writing ca. 1330 A.D., stated that

> The principles of alchemy are twofold, natural and artificial. The natural principles are the causes of the four elements, of the metals, and of all that belongs to them. The artificial principles are sublimation, separation, distillation, calcinations, coagulation, fixation, and creation, besides all the tests, signs, and colours by which the artificer can tell whether these operations have been properly performed or not.[21]

In other words, the operations of alchemy itself constitute the human and artificial element of exoteric alchemy. In this, one sees the connection to the esoteric, for in order to perform these operations, the alchemist himself had to transmute himself, with the aid of divine enlightenment, from the "base metal" of sinful humanity to the "pure gold" of the redeemed and enlightened soul.

---

19 E.J. Holmyard, *Alchemy*, pp. 15–16.
20 Ibid., p. 16.
21 Ibid., pp. 143–144.

1. The Ambiguous Nature of the Technical Language of Alchemy

Petrus Bonus also notes that the dual aspect of alchemy is further mirrored in its actual technical terminology and style of diction, for everywhere one turns in conventional alchemical texts, one is confronted by intentionally ambiguous language, i.e., language that is intentionally designed to have more than one level of meaning. Even though the actual operations of exoteric alchemy

> Could be transmitted in a very short time, he goes on to explain that the search for that knowledge is very difficult, partly because the adepts use words not only in their ordinary sense but in allegorical, metaphorical, enigmatical, equivocal, and even ironical ways.[22]

In other words, alchemy, like any other occult or esoteric art, used language deliberately designed both to reveal and to conceal. The result, as Holmyard observes, "is that it is not always possible to decide whether a particular passage refers to an actual practical experiment or is of purely esoteric significance."[23]

This ambiguity may easily be seen by a glance at a typical table of correspondences between alchemical operations and the signs of the zodiac, both of which share common symbols:

Operation	*Zodiacal Symbol*	*Astrological Meaning*
Calcination	♈	Aries, the Ram
Congelation	♉	Taurus, the Bull
Fixation	♊	Gemini, the Twins
Solution	♋	Cancer, the Crab
Digestion	♌	Leo, the Lion
Distillation	♍	Virgo, the Virgin
Sublimation	♎	Libra, the Scales
Separation	♏	Scorpio, the Scorpion
Ceration	♐	Sagittarius, the Archer
Fermentation	♑	Capricornus, the Goat
Multiplication	♒	Aquarius, the Water-Carrier
Projection	♓	Pisces, the Fishes

Table 1: Alchemical Operations and Zodiacal Correspondences [24]

---

22   E.J. Holmyard, *Alchemy*, p. 142.
23   Ibid., p. 17.
24   Ibid., p. 154, citing Pernety, *Dictionnaire Mytho-Hermétique,* no page citation given.

There is a point in this table whose true significance will only become manifest in later chapters, namely, that alchemy associates certain of its processes and results with the positions of celestial bodies.

There are also dual uses of some symbols of the common base metals of alchemy that associate them with particular celestial bodies.

Base Metal	Symbol	Celestial Body
Gold	☉	The Sun
Silver	☾	The Moon
Copper	♀	Venus
Iron	♂	Mars
Mercury	☿	Mercury
Lead	♄	Saturn
Tin	♃	Jupiter

Table 2: Alchemical Base Metals and Planetary Associations[25]

This last table highlights yet another aspect of alchemy's dualism, and it is a connection readers of my previous books have encountered before. In my book *The Cosmic War: Interplanetary Warfare, Modern Physics, and Ancient Texts*, I noted that one aspect of the most ancient Egyptian and Sumerian astrology was its connection of particular planets and celestial bodies with particular *crystals and precious gems:*

> Most modern people only encounter astrology, if they encounter it at all, in the "horoscope" page of the local newspaper, or in little booklets of sun signs in the grocery store aisle. Because of this type of exposure, most people think of astrology as having only to do with the subtle influences of the stars and planets on human life. But there is most decidedly more to the ancient view, as Budge observes:
>
> "The old astrologers believed that precious and semi-precious stones were bearers of the influences of the Seven Astrological Stars or Planets. Thus they associated with the–
>
> "SUN, yellowish or gold-coloured stones, e.g. amber, hyacinth, topaz, chrysolite.
>
> "With the MOON, whitish stones, e.g. the diamond, crystal, opal, beryl, mother-of-pearl.

---

25 Ibid., p. 154, citing Pernety, *Dictionnaire Mytho-Hermétique,* no page citation given.

"With MARS, red stones, e.g. ruby, haematite, jasper, blood-stone.

"With MERCURY, stones of neutral tints, e.g., agate, carnelian, chalcedony, sardonyx.

"With JUPITER, blue stones, e.g. amethyst, turquoise, sapphire, jasper, blue diamond.

"With VENUS, green stones, e.g. the emerald and some kinds of sapphires.

"With SATURN, black stones, e.g. jet, onyx, obsidian, diamond, and black coral."[26]

As also noted there, while no one really knows the exact origins of astrology, it *is* known that it was present from the inceptions of the "sciences" of the most advanced civilizations of antiquity: Egypt and Sumer. In particular, it is from Sumeria that most contemporary Western astrology stems, for the Sumerians recorded their astronomical and astrological observations on clay tablets. These

> they then interpreted from a magical and not astronomical point of view, and these observations and their comments on them, and interpretations of them, have formed the foundations of the astrology in use in the world for the last 5,000 years.[27]

But that was not all that was claimed for astrology. Budge continues:

> *According to ancient traditions preserved by Greek writers, the Babylonians made these observations for some hundreds of thousands of years, and though we must reject such fabulous statements, we are bound to believe that the period during which observations of the heavens were made on the plains of Babylonia comprised many thousands of years.* [28]

There is ample evidence to suggest, however, that such views were declined scientific legacies of a much more ancient, and much more sophisticated, civilization.[29] It is thus not outside the bounds of possibility that, indeed,

---

26 Joseph P. Farrell, *The Cosmic War: Interplanetary Warfare, Modern Physics, and Ancient Texts,* pp. 243–244, citing Sir E.A. Wallis Budge, *Amulets and Superstitions: The Original Texts with Translations and Descriptions of a Long Series of Egyptian, Sumerian, Assyrian, Hebrew, Christian, Gnostic and Muslim Amulets and Talismans and Magical Figures, with Chapters on the Evil Eye, The Origin of the Amulet, the Pentagon, the Swastika, the Cross (Pagan and Christian), the Properties of Stones, Rings, Divination, Numbers, The Kabbalah, Ancient Astrology, etc.* (Oxford University Press, 1930), p. 406.

27 Budge, op. cit., p. 406.

28 Ibid., emphasis added, cited in my *The Cosmic War: Interplanetary Warfare, Modern Physics, and Ancient Texts,* p. 241.

29 Q.v. *The Giza Death Star,* pp. 38–110; *The Giza Death Star Destroyed,* pp. 99–245; *The Cosmic*

Sumerian astrology has origins that date back "some hundreds of thousands of years."

So one is presented with an interesting picture: on the one hand, from Egypt and Sumer, one encounters the association of planets and stars with certain *crystals* and their color properties, that is to say, with certain *electromagnetic and spectrographical properties,* and on the other hand, from alchemy, one has the association of the *same* celestial bodies not only with certain types of alchemical *operations* but with certain types of *metals* as well. And metals, as everyone knows, have, like crystals, their own unique "lattice" properties of molecular bonding. *In short, one has, from two distinct types of esoteric arts, the association of celestial bodies, with certain materials that in turn possess certain lattice and spectrographical properties. This will become a crucial key into prying open yet another aspect of a paleoancient and very sophisticated physics that may once have underlay the declined astrological and alchemical legacies of Sumer and Egypt.*

So why associate materials with celestial bodies to begin with? Why should the exoteric and esoteric aspects have come to be intertwined in the first place? Why is there such an association of alchemy and astrology?

C. The Goal and Quest of Alchemy: The Transmutative Medium of the Materia Prima

The answer to these questions lies once again in the Egyptian roots of alchemy and in what those roots in turn imply. The basic Egyptian view of creation, as pointed out by the celebrated esotericist and "alternative Egyptologist," René Schwaller De Lubicz, was that all of the existing diversity of the universe stemmed from one underlying "prime matter" or *materia prima,* an absolutely undifferentiated substrate, an "aether" or medium which then began to undergo differentiation. This initial process of "hyper-differentiation" of an undifferentiated medium Schwaller called the "primary scission."[30] Further differentiations are in turn performed upon these initial derivations from the medium, until at last the entire diversity of creation arises. While all this sounds rather fanciful, it is in fact capable of a profoundly sophisticated interpretation from the point of view of certain aspects of modern physics, for an absolutely undifferentiated substrate in fact is physically *non-observable;* it is therefore, as far as physics is concerned, a *nothing,* even though it may be said that this *materia prima* has some sort of "existence." The whole of Egyptian religion and magical practice, then, stem from this viewpoint, for if all arises from this *materia prima,* then everything that exists, by dint of

---

War, pp. 1–66, 100–131, 234–273.

30   Schwaller's views and their profound topological correlations in ancient texts are discussed at length in my *The Giza Death Star Destroyed,* pp.99–129.

its existence, can be described in terms of its "topological descent" from that substrate. In short, every existing thing is connected with every other existing thing by virtue of its creation from the same underlying "stuff" or substrate. The substrate exists in every thing, since every thing is but a particular differentiated manifestation of that substrate.

Consequently, this underlying substrate or medium was *transmutative* in its very nature; it was, so to speak, a "pure potential," *capable* of undergoing differentiation and diversification. To put it succinctly, the medium was *the* Philosophers' Stone par excellence. Moreover, since it was an *undifferentiated* medium, it was above the concepts of space and time themselves. It was, in a word, *non-local.* And hence one can see the connection to Egyptian sympathetic magic and alchemical practice, for if everything is connected via this non-local medium, then one could manipulate and influence another object (or person!) via the medium from which they are descended. This was supposedly accomplished by reconstructing *as exact an analogue of that object's "descent" or process of differentiation from the medium itself.* One had, so to speak, to "back engineer" the whole process of differentiations.

And with this, one perceives the connection to alchemy, to what its real goal was, and to why it connected the exoteric and esoteric aspects of its practice, for if the exoteric operations were to work, the operator performing them had *himself* to ascend back *up* the path of his own differentiation and "topological descent" from that medium, at least, insofar as it was possible for him to do so. The quest of alchemy, in short, was to literally embody that *materia prima* and its transmutative powers as fully within lower diversified matter as was possible in the earthly Philosophers' Stone.

These views of the underlying substrate or *materia prima* persisted into the Hellenic philosophers, and became, via Aristotle, the common currency of later occidental alchemy. For Aristotle,

> The basis of the material world was a prime or primitive matter, which had, however, only a potential existence, until impressed by 'form.' By form he did not mean shape only, but all that conferred upon a body its specific properties.[31]

In terms of the topological metaphor of alchemy, in other words, Aristotle's "form" is the Egyptians' "primary scission" with all its ensuing "differentiations."

This view of the *materia prima* and the topological descents and differentiations of existing things is the real basis of the famous alchemical and

---

31   E.J. Holmyard, *Alchemy*, p. 21.

esoteric axiom "as above, so below." In fact, in the most prized and famous alchemical text, the *Emerald Tablet of Thoth,* this axiom of the transmutative medium found its most famous expression:

> True it is, without falsehood, certain and most true. That which is above is like to that which is below, and that which is below is like to that which is above, to accomplish the miracles of one thing.
>
> *And as all things were by the contemplation of one, so all things arose from this one thing by a single act of adaptation.* [32]

In the last sentence of the above quotation, one may see encapsulated the whole sum and substance of alchemical "physics": the "one thing" is the undifferentiated substrate, the "arising" of all things from that substrate is the primary scission, and the "single act of adaptation" that occurs "by the contemplation of one" is the act of differentiation itself, brought about by Intelligence and a supreme act of will. It will be noted that the act of differentiation thus also connotes placing the undifferentiated medium, which is in a state of utter *equilibrium,* into a more or less constant state of *non-equilibrium* or *stress.* This too, will become a crucial point in the remainder of this book

With this understanding of the ancient view of the transmutative medium in hand — a view that is decidedly modern once all the residue of metaphysics is boiled out of it — one is finally in a position to assess alchemical accounts of the quest for and composition of the Philosophers' Stone. From the act of that initial "differentiation" of the underlying *materia prima,* one ends up with *three* entities: 1) the underlying medium itself, 2) the differentiated parts of it, and 3) their common properties.[33] Interestingly enough, this tripartite structure becomes one of the properties of the Philosophers' Stone itself, in some alchemical texts, for it is often referred to as the "tripartite Stone." For example,

> According to an anonymous seventeenth-century book entitled *The Sophic Hydrolith,* the Philosophers' Stone, or the ancient secret, incomprehensible, heavenly, blessed, and *triune universal stone* of the sages, is made from a kind of mineral by *grinding it to powder, resolving it into its three elements,* and *recombining these elements into a solid stone* of the fusibility of wax.[34]

---

32 The *Tabula Smaragdina,* trans. R. Steele, D.W. Singer, cited in E.J. Holmyard, *Alchemy,* p. 97, emphasis added.

33 A slightly different version of this "ternary" structure of the medium was given in *The Giza Death Star Destroyed.* The view here has been simplified for exposition's sake, but for the fuller and more accurate view and its connection to a profound topological metaphor in ancient Hermetic texts, see that work, pp. 222–245.

34 Holmyard, *Alchemy,* p. 17, emphasis added.

Note here that there are three essential operations to the successful confection of this "triune universal stone":

1) Subjecting it to stress ("grinding it to powder");
2) Recapitulating the process of differentiation ("resolving it into its three elements"); and
3) "Reverse engineering" its original topological descent from the medium ("recombining these elements into a solid stone").

The direct connection of the Philosophers' Stone to the underlying transmutative *materia prima* is made even more apparent in the following passage from the fourteenth-century alchemist, Peter Bonus:

> In the first sense our Stone is *the leaven of all other metals*, and changes them into its own nature — a small piece of leaven leavening a whole lump. As leaven, though of the same nature with dough, cannot raise it until, from being dough, it has received a new quality which it did not possess before, *so our Stone cannot change metals until it is changed itself, and has added to it a certain virtue which it did not possess before. It cannot change, or colour, unless it has first itself been changed and coloured.* Ordinary leaven receives its fermenting power through the digestive virtue of gentle and hidden heat; and so *our Stone is rendered capable of fermenting, converting, and altering metals by means of a certain digestive heat, which brings out its potential and latent properties,* seeing that without heat neither digestion nor operation is possible.[35]

This is a very revealing passage, and for several reasons that will preoccupy us throughout the remainder of this book.

We have the following assertions by Bonus:

1) The Philosophers' Stone is "the leaven of all other metals," in other words, the "leaven" metaphor is employed to denote the fact that in some sense the Philosophers' Stone partakes of the properties of the underlying transmutative medium *directly:* as leaven is present throughout a whole mass of dough, so the medium is present throughout all the differentiations of created things within it, and that are comprised of it. That medium has, so to speak, literally been *"em-bodied"* within the Philosophers' Stone, a metaphor that, as we shall shortly see, Bonus himself employs;
2) The Philosophers' Stone can effect no change or transmutation until it itself has undergone change and transmutation. This is

---
35 Ibid., pp. 146–147, emphasis added.

most likely to be understood in connection with the first point immediately above. But note also that the *ability* to change is connected with *color*. As we shall see in the remainder of this chapter, and throughout the next section of the book, this reference to *color* will assume great significance as a sign of a genuinely alchemical transmutation, *even for modern physics;*

3) This change in turn is accomplished by *heat*, for "our Stone is rendered capable of fermenting, converting, and altering metals by means of a certain digestive heat, which brings out its potential and latent properties."

Thus, the Philosophers' Stone is confected by:

1) Heat, or, once again, a *stress,* which makes it undergo a
2) Change, or once again, *differentiation* which "Brings out its potential and latent properties," which are those of
3) The transmutative medium itself.

In other words, some of the properties of that transmutative medium are literally "em-bodied" in the Philosophers' Stone by dint of some process involving *heat and color.*

Bonus expands on this "embodiment" metaphor by drawing an analogy to the body and the soul as follows:

> It is the body which retains the soul, and the soul can shew its power only when it is united to the body. *Therefore when the artist sees the white soul arise, he should join it to its body in the same instant, for no soul can be retained without its body.* This union takes place through the mediation of the spirit, for the soul cannot abide in the body except through the spirit, which gives permanence to their union, *and this conjunction is the end of the work. Now, the body is nothing new or foreign; but that which was before hidden becomes manifest and that which was manifest becomes hidden.* The body is stronger than soul and spirit, and if they are to be retained it must be by means of the body. *The body is the form, and the ferment, and the Tincture of which the sages are in search. It is white actually and red potentially; while it is white it is still imperfect, but it is perfected when it becomes red.*[36]

In typical fashion, Bonus both reveals, and conceals, much about the Philosophers' Stone in this passage.

---

36 E.J. Holmyard, *Alchemy*, p. 147, emphasis added.

We may summarize these points in connection with the italicized portions of the quotation just cited.

1) Note first of all that the Stone is now a *"tincture,"* implying that it is not a "stone" at all, but a liquid;
2) Observe the important use made of the "body-soul" analogy to "em-body" something. In this case, it is rather clear what Bonus means: the transmutative properties of the medium itself are the "soul," which requires a "body," the Stone itself, through which to work. In his own words, it is "this conjunction" that "is the end of the work" or its goal.
3) Thus, the "body" or material that undergoes the change and acquires its new transmutative properties is "nothing new or foreign," i.e., it remains what it was before in terms of its being a mineral. *But...*
4) ... some "hidden" properties have now become "manifest." This cryptic remark can be explained by reference to the transmutative medium once again. As was previously mentioned, everything differentiated from that medium *inevitably retains to varying degrees* the transmutative properties of that medium by virtue of their descent from it. Thus, in alchemical thinking, these properties remain latent in any substance, particularly in its "pure" or alchemically "refined" form. Thus, the goal of the operations of exoteric alchemy is to make these hidden and latent properties manifest; it is to *sharpen or intensify the latent properties of the transmutative medium that remain in any element.* Thus, the alchemist is really seeking *an altered state of ordinary matter*, so that the "body" can become, in Bonus' words, "the form, and the ferment, and the Tincture of which the sages are in search," that is, so that "ordinary matter" can embody the transmutative properties of the medium itself in this world.
5) And finally, we have a most important clue from Bonus: there is a distinct spectrographic sequence of colors that allows the alchemist to know when he is getting "close": in its manifest form it is "white actually," and when processed through its final stage of refinement to bring out its latent potential, it is "red potentially." Thus, "while it is white it is still imperfect, but it is *perfected when it becomes red.*"

White, and red.

These are the two colors that we shall see recur over and over again in parts one, two, and three of this book. They are the strongest indicators, as we shall see, that someone, somewhere in the secret laboratories of the world's modern "alchemical" powers, has done their homework, and that they are drawing their inspirations from people other than just Max Planck and Albert Einstein!

Thus, the whole goal of the alchemical art lay in "the general idea that the powers of the cosmic soul must somehow be concentrated in a solid, the philosophers' stone or elixir, which would then be able to carry out the transmutations that the alchemists desired."[37] It was an attempt to reconstruct, for specific cases, the descent of specific minerals from that undifferentiated "cosmic soul" or medium as exactly as possible, in order to incorporate that "cosmic soul's" very powers of transmutation in ordinary matter. As Holmyard states it, "The underlying idea seems to have been that since the prime matter was the same in all substances, an approximation to this prime matter should be the first quest of alchemy."[38] The ability to confect such a "philosophical gold" would indeed give its possessors an awesome power.

It is important to view this quest for the Philosophers' Stone in a much wider context in order to reveal its true significance. As I noted first in my book *The Giza Death Star*, and then my books *The Giza Death Star Destroyed*, and *The Cosmic War*, stones played an important if not central role in ancient mythological descriptions of an actual interplanetary war. Stones — *crystals* — were a central component in the awesome technologies of destruction that allowed that war to be waged. And as I noted in *The Cosmic War*, these stones were directly tied by the texts themselves to the geometries of local celestial space and to the fabric of space-time, the transmutative medium, itself. After the war, some of these stones — the infamous Sumerian "Tablets of Destinies" — were destroyed, others were taken and used for other purposes, and a few, which *could not be destroyed*, were secreted away.[39] *In this sense, one may perhaps also view the alchemical quest as the attempt, by technological means, to reconstitute the technology represented by the lost Tablets of Destinies.*

But there is yet another esoteric connection to alchemy and this quest to confect the Philosophers' Stone, and we have already encountered it: the Emerald Tablet of Thoth, or, in his Hellenistic incarnation, Hermes Trismegistus, the Thrice Great Hermes, the grand *magister* of alchemy, the patron of all alchemical adepts. By its constant reference to the Emerald Tablet, alchemy itself acknowledges that the *goal* of its craft is precisely the reconstitution of this ancient, lost power, the power to manipulate the transmutative medium itself.

---

37 Holmyard, *Alchemy*, p. 98.
38 Ibid., p. 26.
39 Q.v. my *The Cosmic War: Interplanetary Warfare, Modern Physics, and Ancient Texts*, pp. 204–233.

3. *The Problem of Alchemy's Survival:*
The Triune Stone and the Augustinian Trinity

As was suggested previously, alchemy was the principal mechanism by which esoteric and occult studies survived in a continuous stream from the death of the last Neoplatonist magician, philosopher, and theurgist — Iamblichus — to the rise of the Templars. As was also suggested, this survival in part depended upon royal, imperial, and even occasionally Lateran patronage.

But there is another mechanism at work in alchemy's survival during this period, particularly in the Latin Christian West, and it is a rather surprising one, and to understand it, one must understand the strange and strong relationship between the "triune Philosophers' Stone" and the Christian West's Augustinized doctrine of the Holy Trinity. To understand it, in other words, one must do a little "theology." And it may come as a surprise to many people, the doctrine which came to prevail in the mediaeval Latin Church was *not* the original Christian doctrine of the Trinity, which survived only in the Orthodox Catholic Churches of the East. Indeed, the Eastern Orthodox Churches to this day regard the doctrine that came to prevail throughout the West and even at Rome itself as a formal heresy of the highest order, which played no small role in the formal severing of communion between Rome and Constantinople in 1014, and outright mutual excommunication and schism later in 1054. While this is not the appropriate place to delve into these issues, an understanding of the resemblance of the Augustinized doctrine of the Christian West *is* essential to understand the enormity of alchemy's view of its "triune Philosophers' Stone."[40]

The clue to deciphering this strange connection first occurred to me when translating the work of an eastern Patriarch of Constantinople, who, upon learning of the exact content and nature of the formulation of Trinitarian doctrine that was increasingly accepted throughout the Christian West, wrote that the doctrine was more appropriate to a formulation of "sensory things," and not as a formulation of theological doctrine. In other words, the doctrine of the Trinity as formulated in the Latin West was more appropriate to *physics* than to *theology*.[41] It was an echo of other comments one often finds in works of earlier Greek patristic authors addressing similar issues.

This indeed was a kind of Rosetta Stone to unlock the possible physics meaning of ancient Hermetic texts, for what this ninth-century Christian patriarch was suggesting was something truly revolutionary: the whole philosophy of Hermeticism and Neoplatonism as outlined in so many texts was

---

40   I have written extensively elsewhere about the specifically theological issues at stake in the Western Augustinian doctrine of the Trinity.

41   The work in question is Patriarch St. Photius the Great's *Mystagogy of the Holy Spirit*.

less about metaphysics in the standard academic sense, and more about the physics of the underlying medium.

Intrigued with this idea, I decided to see if, in fact, certain Neoplatonic and Hermetic texts could indeed be interpreted as containing nothing but an encoded physics of the *materia prima*, and I devoted a whole chapter appendix to the subject in my book *The Giza Death Star Destroyed*. In doing so, one particular passage from the *Hermetica* becomes particularly significant to the discussion here, and it is best to cite what I wrote in that book before exploring the alchemical connection more fully. The passage in question was about a "topological metaphor in Hermes Trismegistus' conception of God, Space, and Kosmos (Θεος, Τομος, Κοσμος)."[42] And in its own very suggestive way, it contained its own insight on the "triune Philosophers' Stone," by way of a peculiar metaphor of topological triangulation:

> A very different and in some respects more sophisticated version of the metaphor of topological non-equilibrium is found in the Hermetica of Hermes Trismegistus...here an attempt will be made to render the implied topological metaphor of one particular passage formally explicit. This passage is the *Libellus II:1-6b*, a short dialogue between Hermes and his disciple, Asclepius:
>
> "Of what magnitude then must be that Space in which the Kosmos is moved? And of what nature? Must not that Space be far greater, that it may be able to contain the continuous motion of the Kosmos, and that the thing moved may not be cramped through want of room, and cease to move? — (*Asclepius*): Great indeed must be that Space, Trismegistus. — (*Hermes*): And of what nature must it be, Asclepius? Must it not be of opposite nature to Kosmos? And of opposite nature to body is the incorporeal.... Space is an object of thought, but not in the same sense that God is, for God is an object of thought primarily to himself, but Space is an object of thought to us, not to itself."[43]

I then commented as follows:

> This passage thus evidences the type of "ternary" thinking encountered in Plotinus, and a kind of metaphysical and dialectical version of topological triangulation... But there is a notable distinction between Plotinus' ternary structure, and that of the *Hermetica*: whereas in Plotinus' system the three principal objects in view are the One, the Intellect, and the World Soul,

---

42 Joseph P. Farrell, *The Giza Death Star Destroyed* (Kempton, Illinois: Adventures Unlimited Press, 2006), p. 239.

43 Joseph P. Farrell, *The Giza Death Star Destroyed*, p. 239, citing *Libellus II: 1-6b, Hermetica*, trans. Walter Scott, Vol. 1, pp. 135, 137.

here the principal objects in view are the triad of Theos, Topos, and Kosmos (Θεος, Τοπος, Κοσμος), or God, Space,[44] and Kosmos.

These three – God, Space, and Kosmos – *are in turn distinguished by a dialectic of opposition based on three elemental functions, each of which implies its own functional opposite:*

f_1: self-knowledge ⇔ $-f_1$: ignorance
f_2: rest (στασις) ⇔ $-f_2$: motion (κινησις)
f_3: incorporeality ⇔ $-f_3$: corporeality.

So in Hermes' version of the metaphor, the following "triangulation" occurs, with the terms "God, Space, Kosmos" becoming the names or symbols for each vertex or region.[45]

I then reproduced the following diagram to exhibit the functional dialectical oppositions of each of these three vertices:

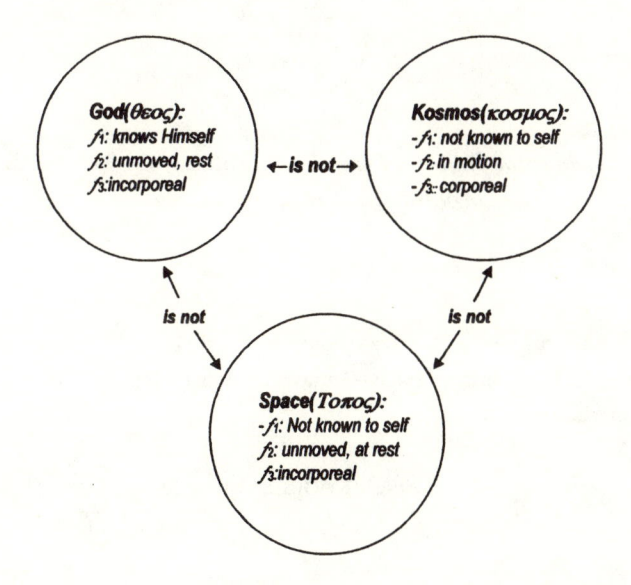

God(Θεος):
f_1: knows Himself
f_2: unmoved, rest
f_3: incorporeal

←is not→

Kosmos(κοσμος):
$-f_1$: not known to self
$-f_2$: in motion
$-f_3$: corporeal

is not

is not

Space(Τοπος):
$-f_1$: Not known to self
f_2: unmoved, at rest
f_3: incorporeal

The "Triune Stone" in the Hermetica of Hermes Trismegistus

I then continued my commentary as follows:

This diagram is significant for a variety of reasons. For one thing, theologically informed readers will find it paralleled in the so-called Carolingian

---

44 The word τοπος can also mean "place."
45 Farrell, op. cit., p. 239, emphasis added.

THE PHILOSOPHERS' STONE | 45

"Trinitarian shield," a pictogram used to describe the doctrine of the Trinity as it emerged in the Neoplatonically influenced Augustianian Christianity of the Mediaeval Latin Church.

The "shield of the faith" I referred to looks like this:

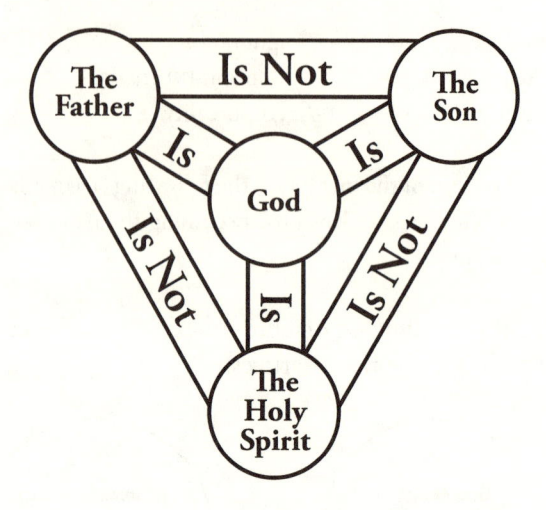

The "Shield of the Faith": A Common Pictogram of the Western Church's Augustinized Doctrine of the Trinity

Continuing the commentary:

> More importantly in this context, however, the diagram illustrates how each vertex — God, Kosmos, Space — may be described as *a set of functions or their opposites*:
>
> *God (Θεος)*   *Kosmos (Κοσμος)*   *Space (Τοπος)*
> $\{f_1, f_2, f_3\}$ $\{-f_1, -f_2, -f_3\}$ $\{-f_1, f_2, f_3\}$
>
> f_1: knowledge $-f_1$: ignorance $-f_1$: ignorance
> f_2: unmoved f_2: in motion f_2: unmoved
> f_3: incorporeal $-f_3$: corporeal f_3: incorporeal
>
> Hermes' version of the metaphor thus lends itself quite neatly to an analysis in terms of Hegelian dialectic, with Space itself forming the synthesis between God, the thesis, and Kosmos, the antithesis, described in terms of the functions f_1, f_2, f_3 or their opposites. But a mere Hegelian analysis would miss the subtlety of the metaphor.

Indeed, the metaphor is almost wholly topological and physical!

To see how, let us extend the formalism by *dispensing* with Hermes' metaphysical descriptions of the functions f_1, f_2, f_3 and take the terms God, Kosmos, and Space as the sigils of distinct topological regions in the neighborhood of each vertex, and model them as empty sets. Since it is possible for combinatorial functions, e.g., ∪ and ∩, to be members of empty sets, then letting $\varnothing_G$, $\varnothing_K$, $\varnothing_S$ stand for God, Kosmos, and Space respectively, one may quickly see that a latticework (may result) from entirely different sets of functional signatures, exactly as was the case in Plotinus, but via a very different route:

$\varnothing_G = \{f_1, f_2, f_3\}$
$\varnothing_K = \{-f_1, -f_2, -f_3\}$
$\varnothing_S = \{-f_1, f_2, f_3\}$.

From this the non-equilibrium, non-equivalence paradigm (of ancient texts embodying the topological metaphor) is evident, for while the sets are equipotent in that they have the same number of members, they are not functionally identical.[46]

Here it is best to pause and comment a bit further than I did in *The Giza Death Star Destroyed*.

Since what is being described here are three regions or what topologists would call "neighborhoods," what is actually being modeled, via this ancient metaphor, is not only a kind of "topological triangulation" but that very triangulation is being accomplished by functional distinctions between each region, each of which is not only a "physical nothing" but more importantly, a *differentiated nothing*. The physical medium on this ancient model thus *creates information*,[47] *and in so doing, transmutes itself*. This conception is the basis, of course, of alchemy.

The resemblance of this Hermetic topological metaphor to the Augustinian Trinitarian shield, with its own dialectics of oppositions between the three divine Persons, will immediately be evident, and this suggests that one of the strongest reasons for alchemy's survival throughout the Western Middle Ages and beyond is that there was a fertile, widely accepted cultural matrix in which it could thrive, for that supposedly theological doctrine, with its own roots

---

46   Farrell, *The Giza Death Star Destroyed*, pp. 239–241, emphasis added.
47   It creates information for the precise reason that if one performs unions and intersections on these sets, it will be evident that many more "regions" may result, each with their own peculiar functional signature.

deep in Neoplatonism, and therefore with its own deep but *largely unsuspected* roots in Egyptian hermeticism, was really about physics and topology in a proper sense.[48]

2. Alchemical References and Analogues to the Augustinized Trinity

A survey of alchemical references to the "triune Philosophers' Stone" will only bring the fact of the West's widespread Neoplatonic and Augustinian cultural matrix, and alchemy's relationship to it, into even sharper focus. For example, the *Rosarium Philosophorum* (*The Rosary of the Philosophers*) refers to the animal, vegetable, and mineral nature of the Philosophers' Stone, as do many alchemical texts.[49] More importantly, it cites the alchemist Arnoldus as stating "Let the Artificers of Alchemy know this, that the forms of metals cannot be transmuted *unless they be reduced to their first matter*, and then they are transmuted into another form than that which they had before."[50] Arnoldus is implying precisely the type of "back-engineering" of matter in its topological descent from the *materia prima* as being essential to the successful confection of the Philosophers' Stone.

Similarly, the *Rosarium Philosophorum* cites an intriguing quotation of Rosinus, in which the doctrinal formularies of the Trinity are very much in evidence:

> We use true nature because nature does not amend nature, unless it be into his own nature. There are three principal Stones of Philosophers. That is mineral, animal, and vegetable. A mineral Stone, a vegetable Stone, and an animal Stone, *three in name but one in essence*. The Spirit is double, that is, tincturing and preparing.[51]

The resonance of this formulary to the standard Trinitarian doctrinal and hymnographical expression — three in persons, one in essence — is astonishing. While there is nothing peculiarly Augustinian about this formulary in and of

---

48 Lest there be *any* misunderstanding here, these remarks are *not* to be construed as the author's rejection of the doctrine of the Trinity, but merely as his rejection of the *Augustinized formulation* of it, that is, of his rejection of that common Augustinized theological inheritance of the Christian West!

Similarly, such a connection is meant only to explain the relative persistence and widespread nature of alchemy in the Christian West. While it continued to be practiced in Byzantium and its satellites, it was never quite as pervasive as it was in the West.

49 *Rosarium Philosophorum,* Tomus II, *De Alchemica Opuscula complura veterum philosophorum* (Frankfurt, 1550, 18th-century English translation in Ferguson 210).

50 Ibid., emphasis added.

51 *Rosarium Philosophorum,* Tomus II, *De Alchemica Opuscula complura veterum philosophorum* (Frankfurt, 1550, 18th-century English translation in Ferguson MS 210 in the Bodelian Library, Oxford), emphasis added.

itself, the last sentence concerning the Spirit being "double" is strongly suggestive of the Augustinian formulation of the Trinity, since in the Augustinian formulation the Spirit is made to take His personal origin from the Father *and the Son,* as the interpolated version of the Nicene creed in use throughout Western Christianity states, and as a glance at the Trinitarian shield cited previously will disclose. Indeed, so strongly is this idea connected with that formulation that the Augustinian doctrine of the Trinity is sometimes simply referred to as the doctrine of the "double procession" of the Holy Spirit. The alchemical reference is therefore suggestive, but hardly conclusive.

Arnoldus de Nova Villa makes yet another reference to the triune Stone in connection with the Christian Trinity in his *Cymicall Treatise:*

> In the beginning of this labour, I'le [sic] say, that the most excellent Hermes teaches the way in plain words to rationall [sic] men, but in occult and hid speeches to the unwise and fools. I say that the father son and holy ghost are one, and yet three, so speaking of our Stone I say three are one, and yet are divided.[52]

The Trinitarian reference again is quite clear, but again, there is nothing distinctly Augustinian about it. Arnoldus additionally makes the comment that the "unifying substance" of the triune Stone is Mercury, for "out of Mercury is everything made."[53] This type of reference causes many who are unfamiliar with alchemical texts to misinterpret what is actually being said. The term "mercury" is often used in two senses, the first in its literal prosaic sense, referring to the chemical element itself. The second usage, often coupled with other designators such as "Our Mercury" or "the Philosophical Mercury" in a kind of code name, simply means the underlying transmutative medium or *materia prima* itself. It is, in short, a code name for the Philosophers' Stone itself, and for the underlying substance that gives it its power.

In context, then, Arnoldus' remarks exhibit their character as disguising the "triune Stone" in the context of a strictly Augustinian formulary of the Trinity, since the unity of that Trinity is seen to lie less in the Person of the Father, and more in the impersonal unity of the circle of the divine substance in the middle of the pictogram. Arnoldus is saying much the same thing. The union of the triune Stone in its animal, vegetable, and mineral parts, is the "philosophical Mercury," the *materia prima itself:* "the Mercury of the sages,"

---

[52] *A Chymicall Treatise of the Ancient and highly illuminated Philosopher, Devine and Physitian, Arnoldus de Nova Villa who lived 400 years agoe, never seene in print before, but now by a Lover of the Spagyrick art made publick for the use of Learners,* printed in the year *1611.* Bodleian Library, MS Ashmole 1415, pp. 130–146, transcribed by Hereward Tilton.

[53] Ibid.

he says, "is not that common Mercury, call'd by the Philosophers prima materia."[54] The connection may easily be seen by reproducing the Trinitarian shield and substituting alchemical references for the Trinitarian ones:

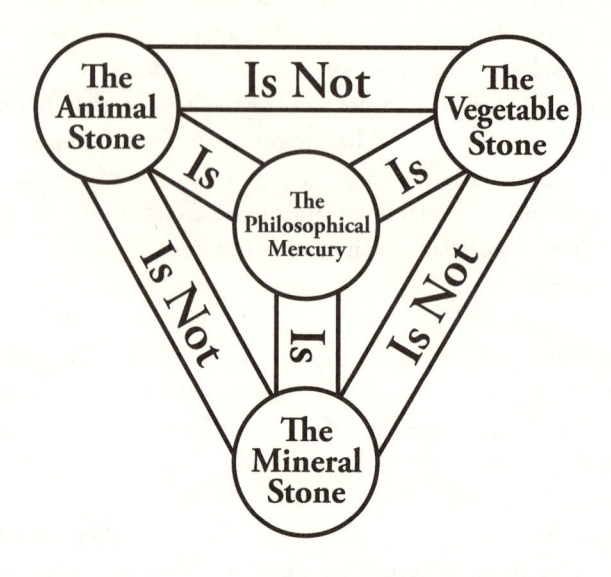

```
         The                Is Not              The
       Animal   ─────────────────────────────  Vegetable
        Stone                                    Stone
           \     Is                    Is      /
            \         The Philosophical       /
        Is Not        Mercury              Is Not
              \                            /
               \        Is       Is       /
                \                        /
                 \                      /
                        The Mineral
                           Stone
```

The "Alchemical Augustine": Augustinian Trinitarian Shield with The Triune Stone In the Vertices, and The Philosophical Mercury or Materia Prima As the Unifying Substance

In other words Arnoldus, if one closely scrutinizes his remarks, reproduces the "topological triangulation" we discovered in the much earlier Hermetic texts.

The transmutative medium and *materia prima* itself receives its own biblical treatment at the hands of some alchemists. For example, Simon Forman interprets the initial creation account of Genesis 1 as an alchemical work, with the primordial waters representing the undifferentiated medium:

> Into the darkness then did descend the spirit of God, Upon the watery chaos, whereon he made his abode. Which darkness then was on the face of the deep, In which rested the Chaos, and in it all things asleep. Rude, unformed, without shape, form or any good, Out of which God created all things as it stood…. Then out of this Chaos, the four elements were made… The quintessence (that some men it call) Was taken out of the Chaos before the four elements all.[55]

---

54 Ibid.
55 Simon Forman, alchemical poem *Of the Division of Chaos*, Bodeleian Library, Oxford, MS Ashmole 240.

Chaos is the "physical nothing" out of which God fashioned, by differentiating within it, the things of the world. Note also that term "quintessence," for it will become quite important to the case in part three that alchemy *may* have formed some of the conceptual basis behind one of physics' most brilliant, and most unknown, thinkers. Forman expands on what the "quintessence" is in the following manner:

> And into every specific thing do put quintessence, To reap such seed thereof; as men do sow, But of themselves. As they are simple and pure in kind.[56]

To put it succinctly, "quintessence" is yet another code name for the Philosopher's Stone itself, and for the alchemical operation of "embodying" it in a natural substance in order to utilize its powers. This, too, as will be seen much later in part three, is yet another suggestive link to the breathtaking thought of a modern physicist.

a. Philippus Theophrastus Areolas Bombastus von Hohenheim, a.k.a. Paracelsus

No survey of alchemy in general nor of alchemy's peculiar relationship to the Augustinized formulation of the Trinity in the Christian West would be complete without noting the many statements made in its regard by perhaps the most famous and controversial of the late Mediaeval and early Renaissance alchemists, Philippus Theophrastus Areolus Bombastus von Hohenheim, better known as Paracelsus (1493–1541). One gets some measure of the man in the fact that he was born simply "Philippus von Hohenheim" but later took on all the other names himself. Widely noted in his day as an extraordinary physician and alchemist, much of his written output does in fact concern alchemy, and his own disagreements with other practitioners of the craft, no matter how ancient or venerable:

> From the middle of this age the Monarchy of all the Arts has been at length derived and conferred on me, Theophrastus Paracelsus, prince of Philosophy and of Medicine. For this purpose I have been chosen by God to extinguish and blot out all the phantasies of elaborate words, be they the words of Aristotle, Galen, Avicenna, Mesva, or the dogmas of any among their followers.[57]

---

56 Ibid.
57 Paracelsus, *Concerning the Tincture of the Philosophers,* compiled and transcribed by Dusan Djordjevic Mileusnic from *Paracelsus, his Archidoxis: Comprised in Ten Books, Disclosing the Genuine Way of making Quintessences, Arcanums, Magisteries, Elixirs, &c., Together with his Books Of Renovation & Res-*

Paracelsus was anything but modest. To drive his point home, he points a definite finger to those whom he regards as the principal corruptors of the ancient craft, and they are the two patrons we have already encountered: "…that sophistical science has to have its ineptitude propped up and fortified by papal and imperial privileges."[58] Briefly put, Paracelsus knew full well where the main mechanisms that accounted for alchemy's survival, and corruption, during the Middle Ages ultimately lied.

(1) On the Paleoancient Very High Civilization and Egypt

Part of the uniqueness of Paracelsus is that he was quite aware, and made no secret, of the implications of alchemy for human history and culture. In his book *Concerning the Tincture of the Philosophers,* he writes that "I have proposed by means of this treatise to disclose to the ignorant and inexperienced: *what good arts existed in the first age…*"[59] That Paracelsus means by "first age" something predating Egypt and Sumer will be apparent in a moment.

In any case it is to its possession of the secrets of alchemy that Paracelsus ascribes ancient Egypt's power: "If you do not yet understand from the aforesaid facts, what and how great treasures these are, tell me why no prince or king was ever able to subdue the Egyptians."[60] But the alchemical science itself he clearly states came from Adam, whom he calls "the first inventor of arts, because he had knowledge of all things as well after the Fall as before."[61] Because of this plenitude of knowledge, Adam was also able on Paracelsus' view to predict

> the world's destruction by water. From this cause, too, it came about that his successors erected two tablets of stone, on which they engraved all natural arts in hieroglyphical (sic) characters, in order that their posterity might also become acquainted with this prediction, that so it might be heeded, and provision made in the time of danger. Subsequently, Noah found one of these tables under Mount Araroth, after the Deluge. In this table were described the courses of the upper firmament and of the lower globe, and

---

tauration, of the Tincture of the Philosophers; Of the Manual of the Philosophical Medicinal Stone; Of the Virtues of the Members; Of the Three Principles; and Finally his Seven Books; of the Degrees and Compositions of Receipts, and Natural Things, Faithfully and plainly Englished, and Published by J.H. Oxon, London, 1660.

58   Ibid.
59   Ibid.
60   Ibid.
61   Paracelsus, *The Aurora of the Philosophers,* from *Paracelsus, his Aurora & Treasure of the Philosophers, As also The Water-Stone of the Wise Men; Describing the matter of, and manner hot to attain the universal Tincture. Faithfully Englished, and Published by J.H. Oxon.* London, Giles Calvert, 1659. Transcribed by Dusan Djordevic Mileusnic.

also of the planets. *At length this universal knowledge was divided into several parts, and lessened in its vigour and power. By means of this separation, one man became an astronomer, another a magician, another a cabalist, and a fourth an alchemist.* Abraham, that Vulcanic Tubalcain, a consummate astrologer and arithmetician, *carried the Art out of the land of Canaan into Egypt, whereupon the Egyptians rose to so great a height* and dignity that this wisdom was derived from them by other nations.[62]

Paracelsus makes two very crucial and significant observations here, whose subtle importance may be overlooked since they are so apparently obvious, but only *apparently*:

1) That alchemy itself is only a *fragment* of a once larger, highly unified body of knowledge and science that also encompassed astronomy and, for want of a better word, "metaphysics," or, to put it differently, *hyper-dimensional* physics, a physics *beyond* (meta, from the Greek μετα, meaning "beyond") the ordinary physical and natural world (physics, from the Greek φυσις, meaning "nature" or, in this case, the natural world). In other words, Paracelsus views the Deluge itself as yet another "Tower of Babel Moment" in which an ancient highly unified scientific-religio-philosophical world-view was further fragmented in what could be taken as a classical guerrilla warfare operation, just like the Tower of Babel: massive interference with an enemy's communications and science and decision-making processes.[63] Paracelsus is, in other words, thoroughly familiar with the esoteric traditions on this point, and more aware than most of their massive implications, for he states them clearly: prior to the high civilizations of classical antiquity — Egypt and Sumer — there was something even higher and much more sophisticated. A little later on Paracelsus even elaborates this concept a bit more fully, attributing the fragmentation of knowledge to the political fragmentation that occurred after the Deluge: "When a son

---

62   Paracelsus, *The Aurora of the Philosophers*, from *Paracelsus, his Aurora & Treasure of the Philosophers, As also The Water-Stone of the Wise Men; Describing the matter of, and manner hot to attain the universal Tincture. Faithfully Englished, and Published by J.H. Oxon.* London, Giles Calvert, 1659. Transcribed by Dusan Djordevic Mileusnic, emphasis added.

63   For more on the Tower of Babel Moment viewed in this "paleophysical" manner, see my books *The Giza Death Star Destroyed: The Ancient War for Future Science* (Kempton, Illinois: Adventures Unlimited Press, 2005), pp. 77–78; *The Cosmic War: Interplanetary Warfare, Modern Physics, and Ancient Texts* (Kempton, Illinois: Adventures Unlimited Press), pp. 210–212. For the Deluge and its centrality in the esoteric tradition's view of ancient lost science, see my *Giza Death Star Destroyed*, pp. 53–96.

of Noah possessed the third part of the world after the Flood, this Art broke into Chaldaea and Persia, and thence spread into Egypt."[64]

2) That Egypt *gains* its knowledge from Sumer, that is, that there is a relationship between the two *to* that paleoancient Very High Civilization. Paracelsus, of course, disguises this relationship by a "biblical" reference to Abraham and his journey from Ur in Sumer to Canaan, and thence, via his descendants, into Egypt. But of course, the biblical reference is a mere pietism, for Paracelsus cannot have missed the fact that Egypt was already *there* by the time of their arrival, and cannot have missed the significance of Moses' growth in the Egyptian royal court. As indicated above, Paracelsus really attributes the further fragmentation of knowledge to the post-flood political fragmentation of the world. *He is suggesting, in other words, that the fragmentation of knowledge is threefold: the Chaldaeans (Sumerians) preserving the predominantly astrological and astronomical component, the Hebrews preserving the predominantly Cabalistic component, and the Egyptians preserving the alchemical component.* Thus, the "triune Stone" is also a triune Stone of a lost science, fragmented into three scientific and political cultures of the ancient classical world.

Consequently, Paracelsus uses the alchemical symbol of the "triune Stone" as a sigil of his entire philosophy of a hidden history of science and a paleoancient Very High Civilization predating Egypt and Sumer.

(2) On the Relation Between Astronomy and Alchemy

With Paracelsus' views on ancient history and science in hand, one may more readily appreciate why he not only insists that there are many valid methods to achieving the alchemical goal,[65] but also why he insists upon "the agreement of Astronomy and Alchemy."[66] Indeed, Paracelsus was to note — with some frustration — that his many attempts to perform the same experiment were sometimes successful, and sometimes not, *depending upon the season or other temporal factors.* Bear this point in mind, for it will become

---

64 Paracelsus, *The Aurora of the Philosophers,* from *Paracelsus, his Aurora & Treasure of the Philosophers, As also The Water-Stone of the Wise Men; Describing the matter of, and manner hot to attain the universal Tincture. Faithfully Englished, and Published by J.H. Oxon.* London, Giles Calvert, 1659. Transcribed by Dusan Djordevic Mileusnic, emphasis added.

65 Paracelsus, *Concerning the Tincture of the Philosophers,* compiled and transcribed by Dusan Djordjevic Mileusnic.

66 Ibid.

extraordinarily crucial in part two, for he anticipated, by no less than four centuries, similar observations made meticulously over several years and in several ways, by a brilliant Russian physicist.

(3) Crystals and Metals: Sapphires and Mercury

So precisely does Paracelsus echo some ancient views of space — and anticipate some very modern ones — that some of his words will seem, from any standpoint, ancient *or* modern, rather breathtaking:

> To conjure is nothing else than to observe anything rightly, to know and to understand what it is. *The crystal is a figure of the air.* Whatever appears in the air, movable or immovable, *the same appears also in the speculum or crystal as a wave.* For the air, the water, and the crystal, so far as vision is concerned are one, like a mirror in which an inverted copy of an object is seen.[67]

At first reading it would appear that Paracelsus is talking about nothing more than eyeglasses and mirrors, and indeed, on one level, he is. Moreover, he is advancing a wave theory of light that at that time is beginning to become part of optical study.

But such a prosaic reading would miss the true significance of his observations, for "air" oftentimes functions, particularly in alchemical literature, as yet another code name for "space" itself, and sometimes even for the *materia prima*. Consequently, his association of crystals with their implied latticework to *air*, that is to *space itself*, implies a very modern, and indeed topological, view of space being advanced by some modern physicists and topologists.[68] Given his familiarity with biblical texts, and his penchant for interpreting them as referring to a lost paleophysics, Paracelsus could hardly have been oblivious to the biblical reference in Ezekiel which seems to refer to space itself as a crystal, as a lattice structure: "And the likeness of the firmament upon the heads of the living creature was as the colour of the terrible crystal, stretched forth over their heads above."[69]

Paracelsus also, somewhat unusually, couples the idea of crystals and gemstones with metals.[70] Metals, like the more familiar gemstones and crystals, also possess a regular lattice structure, so once again, Paracelsus' views are, in their own way, very advanced for the day. Most importantly, in yet another rather obscure

---

67 Paracelsus, *Coelum Philosophorum*.
68 See my *Giza Death Star Destroyed*, pp. 196–221, referring to the work of Ukrainian physicist Volodymyr Krasnoholovets and topologist Michel Bounias. Krasnoholovets, let it be noted, was also involved in the former Soviet Union's secret work on studying the mysterious power of pyramids.
69 Ezekiel 1:22.
70 Paracelsus, *Coelum Philosophorum*, 12.

work, the *De Elemente Aquae,* Paracelsus states that "in the matter of body and colour the Sapphire is generated from Mercury (the prime principle)."[71] This mention of Sapphire in connection with the philosophical Mercury or *materia prima* is not without its own contemporary significance, for the former Soviet Union undertook experiments to measure minute fluctuations of the Earth's gravitational acceleration by means of a large artificial sapphire.[72]

By associating crystals with the Philosophical Mercury and the *materia prima* and implying that alchemy and the Philosophers' Stone is a technology with the power to manipulate the latter, Paracelsus is also implying that the lattice structure of crystals and of space itself is intimately related to physical forces.

So what do we have?

Paracelsus has outlined the following revolutionary ideas:

1) That an ancient and highly unified science was fragmented by various means, including political, as a consequence of the Deluge which he interprets as yet another "Tower of Babel Moment," and that these fragmentations went, in their astrological-astronomical, cabalistic, and alchemical components to Sumer, the Hebrews, and Egypt, respectively;
2) That some alchemical operations are successfully performed only at certain times and seasons, anticipating by some four centuries the views that will be more fully explored in a modern physics context in part two;
3) That there is an intimate relationship between the lattice structure of crystals and metals and the primary transmutative medium itself, implying that lattice structures and physical forces are intimately connected, a very modern idea more fully explored in a modern physics connection in part three; and,
4) That there is a metaphysical, or to put it into modern terms, a hyper-dimensional physics component always at work throughout all of the above.

But there's more, and in uncovering it, we once again see the profound relationship between the Augustinized formulary of the Trinity, its deep

---

71 Paracelsus, *De Elemente Aquae,* Lib IV, Tract IC, c. 15.
72 See my *Giza Death Star* (Kempton, Illinois: Adventures Unlimited Press, 2001) pp. 267–271. The sapphire recurs in peculiar contexts throughout the esoteric tradition, often in contexts suggesting precisely the connection to space-time itself and the ability to manipulate them. Cf. also my *Giza Death Star Destroyed,* pp. 65–68, 247–263, particularly pp. 259–260. For the association of stones and crystals with a wider weapons technology and the ancient cosmic war myths, see my *The Cosmic War: Interplanetary Warfare, Modern Physics, and Ancient Texts* (Kempton, Illinois: Adventures Unlimited Press, 2007), pp. 204–273.

and unsuspected roots in ancient Hermeticism (and therefore in an ancient paleophysics), the textual metaphors of topological triangulation in ancient Hermetic texts, and the triune Philosophers' Stone.

(4) And the Augustinized Trinity and Alchemy

The connection begins with Paracelsus' statement that he "will teach you the tincture, the Arcanum 1, the *quintessence*, wherein lie hid the foundations of all mysteries and of all works."[73] Again, file that reference to "quintessence" in the back of your mind, for it will play a significant role in deciphering the possible influences at work in the thinking of a brilliant modern theoretical physicist in part three. According to Paracelsus, this arcane operation is to be accomplished by means of the Holy Spirit. But once one delves into what he says about this Holy Spirit, one perceives the connection:

> This is the Spirit of Truth, which the world cannot comprehend without the interposition of the Holy Ghost, or without the instruction of those who know *it*. The same is of a mysterious nature, wondrous strength, boundless power. The Saints, from the beginning of the world, have desired to behold *its* face. By Avicenna this Spirit is named *The Soul of the World*. For, as the Soul moves in all the limbs of the body, so also does this Spirit move all bodies. And as the Soul is in all the limbs of the Body, so also is this Spirit in all elementary created things.[74]

The reference to the Holy Spirit as an *It* rather than a fully fledged Person, a *Him,* is one of those consequences of the Augustinized Trinitarian formulary that began to influence Western Christian piety during the Middle Ages and down to our own day.[75]

But the real clue lies in the identification of the Holy Spirit of Christian doctrine with the World Soul of Neoplatonism, for in the latter conception, this World Soul is made to take its origin from the Intellect, which in turn takes its origin from the One. The World Soul is thus caused by two classes of causes, the Uncaused Cause (the One), and the Caused Cause (the Intellect), much like the Holy Spirit, on the Augustinian formulation, takes His origin from an Uncaused Cause (the Father) and a caused cause (the Son). And note, in the Paracelsan version, it is this World Soul-Holy Spirit that is essential to

---

73   Paracelsus, *The Book Concerning the Tincture of the Philosophers*, 4.
74   Paracelsus, *The Apocalypse of Hermes*, 14.
75   Consider only the fact that, in spite of their professed "trinitarianism" most Western Christians only invoke the Father and the Son in their popular prayers and piety, in a kind of de facto functional *binitarianism*. Such a manner of prayer is quite foreign to the Orthodox Christian East, which is consistently Trinitarian in its pieties.

a purely physical alchemical work. In other words, in Paracelsus' hands the Augustinized Trinity, which began as was seen in a purely physical and topological metaphor in the *Hermetica,* and after its peregrinations throughout the Western Middle Ages as a theological doctrine and even as a dialectical interpretation of history itself, ends once again by being reduced back into a purely physics-related phenomenon. No better words can summarize this historical process than the words of Thomas Aquinas himself on the doctrine of the Holy Spirit's double procession within the Western Trinitarian doctrine, for in the hands of Paracelsus, the historical "cycle has concluded when it returns to the very same substance from which the proceeding began,"[76]

This is made more apparent in the following passage: "Eye hath not seen," says the alchemical author of the *Apocalypse of Hermes,* "nor hath ear heard, nor hath the heart of man understood what Heaven hath *naturally incorporated with this Spirit.*"[77] In other words, once reduced to an *It,* the "Spirit" becomes merely the sigil for the Philosophers' Stone itself, and *its embodiment into physical matter*, into the Stone, which has both physical and metaphysical or hyper-dimensional components. One can see the logic of Paracelsus (and other Western Christian alchemists) at work, for the alchemical work of confecting the Philosophers' Stone by embodying "the Spirit" into matter is the alchemical analogue to the Incarnation, where the Son becomes incarnate. The Son becomes human, the Spirit becomes the matter of the Stone.[78]

Thus, since this "Spirit"-as-transmutative physical medium is the quintessence of the Philosophers' Stone, the Stone in turn confers not only longevity and even a kind of immortality, but it is also *indestructible.*[79]

That Paracelsus seems to be quite knowledgeable of the ultimately Hermetic roots of the Augustinized Trinitarian formulary is made abundantly clear in *The Aurora of the Philosophers:*

> Magic, it is true, had its origin in the Divine Ternary and arose from the Trinity of God. For God marked all His creatures with this Ternary and engraved its hieroglyph on them with His own finger. Nothing in the nature of things can be assigned or produced that lacks this magistery of the Divine Ternary, or that does not even ocularly prove it. The creature teaches us to

---

76 Thomas Aquinas, *Summa Contra Gentiles,* IV *Salvation.*
77 Paracelsus, *The Apocalypse of Hermes,* 15.
78 That this is precisely the analogy is made clear from the passage which immediately follows that just cited, for it states "Therefore have I briefly enumerated some of the qualities of this Spirit...and I will herewith shew what powers and virtues it possesses in each thing, also its outward appearance, that it may be more readily recognized. In its first state, it appears as an impure earthly body..." and so on. (Paracelsus, *The Apocalypse of Hermes,* 15). Note again the reference to the Spirit as an *it,* i.e. as the transmutative physical medium itself.
79 Paracelsus, *The Apocalypse of Hermes,* 15, 16.

understand and see the Creator Himself, as St. Paul testifies to the Romans.[80] This covenant of *the Divine Ternary, diffused throughout the whole substance of things, is indissoluble*. By this, also, we have the secrets of all Nature from the four elements. For the Ternary, with the magical Quaternary, produces a perfect Septenary, endowed with many arcane and demonstrated by things which are known.[81]

Even the reference to the "divine quaternary" has its echo in the Augustinized Trinitarian formulation, for in the Trinitarian shield, one counts not three, but *four* circles, with the center circle representing the undifferentiated medium, the absolute simplicity, of the divine essence itself, which in the hands of the alchemists, has become a symbol of the Philosophical Mercury, the *materia prima* itself.

The previous reference to Neoplatonism's World Soul recalls yet another Neoplatonic theme that enters heavily into alchemical texts, and Paracelsus in particular. This is the doctrine that as all things come from the One, a process called emanation (πρooδoς in the Greek), so everything returns to It (περιαγωγη). The alchemical text, *The Aurora of the Philosophers*, having outlined the derivation of all from the Divine Ternary, thus reverses the process:

> Here also it refers to the virtues and operations of all creatures, and to their use, since they are stamped and marked with their arcane, signs, characters, and figures, so that there is left in them scarcely the smallest occult point which is not made clear on examination. Then when the Quaternary and the Ternary mount to the Denary *is accomplished their retrogression or reduction to unity.*[82]

But note that the Neoplatonic metaphor of return occurs in an alchemical context. In other words, the conception of return is itself in turn a metaphor for the alchemical operation of confecting the Philosophers' Stone, for *back-engineering the "topological descent" from the primary physical medium.* Paracelsus has seen through all the centuries' misunderstandings of such metaphysical texts, and their misunderstanding precisely as "philosophical" texts, to the underlying metaphor of a hidden and occulted *physics*. If this seems a constrained reading of the passage, he himself makes it abundantly clear a little further on:

---

80 Romans 1:21. Thus, Paracelsus understands the verse, not even in the sense of the Mediaeval scholastics, as referring to a program of natural *theology*, but to a program of natural alchemy and *magic*.
81 Paracelsus, *The Aurora of the Philosophers,* 6.
82 Paracelsus, *The Aurora of the Philosophers,* 6, emphasis added.

> The Magi in their wisdom asserted that all creatures might be brought to one unified substance...by many industrious and prolonged preparations, exalted and raised up above the range of vegetable substances into mineral, above mineral into metallic, and above perfect metallic substances into a perpetual and divine Quintessence.[83]

Thus, as "emanation" functions as a metaphor for the differentiation of the *materia prima* giving rise to and accounting for the diversity of creatures, so "return" functions, for the alchemist "in the know," as a metaphor for the process of "back-engineering" that topological descent in order to confect and embody the medium more immediately itself in a creature, in the Philosophers' Stone.

Whatever else must be said of Paracelsus, one thing perhaps goes frequently unnoticed. If the old Mediaeval adage be true that "philosophy is the handmaiden of theology," then Paracelsus has seen through the Augustinized Trinitarian formulary to its ultimately Hermetic roots, and alchemically reduced it once again to its original basis in physics and "sensory things" shorn of the theological associations it subsequently came to hold. With that, he has performed the irrevocable divorce of theology and philosophy as it obtained in the Christian West.

b. The Ultimate Reductio and the Clincher

If all this is not enough to convince the most resistant skeptic, there is an alchemical text where the identification of the Christian Trinity, in its Augustinzed form, with the topological metaphor of the *Hermetica* of Hermes Trismegistus is clearly made, and the reduction to physics is accomplished in no uncertain terms. This is a small manuscript entitled simply "*Place in Space, the Residence of Motion.*" The subtitle of this is even more provocative: "*The Secret Mystery of Nature's progress, being an Elucidation of the Blessed Trinity, Father — Son — and Holy Ghost. Space — Place — and Motion.*"[84] Notably, the same sort of functional sets of dialectical oppositions are used to distinguish each vertex as were previously shown in the citations from the *Libellus II* of the *Hermetica* of Hermes Trismegistus:

> Space, Place, Father & Son are inseparable fixed & immoveable. Motion ye Holy Ghost is that which brings all things to the Blessed determination of the Dei, as in the Floria Patri, Filii & Spiriti Sancti [sic], etc.[85]

---

83 Ibid., 7.
84 Transcribed by Adam McLean, Sloane MS 3797, folios 3–5, at www.levity.com/alchemy/place_in_space.html, p. 1.
85 Ibid.

In other words, the Spirit is distinguished by functional opposition — motion — from the Father and the Son, each of whom are symbols of Space and Place. The reduction to physics is once again in evidence.

But there is yet more to the mystery of alchemy...

The Indestructible Stone
The Persistence of Alchemy and the Powers of the Philosophers' Stone

•••

"No fire or other element can destroy it. It is also no stone, because it is fluid, can be smelted and melted."
Ruland[1]

A. Stories of Alchemical Success

The persisting mystery of alchemy is precisely its persistence. That is, if the possession of the tremendous powers of the Philosophers' Stone was the goal of alchemy, why then did no one notice, over the several centuries if not thousands of years of its practice, that no one ever actually *did* it? How does one account for the *persistence of the practice* in the face of, presumably, the centuries' accumulated mountain of failures? Ancient and mediaeval man was no less rational than modern man, and repeated failures would, eventually, have led rational people to abandon the whole enterprise as a futility. How does one account for *the persistence* of alchemy through the centuries when faced with the enormity, and the likely *futility*, of its quest to confect the Philosophers' Stone?

Should not the whole enterprise have been abandoned much sooner than actually was the case? One possibility of explaining this persistence is,

---

[1] Ruland, *On the Materia Prima*, from *Lexicon alchemiae sive dictionarium alchemisticum, cum obscuriorum verborum, et rerum Hermeticarum, tum Thephrast-Paracelsicarum phrasium, planam explicationem coninens* (Frankfurt, 1612), transcribed by John Glenn, www.levity.com/alchemy/ruland_e/html, p. 1

of course, that the extremity of the power of the Stone itself was a sufficient motivation to continue pursuing it in spite of the presumably large amount of data that said it could not be done.

But there is another possibility, before which one hesitates, given our inbred modern skepticism concerning all that is "extraordinary." That possibility is that, on occasion, for whatever reason, there *were* some successes, that they actually *did it*. In fact, this is what is actually claimed in the historical record, if one reads such accounts at face value. Only a few of the many examples that Holmyard cites in his book are reproduced here.

1. *The Swedish General*

An intriguing story comes out of eighteenth-century Sweden, when the Scandinavian kingdom was at the height of her power. In 1705 the Swedish general Paykhull

> had been convicted of treason and sentenced to death. In an attempt to avert this punishment, he offered the king, Charles XII, a million crowns of gold annually, saying that he could make it alchemically; he claimed to have received the secret from a Polish officer named Lubinski, who had himself obtained it from a Corinthian priest. Charles accepted the offer, and a preliminary test was arranged under the supervision of a British officer, General Hamilton of the Royal Artillery, as an independent observer. All the materials were prepared with great care, in order to prevent the possibility of fraud, then Paykhull added his elixir and a little lead, and a mass of gold resulted which was coined into 147 ducats. A medal struck at the same time bore the inscription (in Latin as usual): 'O.A. von Paykhull cast this gold by chemical art in Stockholm, 1706."[2]

One has difficulty imagining any alchemical process as being successful. But equally, one has difficulty imagining any king, especially King Charles XII, and a British artillery general, being taken in by a charlatan, especially when the charlatan is on trial for his life, and when Charles XII had potential riches to gain. The Swedish royal courtesans and ministers would have literally hovered over General von Paykhull, scrutinizing his every move.

2. *A Provincial Frenchman*

Another interesting story is mentioned by Holmyard, this time from provincial France, and again, from the early eighteenth century. On this occasion

---
2 Holmyard, *Alchemy*, pp. 132–133.

An ignorant Provencal rustic named Delisle caused a sensation by claiming, with apparent good reason, that he could transform iron and steel into gold. The news came to the ears of the Bishop of Senez, who after witnessing one of Delisle's experiments wrote to the Minister of State and Comptroller-General of the Treasury in Paris that he could not resist the evidence of his senses. In 1710 Delisle was summoned to Lyons, where, in the presence of the Master of the Lyons Mint, he made much show of distilling some unknown yellow liquid. He then projected two drops of the liquid upon three ounces of pistol bullets fused with saltpeter and alum, and poured the molten mass out on to a piece of iron armour, where it appeared as pure gold, withstanding all tests. The gold thus obtained was coined by the Master of the Mint into medals inscribed *Aurum arte Factum*, 'Gold made by Art,' and these were deposited in the museum at Versailles. Of Delisle's subsequent life, history has nothing to relate.[3]

Again, we find the association with a Royal government, this time that of France, and in connection with its Treasury and Royal Mint. And again, reason pauses to make us hesitate over the alleged success. But similarly, reason also forces one to consider that the possibility of a fraud being perpetrated under the watchful eyes of the Master of the Lyon mint is rather slim, especially since the gold that was produced was subjected to "all tests." Was Delisle's success the reason for his subsequent disappearance? Or, conversely, did the Versailles Palace, subsequently determining that it had been defrauded, "disappear" Delisle into the bowels of the Bastille? We will never know.

3. The Hapsburg Emperors Ferdinand III and Leopold I

One of the best known examples of royal or imperial patronage of alchemical practice was the Holy Roman Hapsburg Emperor Ferdinand III, of which no less than four examples are known. The first concerns an incident in 1647, when an alchemical adept name J.P. Hofmann allegedly successfully performed a transmutation in the presence of Ferdinand himself in the city of Nuremberg.

From this hermetic gold the emperor caused a medal of rare beauty to be struck. It bears on the obverse two shields, in one of which are eight fleurs-de-lys, and in the other is a crowned lion. The Latin inscriptions signify, 'The yellow lilies lie down with the snow-white lion; thus the lion will be tamed, thus the yellow lilies will flourish; and that the metal was made by

---
3   Holmyard, *Alchemy*, p. 133.

Hofmann. A further inscription reads *Tincturae Guttae V Libram*, denoting that five drops of the tincture of elixir transmuted a whole pound of the base metal. On the reverse is a central circle with Mars in it, holding the symbol ♂ in one hand and a sword in the other. Around this central circle are six smaller ones, containing the signs of gold, silver, copper, lead, tin, and mercury, with an inscription claiming that in this case the active agent in transmutation was made from iron.[4]

While this is not the last time we shall encounter an alchemical connection to Mars, it is important to note here that once again, the Philosophers' Stone is not so much an actual *stone* as it is an elixir or tincture, a *liquid*.

The next year Ferdinand was at it again, this time in connection with

> A certain Richthausen, who claimed to have received the secret of the Art from an adept now dead, (who) performed a transmutation in the presence of the emperor and of the Count von Rutz, director of mines. All precautions against fraud were taken, yet with one grain *of the powder* provided by Richthausen two and a half pounds of mercury were changed into gold.[5]

Once again, the Emperor Ferdinand had a medallion struck with a value of "300 ducats," and whose Latin inscription stated that "the Divine Metamorphoses" had been "exhibited at Prague, 15 January 1648, in the presence of his Imperial Majesty Ferdinand III."[6]

The mysterious Richthausen was active again in 1650, for in that year Ferdinand apparently performed his own transmutation with some of the adept's powder, and once again struck a medallion indicating that lead had been successfully transformed into gold.[7] The Emperor's patronage became "official" when, following another successful transmutation in 1658 when Richthausen gave the Elector of Mainz some of the Stone, a transformation of mercury into gold occurred. In gratitude, Ferdinand raised Richthausen to the ranks of the nobility.[8]

Nor did the Hapsburg interest end with Ferdinand. Ferdinand's son Leopold I was visited in Vienna by an Augustinian monk in 1675, where, again, a "copper vessel" was transmuted into gold, along with some tin. In commemoration of the occasion, gold ducats were struck from the tin. The monk, Wenzel Seyler, had accomplished the whole operation with a powder, for on the obverse of these coins was a bit of poetry:

---

4   Holmyard, *Alchemy*, p. 129.
5   Ibid., emphasis added.
6   Holmyard, *Alchemy*, p. 129.
7   Ibid.
8   Ibid.

Aus Wenzel Seyler's Pulvers Macht
Bin ich von Zinn zu Gold Gemacht.

That is, 'By the power of Wenzel's powder I was made into gold from tin.'[9]

Beyond the patronage and consistent interest in alchemy of the powerful von Hapsburg dynasty, what is of interest in these accounts is that the Stone appears in two now familiar guises: as a *powder,* and as a *liquid.* As will be seen subsequently, these are important clues as to what the substance may actually have been. What also interests us is the apparent frequency with which Ferdinand's adept, (von) Richthausen, was supposedly able to effect the transformation. Not just once, but no less than four times, and on one occasion with the Emperor himself performing the operation. Again, one is confronted with two horns of a very significant problematic: 1) either the Emperor was an incredibly stupid man and easily duped, a very unlikely possibility since, again, Richthausen's actions would have been carefully monitored and scrutinized by the imperial court, or 2) they actually occurred, which is problematical for obvious reasons.

B. The Alchemical Reading of Particular Texts

Throughout the foregoing stories of alchemical successes, we were once again confronted with another aspect of the "dual" nature of alchemy and its ambiguous claims and language, for at almost every turn two possible explanations — fraud or actual transmutation — presented themselves, *each* with its own set of irrationalities and rationalities. Not only can *we* read these accounts in two ways, *so could the alchemists themselves.* In fact, they could read almost *any* text concerning transformations in an alchemical way.

1. Ovid's Metamorphoses

We have already repeatedly encountered numerous biblical allusions and references in our alchemical survey. But these are not the only kinds of texts in which alchemists were able to discover metaphors for their art.

A famous case in point is the classical Roman poet Ovid, and his celebrated epic poem *The Metamorphoses.* The fourteenth-century alchemist Petrus Bonus, whom we have encountered before, "claimed that Ovid's *Metamorphoses* dealt esoterically with the philosophers' stone, and that many other ancient poems and myths had hidden alchemical meanings."[10] In other words, *the alchemists*

---

9 Ibid., p. 130.
10 Holmyard, *Alchemy*, p. 148.

themselves believed that ancient texts of a certain type contained a hidden and encoded physics, the hidden and encoded physics of alchemy itself, which was thus an attempt to recover a lost technology and science.

2. The Nine Philosophers, The Enneads, The Council of the Gods, and the Egyptian Neters

A more complex example of the alchemical reading of ancient texts *and motifs* occurs in a tenth-century Muslim alchemical text known as *The Book of the Controversies and Conferences of the Philosophers*, compiled by the Muslim alchemist Uthman ibn Suwaid of the town of Akhimm, in other words, *Panopolis in Egypt*.[11] In the Latin version of this work,

> Nine philosophers take part in the preliminary discussion, with the names Iximidrus, Exumdrus, Anaxagoras, Pandulfus, Arisleus, Lucas, Locustor, Pitagoras, and Eximenus. 'Anaxagoras' and 'Pitagoras' seem to indicate that the remaning seven names are mistransliterations of Greek names, and by transcribing them back into Arabic characters Plessner was able to show that the list should read Anaximander, Anaximenes, Anaxagoras, Empedocles, Archelaus, Leucippus, Echpantus, Pythagoras, and Xenophanes — thus solving an age-old mystery.
>
> These nine philosophers are all pre-Socratic, and Plessner demonstrates that in their speeches (in the work) they are reciting theories that, from classical sources, they are known to have held.[12]

But why *nine* philosophers?

Beyond the work's own obvious Egyptian provenance, there are deeper Egyptian esoteric connections to be considered, for one has only to recall that the famous Neoplatonic philosopher Plotinus, regarded by many as the actual *founder* of the school,[13] wrote his version of the unfolding differentiation of creation from an underlying undifferentiated, or "simple" (απλως) medium of The One (το εν) and The All (το παν) in his well-known philosophical treatise *The Enneads*, or quite literally, "The Nines." And this in turn, according to many interpreters, is but a "public popularization" of the ancient nine primordial Egyptian *neters* or "gods,"[14] the first nine differentiated entities that, in the Egyptian cosmogenesis, unfold from the first act of differentiation

---

11   Holmyard, *Alchemy*, p. 83.
12   Ibid., p. 84.
13   Ammonius Saccas, teacher of Plotinus and of the early Christian writer Origen, being the other.
14   The term "gods" here is often used to translate the Egyptian term *neter*, though it is more likely the case that the term should be translated by the English word "natures" or even by the phrase "differentiated entity."

of the prime matter in the primary scission.[15] In point of fact, the first nine Egyptian *neters*, and indeed the whole *concept* of *neters* in the Egyptian cosmology is rather more involved than the mere translation of the term as "gods" would suggest, for as the celebrated "alternative Egyptologist" René Schwaller De Lubicz pointed out, the term actually implies an exact science of the nature of cause and effect:

> *There exists a bond between cause and effect, and that bond is called the Neter.* It is this *Neter* which the believer hopes to spur into action through his appeal. Further, what this *Neter* represents **as energetic and harmonic activity** is consciously evoked by the sage. This conscious evocation must necessarily be a gesture or a word of the same nature as that of the *Neter* summoned, and by that fact the evocation becomes *the cause of the magical effect.*[16]

Bear in mind this idea of an exact science of cause and effect, as "energetic and harmonic activity," for it will assume great importance not only in our subsequent explorations of alchemy in this chapter, but also in our examination of the modern Russian "alchemical" work in Part Two.

In any case, it is now apparent that the whole alchemical enterprise is suffused through and through with connections to ancient Egyptian cosmological views, views which were operative in their understanding of sympathetic magic, and that gave rise to the more sophisticated, "scientific," and "technological" attempts of alchemy to recover the science of the ability to manipulate the *materia prima*.

3. *The Alchemical Reading of the Episode of the Golden Calf in the Old Testament*

But there is one more alchemical association with, and reading of, an ancient text that in turn has an Egyptian connection, and that, of course, is the Episode of the Golden Calf from the Old Testament biblical book of Exodus. I first mentioned this connection in my book *The Giza Death Star Destroyed,* and it is appropriate to recall here what I said:

> There are numerous biblical references to being "purified" as for example, phrases that attribute a purifying action "like a refiner's fire" to God.... Alchemical "transmutation" of lead into gold, the "Philosophers' Stone" was also understood to be a purification process for the soul. Thus, such references in the Bible often have a clear though esoteric meaning within

---

15 Q.v. R.A. Schwaller De Lubicz, *Sacred Science: The King of Pharaonic Theocracy* (Rochester, Vermont: Inner Traditions International, 1988), pp. 153-158.
16 Ibid., p. 153, italicized emphasis in the original, bold and italicized emphasis added.

alchemy. But is there any evidence to suggest that the material aspect of this transformation, the *technology* of transmutation itself, may once have existed?

For (Sir Laurence Gardner), the answer is an unequivocal "yes," and he points to the Book of Exodus as his evidence. The story is familiar to most of us. In Sinai, while Moses is on the Mount communing with God, Aaron melts down the gold of the Israelites and molds an idol, a "golden calf." Upon his return from the Mount, Moses became angry with this idolatrous act, and then, according to Gardner, "Performs a most extraordinary transformation."

"Exodus 32:20 explains: 'And he took the calf which they had made, and burnt it in the fire, *and ground it to powder, and strawed it upon the water, and made the children of Israel drink of it.*'

"In practice, this sounds rather more like a ritual than a punishment, even though the latter is how the story is conveyed. Aaron had previously melted the gold in the fire to mold the image, *but what Moses did was plainly different because firing gold produced molten gold, not powder.*"[17]

In the Septuagint Greek Old Testament version of the account, it is stated that Moses "consumed the gold with fire." This implies "A more fragmentary process than heating and melting. The *Oxford English Dictionary* defines 'to consume' as 'to reduce to nothing or to tiny particles.' So what is this process that, through the use of fire, can reduce gold to powder?"[18] Put differently, some very extreme heat must be applied beyond the melting temperature of gold in order for it to actually *burn* and be reduced to a particulate.[19]

Clearly Gardner is correct, and there is more involved in Moses' melting of the Golden Calf into a *powder* and consumable *liquid* than meets the eye of the normal pulpit exegete or standard biblical commentary, not the least of which is the fact that in order to reduce gold to a particulate would require *extremities of heat* that could only be obtained by some technology. More on this point below, for as will be seen, extreme heat was precisely one of the obsessions of the alchemists.

It is worth noting that according to Gardner, a third-century alchemical treatise entitled *The Domestic Chemistry of Moses* was actually the origin of the mediaeval alchemical practice of referring to their craft as "the Mosaico-hermetic art," and that Moses himself was considered to be an alchemist by mediaeval alchemical texts.[20]

---

17   Citing Sir Laurence Gardner, *Lost Secrets of the Sacred Ark*, pp. 13–14, emphasis added.
18   Citing Sir Laurence Gardner, *Lost Secrets of the Sacred Ark*, p. 14.
19   Joseph P. Farrell, *The Giza Death Star Destroyed*, pp. 159–160.
20   Gardner, op. cit., p. 84, cited in *The Giza Death Star Destroyed*, p. 161.

Finally, a curious reference to the powdered form of the Philosophers' Stone occurs in a short treatise of the mediaeval alchemist Glauber entitled "How to Make and Prepare the 'Atoms' of Gold," wherein the "atoms" referred to are said to be "most fine and subtle," and capable of medical uses.[21] This fine powder, which Glauber does not hesitate to call "the whole secret of the thing,"[22] is also of a transparent ruby color.[23] Bear this fine powder in mind, for it will assume great significance in the next chapter.

Yet another alchemical treatise likewise mentions the "ruby-red transparent glass" and that the production of the "calx" or powder of gold "is quite essential to success in the prosecution of our Art."[24]

C. Stone, Powder, Elixir, and the Sequence of Colors

Thus far we have encountered references to the Philosophers' Stone being of a threefold material nature; it exists as

1) a stone or mineral or metal of some sort;
2) a tincture, elixir, or liquid of some sort; and,
3) a powder or particulate material of some sort.

Additionally, the previous biblical references indicate that

4) high heat was involved in its confection;
5) some sort of sequence of colors — from white to red — was the signal of a successful operation.

More importantly, it has also been shown that

6) the Stone involved a manifest "bodily" aspect, and a latent or hidden aspect, the "soul" aspect, that appeared to tie it directly to the assumed properties of the underlying transmutative substrate or prime matter.

But, notes Holmyard, the details of such a simple outline of its properties

are scarcely as simple as this outline would suggest. It is first necessary to purge the original material of all that is thick, nebulous, opaque, and

---

21   *A Compendium of Alchemical Processes Extracted from the Writings of Glauber, Basil Valentine, and Other Adepts* (Kessinger, no date), pp. 33, 34.
22   Ibid., p. 37.
23   Ibid.
24   Ibid., p. 59.

dark in it... Then the extracted body, soul, and spirit must be distilled and condensed together by their own proper salt, yielding an aqueous liquid with a pleasant, penetrating smell, and very volatile. This liquid is known as *mercurial water or water of the Sun*. It should be divided into five portions, of which two are reserved while the other three are mixed together and added to one-twelfth their weight of the divinely endowed body of gold. Ordinary gold is useless in this connexion, having been defiled by daily use.

When the water and the gold have been combined in a solutory alembic they form a solid amalgam, which should be exposed to gentle heat for six or seven days. Meanwhile one of the two reserved fifths of the mercurial water is placed in an egg-shaped phial and the amalgam is added to it. Combination will slowly take place, and one will mingle with the other gently and imperceptibly as ice with warm water. This union the sages have compared to the union of a bride and bridegroom. When it is complete the remaining fifth of the water is added a little at a time, in seven instalments [sic]; the phial is then sealed, to prevent the product from evaporating or losing its odour, and maintained at hatching-temperature. The adept should now be on the alert for various changes. At the end of forty days the contents of the phial will be as *black as charcoal*: this stage is known as the raven's head. After seven days more, at a somewhat higher temperature, there appear granular bodies, like fishes' eyes, then a circle round the substance, which is first *reddish, then white, green, and yellow, like a peacock's tail, a dazzling white, and finally a deep red*. That marks the climax, for now, under the rarefying influence of the fire, soul and spirit combine with their body to form a permanent and indissoluble Essence, an occurrence that cannot be witnessed without admiration and awe. The revivified body is quickened, perfected, and glorified, and *is of a most beautiful purple colour; its tincture has virtue to change, tinge, and cure every imperfect body.* [25]

The process of confecting the Philosophers' Stone is thus a rather lengthy and complicated one, one rather more involved than merely superheating gold. The complexity of the process is disclosed by the fact that the spectrographic response of the material, the color sequence, is more than just a transition from white to red. In fact, in this version of the sequence, one has the following progression of colors:

1) Black, to
2) "reddish," to
3) white, to
4) green, to

---

25 Holmyard, *Alchemy*, pp. 17–18, emphasis added.

5) yellow, to
6) a dazzling white, and
7) deep red, to
8) "a most beautiful purple."

That this sequence is indicative of reactions of *some* sort — whether chemical or otherwise – cannot be doubted.

A slightly different sequence is recorded by Petrus Bonus:

> *At first, indeed, the whole mass is white because quicksilver predominates… when it is ferment, the mass in the second stage of the magistery becomes red in the fullness of the potential sense, while in the third stage, or the second and last decoction, the ferment is actively dominant, and the red color becomes manifest and possesses the whole substance.* Again, we say that this ferment is that strong substance which then turns everything into its own nature. Our ferment is of the same substance as gold; gold is of quicksilver, and our design is to produce gold.[26]

While Bonus' list is not nearly as detailed, nevertheless one again finds the beginning of the sequence denoted by white, and the end by red.

This basic sequence is confirmed by the alchemist Glauber in a treatise entitled "The True Tincture of Gold," once again indicating that the Philosophers' Stone is as much a liquid or powder as it is an actual stone. After taking one part gold, three parts mercury, and one part of silver, and putting them into a "philosophical vessel" wherein they are dissolved, Glauber then notes the color sequence:

> In the space of one-quarter of an hour these mixed metals will be radically dissolved by the mercury, and will give a purple colour. Afterwards increase the heat gradually, and then the colour changes to a very fine green. Hereon, when taken out, pour the "water of the dew," to dissolve (which may be done in half-an-hour). Filter the solution, and abstract the water through a glass alembic in balneo, which pour out again fresh and abstract; repeat this three times. In the meanwhile, that greenness will be turned to blackness, like ink, stinking like a carcass, and therefore odious. It behoves sometimes to take away the water, re-affused and digested when the said black colour and stench will disappear in the space of forty hours, and will give place to a pure and milky whiteness, which appearing, remove the water by evaporation to dryness. Diverse colours now appear,

---

26 Ibid., p. 148, emphasis added.

and the remaining white mass is to be affused with highly rectified spirit of wine, when the dissolved green gold will impart to it a quintessence as red as blood or ruby.[27]

While significantly different in some respects from other sequences, the broad signatures remain, with purple, black, white, and red being the salient stages.

Yet another treatise of Glauber notes that red is not only the signature of the final stage of the Stone's confection, but that in its new, powdered state, it "admits not any more melting, nor does it of itself return into a malleable metallic body."[28] In other words, at the final stage of the alchemical process, the "philosophical gold" has been so consumed and burnt that further melting simply is impossible. Finally, in what is surely one of Glauber's most curious series of statements, the "gold calx" or powdered gold appears to be produced "by the oil of sand" in a crucible exposed to a gentle heat.[29] Bear in mind this connection of the Philosophers' Gold to *soil,* for it will become very important in the future.

The most famous of the alchemists, Nicholas Flammel, however, recorded a sequence in many respects similar to the detailed sequence noted above, with but one or two significant exceptions. Holmyard summarized his efforts as follows:

> Although Flamel now had most of the necessary information, it took him three years of unremitting labour to achieve final success. The penultimate stage was reached: all that was left to be done was to heat the product in a glass flask of 'philosophical egg' set in an athanor. With beating heart, Flamel watched for the revealing colours. They came, *and in the correct sequence: from grey to black, 'the crow's head,' then from black to white, the white first appearing like a halo around the edge of the black, and the halo then shooting out white filaments towards the centre, until the whole mass was of a perfect white. This was the white elixir,* and Flamel could wait no longer: he opened the flask, took out the elixir, called (his wife) Pernelle, and prepared to make the trial. This was on Monday, 17 January 1382. Taking about half a pound of lead, he melted it in a crucible, and to it added a little of the white elixir, whereupon the lead was at once converted into silver, purer than the silver of mines.
>
> Sure at last that he had achieved mastery of the Art, he replaced the rest of the elixir in the flask and continued the heating. Now the rest of the colours appeared one after the other: *the white turned to the iridescence of the peacock's*

---

27   *A Compendium of Alchemical Processes*, pp. 23–24.
28   *A Compendium of Alchemical Processes*, p. 47.
29   Ibid., p. 28.

tail, this to yellow, the yellow to orange, the orange to purple, and finally the purple to red — the red of the Great Elixir. [30]

Here the sequence of colors is slightly different:

1) Gray, to
2) Black, to
3) White, to
4) "the iridescence of the peacock's tail" (blue?), to
5) Yellow, to
6) Orange, to
7) Purple, to
8) Red, the "red of the Great Elixir."

Notably, it is the last two components, the red and purple (in Holmyard's first version), and the purple and red (in Flammel's), which is different, which suggests that the final stage was represented by a reddish-purple, or almost maroon color.

The sequence of colors is, for Flammel as for other mediaeval alchemists, the apotheosis of the Philosophers' Stone. For Flammel, achievement of the Stone denotes the achievement of the key to all the sciences, the *materia prima* itself, for it is that "first agent" of all other creations "which is the *Key*, opening the gates of all *Sciences*."[31] Yet, while in many respects the clearest of the alchemists, Flammel, in the midst of detailing how he achieved the confection of the Philosophers' Stone, once again retreats into obscurity with the following passage:

> I must take heed to represent or write where it is that we hide the *keys* which can open all the doors of the secrets of nature, or to open or cast up the *earth*, in that place contenting myself to show the things which will teach every one to whom *God* shall give permission to know, what property the sign of the *Balance* or *Libra* hath, when it is enlightened by the *Sun* and *Mercury* in the month of *October*.[32]

Flammel seems to be suggesting once again that astrology plays some crucial and significant role in the successful confection of the Philosophers' Stone. *To*

---

30   Holmyard, *Alchemy*, p. 245, emphasis added.
31   Nicholas Flammel, *Hieroglyphical Figures: Concerning both the Theory and Practice of the Philosophers Stone* (Kessinger, reprint of the 1624 Walsey edition), p. 18, emphasis in the original. Flammel interprets the Ouroboros image, the serpent biting its tail, as a symbol of the prime matter and its transmutative properties (p. 21).
32   Ibid., pp. 40–41, emphasis in the original.

put it differently, he seems to be suggesting that celestial geometries are somehow capable of amplifying or damping the effects necessary to achieve the Stone.

The sequence of the colors red and purple is not unique to Mediaeval Latin alchemy. It occurs as well in the writings of the famous Muslim alchemist Al Jabir, who ascribed the power to effect all manner of transmutations to the "grand or master elixir."[33] The Muslim alchemist Dubai ibn Malik, moreover, records the properties of this "red elixir" as being able to effect transmutations of base metal into gold up to the amount of 500 times its own weight.[34]

The *Rosarium Philosophorum* produces a similar color sequence. It states that the Great Tincture "hath a clear and bright tincture in itself, white, red, pure, incombustible, stable and fixed, which neither the fire is able to change nor burnt sulphurs [sic] or sharp corrosives able to corrupt."[35]

D. The Claimed Properties of the Philosophers' Stone
1. Longevity and Healing

Intriguingly, it is when one considers the claimed properties of the Philosophers' Stone or "Great Elixir" that one is confronted with even stranger claims. Some of these have already been encountered, namely, the ability of the Stone to effect "cures" or healing, recalling the biblical account of Moses' requiring the Israelites to *drink* the refined gold from the Golden Calf. This claim is not confined solely to Occidental alchemy, for one finds similar claims for alchemical gold being able to prolong life in Chinese alchemy.[36] The famous Chinese philosopher Lao Tse himself composed a treatise entitled, evocatively enough, *Document Concerning the Three Similars*, whose central subject was "the preparation of the 'pill of immortality,' which was made from gold and which is so extremely efficient that it need be only very tiny."[37] Pills, obviously, suggest once again the other form of the Philosophers' Stone: powder.

Nor is this the only strange parallel to be found in Chinese alchemy, for there too the role of belief and the proper spiritual preparation of the operator was considered essential to the success of the confection. The alchemist Go-Hung put it succinctly: "Disbelief brings failure."[38] Even more suggestively, however, Go-Hung states that the alchemical confection must take place "on a famous great mountain, for even a small mountain is inadequate."[39] Is this yet another version of the association of Mountain with Planets with Pyramids, that

---

33   Holmyard, *Alchemy*, p. 78.
34   Ibid., p. 79.
35   *Rosarium Philosophorum*, 9.
36   Holmyard, op. cit, p. 34.
37   Ibid., p. 36.
38   Ibid., p. 37.
39   Ibid.

occurs so often in the ancient Sumerian texts?[40] It is interesting that China is home to some of the world's largest earthen pyramids, and that the Communist government of that country carefully restricts access to these sites.

Another curious reference to the powdered form of the Philosophers' Stone occurs in none other than Albertus Magnus, famous teacher to his even more celebrated student, the mediaeval scholastic philosopher and theologian Thomas Aquinas. Albertus Magnus stated that he had himself "tested alchemical gold and found that after six or seven ignitions it was converted into powder."[41] Yet another mediaeval monk-cum-alchemist, Roger Bacon, indicated that one goal of alchemy was "How to discover such things as are capable of prolonging human life for much longer periods than can be accomplished by nature."[42] Concisely put: from mediaeval China to mediaeval Europe, there is an alchemical consensus that one effect of the Philosophers' Stone is not only its ability to transmute base metals into gold, but also to effect healing and promote longevity.

2. Occupying No Space

In this respect one of the most unusual claims about the Great Elixir was made by the English alchemist John Dastin, a fourteenth-century alchemist in regular contact with Pope John XXII and Cardinal Orsini, which fixes his period of active writing on the subject rather exactly to the years 1316–1334, the years of John's pontificate, with a terminus *post quem* of 1342, the final year of Cardinal Orsini's life.[43] In Dastin's writings one encounters the usual references to preparation of the elixir in a clear glass flask through which one can observe the sequence of colors.[44] But with him there is a new claim for the Great Elixir: even though it must be "confined in some kind of matter, for otherwise it could not be manipulated, *it nevertheless occupies no space,*"[45] an idea, says Holmyard, that is echoed in the ideas of the great Renaissance alchemist Paracelsus. This claim is tantamount to stating that the Philosophers' Stone has hyper-dimensional properties, i.e., that some part of it exists in a space of higher dimensionality, another manifestation of its non-locality again.

It is not difficult to see where such an idea might have come from if one recalls the fact that the whole preoccupation of alchemy was an attempt

---

40 For this formula and the full implications of its meaning, see my *The Cosmic War: Interplanetary Warfare, Modern Physics, and Ancient Texts*, pp. 83, 232–233, 239–240.
41 Holmyard, *Alchemy*, p. 116.
42 Ibid., p. 120.
43 Ibid., p. 149.
44 Ibid., p. 151.
45 Holmyard, *Alchemy*, p. 151.

to embody the "soul" of the transmutative *materia prima,* or as much of its properties as possible, in the "body" of the Stone itself. It has already been suggested that in such references that alchemy might indeed be implying a kind of hyper-dimensional component to the Philosophers' Stone, as we have also seen, as some of Paracelsus' comments will find breathtaking echoes in parts two and three of the present book.

3. Ability to Affect Action at a Distance

Closely related to this strange assertion are the claims of the seventeenth-century English alchemist Sir Kenelm Digby. Digby claimed to have confected a "sublime remedy" called a "Powder of Sympathy or weapon-salve," whose secret formula he claimed he had learned from a Carmelite monk whom "he had met in Florence in 1622."[46]

> *This remarkable powder acted at a distance*; it was not to be applied to a wound but to a bandage stained with blood from the wound. Digby says that he first employed it to cure James Howell…a Welshman and a prolific author…The powder, which had a great vogue for a time, was nothing but green vitriol (ferrous sulphate) but Digby claimed that a current of particles of the powder and blood somehow found its way to the wound and performed its cure.[47]

While clearly nothing more than an ordinary chemical compound, Digby's powder greatly impressed King James I. So, could it have been an ordinary compound that, subjected to alchemical processes, somehow acquired exotic properties? Digby's own suggestion that a "current of particles" somehow found "its way to the wound" is suggestive enough.

4. Indestructibility:
A Link to The Sumerian Tablets of Destinies?

Since the Philosophers' Stone is at least a partial embodiment of the *materia prima* or the transmutative medium itself, and since the latter is, by alchemical lights, indestructible, very often one finds claims that the Philosophers' Stone itself partakes of that indestructibility. For example, in an alchemical dictionary published in 1612, Ruland has the following curious entry about the *materia prima,* the Philosophers' Stone, and its indestructibility:

---

46 Ibid., p. 213.
47 Ibid., pp. 213–214, emphasis added.

Materia Prima et hujus vocabula — The philosophers have so greatly admired the Creature of God which is called the Primal Matter, especially concerning its efficacy and mystery, that they have given it many names, and almost every possible description, for they have not known how to sufficiently praise it.

1. They originally called it Microcosmos, a small world, wherein heaven, earth, fire, water, and all elements exist, also birth, sickness, death, and dissolution, the creation, resurrection, etc.

2. Afterwards it was called the Philosophical Stone, because it was made of one thing. Even at first it is truly a stone. Also because it is dry and hard, and can be triturated like a stone. But it is more capable of resistance and more solid. *No fire or other element can destroy it.* It is also no stone, because it is fluid, can be smelted and melted.[48]

The indestructibility of the Philosophers' Stone recalls similar statements from the Sumerian-Babylonian epic *The Exploits of Ninurta*, in which an inventory of the stones comprising the Tablets of Destinies — over which a horrifyingly destructive "cosmic war" was fought between factions of the Sumerian pantheon — was made. As I noted in one of my previous books, *The Cosmic War,* some of these stones were deliberately destroyed, and others preserved and secreted away. But as I also observed, there was a special class of a very few stones that *could not be* destroyed. One of them even has the suggestive name of "the Elel Stone," and "El" of course, is the ancient Sumerian and Akkadian word root for "god." The "Elel" stone was, like the alchemical Philosophers' Stone, a *divine stone*, divine perhaps for its association with the technologies of the gods, but divine certainly for its indestructibility.[49]

Yet another alchemical treatise, while not directly invoking Sumer or the Tablets of Destinies, makes its own alchemical speculative interpretation of the Urim and Thummim and the stones on the breastplate of the Hebrew high priests, attributing to them an ultimate origin in whatever paleoancient Very High Civilization that may have existed before the Deluge:

> That Urim and Thummim were given on the Mount cannot be proved, yet they are potential from the Creation may appear, for they were substances whose name and essence did predicate each other, being convertible terms: the name and essence one, the words signify Light and perfection, knowledge and holiness, also manifestation and truth; even as since and essence

---

48 Ruland, *Lexicon alchemiae sive dicttionarium alchemistucm, cum obscuriorum verborum, et rerum Hermeticarum, tum Theophrast-Paracelsicarum phrasium, planam explicationem continens*, Frankfurt, 1612, transcribed by John Glann, "On the *Materia Prima,*" www.levity.com/alchemy/ruland_e.html, p. 1, emphasis added.

49 Q.v. my *The Cosmic War: Interplanetary Warfare, Modern Physics, and Ancient Texts* (Kempton, Illinois: Adventures Unlimited Press, 2007), pp. 204–233, especially pp. 227–232.

make one perfection. *It is likely that they were before the Law given*, for the Almighty commended Noah to make a clear light in the Ark...[50]

The theme of gems or stones of power is not unique to Sumer or the Hebrews, but finds similar echos in esoteric myths concerning the Egyptian wisdom god Thoth, whose *Emerald Tablet* was highly prized by alchemists as the most succinct statement of their philosophy and art.[51]

E. Conclusions

What does one make of all this? What conclusions may be drawn to guide us in the coming pages through the labyrinth of modern science's own alchemical quest for exotic materials able to manipulate the substrate of space-time itself? Recalling alchemy's belief in an underlying prime matter or substrate that is transmutative, and recalling that this implies a fundamental condition of non-equilibrium, the following properties of the Great Elixir or the Philosophers' Stone have been noted:

1) The Stone comes in at least two or, more especially, three "parts," a material "body" and a more ethereal "soul," implying that whatever properties were being observed and referred to by the alchemists, that some of them appeared to exist in a state *beyond that of the three dimensions of normal matter itself;* in short, and in more modern terms, there was a *hyper-dimensional* aspect to the Philosophers' Stone, points which shall be taken up again in parts two and three;
    a) This aspect is made more apparent by the later references in the Middle Ages to the Stone having at least one constituent which "occupies no space," a peculiar quality easily explainable by reference to higher dimensions;
    b) In one very questionable instance, that of the Englishman Digby, it appears to be able to effect action at a distance;
2) The Stone appears to exist in at least three states:
    a) As a stone or mineral;
    b) As a reddish-purple Elixir, Tincture, or Liquid; and,
    c) As a powder or granular particulate;
3) The successful confection of "Alchemical Gold" is signaled by a distinctive progression through the electromagnetic optical spectrum, with the salient points being a progression from white to reddish-purple, the latter of which denotes the acquisition of the Great Elixir;

---

50   *The Glory of Light*, Transcribed by Adam McLean from MS Ashmole 1415, 161–70, p. 2.
51   Q.v. my *Giza Death Star Destroyed*, pp. 53–96.

4) Throughout all versions of the confection of the Stone, an elaborate process is implied, involving the use of heat at every step, and in one case — that of Moses and the Golden Calf — involving very intense heat able actually to *consume or burn* gold itself, an observation that in turn requires a rather sophisticated furnace or forge technology;
5) It is also claimed that the Stone, Great Tincture, or Grand Elixir is able to effect cures, healing, and to prolong human life far beyond its natural lifespan;
6) It is also claimed that the Stone is indestructible, partaking directly of the indestructibility of the underlying primary transmutative medium;
7) In the suggestive references to be found in the Chinese alchemist Go-Hung, the "Pill of Immortality" must be fashioned on a great and famous mountain, an allusion, perhaps, to the distant connection between alchemy and pyramids;
8) Throughout alchemical literature not only are certain minerals associated with certain celestial bodies but also the very processes of alchemical operations are with those bodies as well, implying that these processes are wrought most efficiently when those bodies are in certain alignments. Moreover, the association of astrology with certain gems — with *crystals* — also implies an association of lattice structures of crystals with celestial geometries, a significant insight as we shall see in future chapters; and finally,
9) As noted in the examination of Paracelsus, some alchemical experiments and operations could only successfully be performed at specific times and seasons, a point which will become a primary focus of part two.

While all this seems, at first glance, supremely implausible, subsequent chapters will demonstrate that modern science would seem to be vindicating many of these fundamental aspects of the alchemical art...

...and the parallels, as we shall see now, are far more than coincidental...

Part Two
The American "Gold"

∴

"Dastin seems to endow the elixir with a spiritual nature, so that although it must be confined in some kind of matter, for otherwise it could not be manipulated, it nevertheless occupies no space; an idea to which we find an interesting parallel in the views of Paracelsus."

E.J. Holmyard,
Alchemy, p. 151.

Scorched Gold in the Arizona Desert
David Hudson and the Beginnings of Monatomic "Gold"

•••

"No fire or other element can destroy it. It is also no stone, because it is fluid, can be smelted and melted....it is light and bright, indestructible, and is Heaven in operation."
Ruland, *On the Materia Prima*[1]

Arizona farmer David Hudson seems at first glance to be the least likely candidate to become an adept of alchemy, able to synthesize the Philosophers' Stone, for by his own admission, he came from "an ultra-conservative right wing background" steeped in the values of famous Arizona U.S. Senator Barry Goldwater.[2] And, by his own admission, he was a very wealthy man, the materialist's materialist:

> I was farming about 70 thousand acres in the Phoenix area in the Yuma Valley. I was a very large, materialistic person.... I had a forty man payroll every week. I had a four million line of credit with the bank. I was driving Mercedes-Benz's. I had a 15,000 square foot home.[3]

---

[1] Ruland, *On the Materia Prima, Lexicon alchemiae sive dictionorium alchemisticum, cum obscuriorum verborum, et rerum Hermeticarum, tum Theophrast-Paracelsicarum phrasium, planam explicationem continens,* Frankfurt, 1612. www.levity.com/alchemy/ruland_e.html, pp. 1-2.
[2] David Hudson, "The Chemistry of M-State Elements," www.asc-alchemy.com/hudson.html, p, 1.
[3] Ibid.

But when Hudson decided that he was going to try to make the sodium-rich soil of Arizona suitable for growing crops, little did he know that his life was about to undergo a dramatic change and set him off on a quest that, by any lights, could reasonably be described as alchemical.

It all happened simply enough.

"You have to understand," says Hudson, "that in agriculture in the state of Arizona we have a problem with sodium soil." This soil is so rich in sodium content that it "looks like chocolate ice cream" but is so brittle that "it crunches when you walk on it." Even worse for the prospective farmer, "Water will not penetrate this soil. Water will not leach the sodium out of the ground."[4]

A. Acid Baths for the Soil and an Explosive Discovery

As a result of this condition, Hudson determined that the soil would have to be intensely prepared chemically in order to make it suitable for growing crops. Hudson purchased a very high concentration of sulfuric acid — 93%, compared to a 40–60% concentration in normal automobile battery acid — and shipped it into his farm "in truck and trailer loads," and literally injected "thirty tons to the acre into the soil."[5] At this point Hudson observed that when soil was subjected to this type of extreme chemical treatment to prepare it for planting, that it was vital to know the chemical composition of the soil. It was while determining this that the mystery — and Hudson's alchemical quest — began in earnest:

> In doing the analysis of these natural products we were coming across materials that no one seemed to be able to tell us what they were…. We took the material into chemistry and we dissolved it and got a solution that would be *blood red*. Yet when we precipitated this material out chemically by using a reductant of powdered zinc the material would come out as a *black precipitant just like it was supposed to be if it was a noble element*. A noble element if you chemically bring it out of the acid it won't re-dissolve in the acid.[6]

While he did not know it at the time, Hudson had accidentally discovered two of the essential colors — red and black, or the "crow's head" or "raven's head" — that were seen in the previous chapter to be essential clues to the successful confection of the Philosophers' Stone or the Great Elixir.

After precipitating the unknown noble element out of the unknown substance, Hudson dried it out in the 115-degree heat and low 5% humidity of the

---

4   David Hudson, "The Chemistry of M-State Elements," www.asc-alchemy.com/hudson.html, p. 1.
5   Ibid.
6   Ibid, p. 2, emphasis added.

Arizona desert. What happened next caught Hudson completely by surprise:

> What happened was that after the material dried it exploded. It exploded like no explosion I had ever seen in my life, and I've worked with a lot of explosive materials. There was no explosion and there was no implosion. It was as if somebody had detonated about fifty thousand flash bulbs all at one time....just poof.[7]

Not only had all the material itself been consumed in the "explosions," but so had the paper on which it had been placed.[8]

1. Sunlight and a Pencil

Intrigued by this very peculiar result, Hudson took a new, unsharpened pencil and stood it on end next to another sample that was drying. Once again, the material "detonated," but the pencil was not knocked over. "So this was not an explosion and was not an implosion. It was like a tremendous release of light."[9] Hudson admitted the phenomenon had him "baffled." Moreover, he also discovered that if he dried the material out of the sunlight it would *not* explode. Only when dried in sunlight would it "detonate."

2. Reductions, Beads, and Anomalous Shattering

After several such anomalous "explosions," Hudson decided to try to discover exactly what the strange material was made of. The method he decided upon was that of "crucible reduction," wherein the tested material is mixed in a heating element along with lead. When heated sufficiently, the lead melts, and metals that are heavier than lead stay toward the bottom of the molten compound, while those lighter float to the top and may be poured off. Normally, this will leave a small "bead" of gold and silver at the bottom of the crucible.

When he had completed this process, Hudson took his bead reduction for analysis "to all the commercial laboratories" and was informed that, indeed, there was nothing but gold and silver in it.[10] There was, however, just one "small" little problem:

> ...I could take that bead and set it on a table and hit it with a hammer and it shattered like glass. Now there is no known alloy of gold and silver that is

---

7 www.asc-alchemy.com/hudson.html, p. 2.
8 Ibid.
9 Ibid., p. 3.
10 Ibid.

not soft.... So any alloy of gold and silver if that's all that's there is going to be soft and ductile. You can flatten it out and make a pancake out of it. Yet this material shattered like glass. I said something's going on here that we are not understanding. Something unusual is happening...[11]

While there was no way to determine at this stage if this strangely brittle gold and silver were somehow linked to the anomalous "explosions", Hudson now had four mysteries on his hands: He had discovered a chemical compound that:

1) "exploded" when dried in sunlight but did not do so when it was not dried in it;
2) would not knock over a pencil placed next to it but yet would totally consume the material itself, and burn the pencil when it "exploded";
3) was apparently reducible to an alloy of silver and gold by normal means of reduction; yet,
4) while chemically analyzable as gold and silver alloy, nonetheless shattered like glass when struck by a hammer.

Hudson decided that further tests of his strange silver-gold alloy were in order.

When the gold and silver were further separated out of the little beads, what was left was, in Hudson's own words, "a whole bunch of black stuff."[12] Once again, the "crow" had reared its alchemical black head, for when this new black residue was analyzed at commercial laboratories "they told me it was iron, silica, and aluminum."[13] This mystified Hudson even more:

I said it can't be iron, silica, and aluminum. First of all you can't dissolve it in any acids or bases once it is totally dry. It doesn't dissolve in fuming sulfuric acid; it doesn't dissolve in sulfuric nitric acid; it doesn't dissolve in hydrochloric nitric acid. Even this dissolves gold yet it won't dissolve this black stuff.[14]

As if that were not all, the engineers in the analytical laboratories appeared to be just as baffled as Hudson. "No one could tell me what it was."[15]

3. *The Cornell Labs Episode*

Having reached this impasse, Hudson determined that if he was ever going to solve the mystery of the enigmatic material, he would "have to throw

---

11 Ibid.
12 www.asc-alchemy.com/hudson.html, 3.
13 Ibid.
14 Ibid., pp. 3–4.
15 Ibid., p. 4.

some money" at the problem.[16] Hiring a Ph.D. at Cornell University "who considered himself an expert on precious elements,"[17] the scientist informed him that Cornell had a machine that could analyze the material to within "parts per billion."[18] Unfortunately, the sophisticated machine ultimately told Hudson nothing different than the commercial laboratories had earlier: the enigmatic material was nothing but iron, silica, and aluminum.[19]

Undaunted, Hudson and the scientist separated out the iron, silica, and aluminum from the strange black bead. When this was done, according to Hudson 98% of the sample still remained. And that sample, as Hudson put it, "was pure nothing. I said, 'Look, I can hold this in my hand; I can weigh it; I can perform chemistries with it.' I said 'That is something. I know that is something. It is not nothing.'"[20]

The scientist then informed Hudson that the problem was that the machine, which used spectrographic analysis to determine the chemical composition of compounds, could not make a determination of its composition because the strange black material's emission spectra did not agree with any of the chemical elements programmed into it! [21] Hudson returned from Cornell, totally disillusioned with American academia, but more determined than ever to find out what the substance was composed of. At this point, new players entered the scene...

B. Spectroscopic Analysis: Enter the Russians

Returning to Arizona, Hudson inquired in the Phoenix area for an expert in spectroscopic analysis, and eventually found a man who had been trained in the then West Germany, and who moreover actually designed and built spectroscopic analytical equipment.[22] Hudson approached him with a copy of a book published by the Soviet Academy of Sciences, entitled *The Analytical Chemistry of the Platinum Group Elements*. Hudson noted that "In this book, according to the Soviets, you had to do a 300 second burn on these elements to read them."[23] This posed a rather great difficulty for standard spectroscopic equipment. Hudson explained the difficulty this posed as follows:

> Now for those of you who have never done spectroscopy it involves taking a carbon electrode that is cupped at the top. You put the powder on that

---

16 Ibid.
17 Ibid.
18 Ibid.
19 Ibid.
20 www.asc-alchemy.com/hudson.html, p. 4
21 Ibid.
22 Ibid., pp. 4–5.
23 Ibid., p. 5.

electrode and you bring the other electrode down above it and you strike an arc. In about fifteen seconds the carbon at this high temperature burns away and the electrode's gone and your sample's gone. So all the laboratories in this country are doing fifteen second burns and giving you the results. According to the Soviet Academy of Sciences the boiling temperature of water is to the boiling temperature of iron just like the boiling temperature of iron is to the boiling temperature of these elements.

.... So literally we had to design and build an excitation chamber where argon gas could be put around this electrode so that no oxygen or air could get into the carbon electrode and we could burn it not for fifteen seconds but for three hundred seconds. According to the Soviet Academy of Sciences this is the length of time we have to burn the sample.[24]

Having built the needed equipment, which filled a garage,[25] Hudson and his new scientist advisor performed the burn. What happened next is best told in Hudson's own words:

Anyway when we ran this material during the first fifteen seconds we got iron, silica, aluminum, little traces of calcium, sodium, maybe a little titanium now and then, and then it goes quiet and nothing reads. So at the end of fifteen seconds you are getting nothing. Twenty second, twenty five seconds, thirty seconds, thirty five seconds, forty seconds still got nothing. Forty five seconds, fifty seconds, fifty five seconds, sixty seconds, sixty five seconds but if you look in through the colored glass sitting there on the carbon electrode is this little ball of *white* material.[26]

Note that, just as in the alchemical sequence of colors, Hudson's black material has now turned *white*.

1. *The Platinum Group Metals*

But back to Hudson:

At seventy seconds, exactly when the Soviet Academy of Science said it would read, palladium begins to read. And after the palladium platinum begins to read. And after the platinum I think it was rhodium begins to read. After rhodium ruthenium beings to read. After ruthenium then iridium begins to read and after the iridium osmium begins to read.

---

24 www.asc-alchemy.com/hudson.html, p. 5.
25 Ibid.
26 Ibid., p. 6, emphasis added.

However, Hudson was soon to discover something even more amazing about the platinum group metals found in his small sample:

> Well, we came to find out that rhodium was selling for about three thousand dollars per ounce.... Iridium sells for about eight hundred dollars an ounce and ruthenium sells for one hundred and fifty dollars an ounce.[27]
>
> Then you say gee these are important materials aren't they. They are important materials because in the world the best known deposit is now being mined in South Africa. In this deposit you have to go a half mile into the ground and mine an 18 inch seam of this stuff. When you bring it out it contains one third of one ounce per ton of all the precious elements.
>
> Our analysis, which we ran for two and a half years and we checked over and over; we checked every spectral line, we checked every potential on interference, we checked every aspect of this....
>
> When we were finished the man was able to do quantitative analysis and he said "Dave, you have six to eight ounces per ton of palladium, twelve to thirteen ounces per ton of platinum, one hundred fifty ounces per ton of osmium, two hundred and fifty ounces per ton of ruthenium, six hundred ounces per ton of iridium, and eight hundred ounces per ton of rhodium. Or a total of about 2400 ounces per ton when the best known deposit in the world is one third of one ounce per ton.
>
> As you can see this work wasn't an indicator that these elements were there; these elements were there and they were there in boucoups [sic] amounts.[28]

Hudson's German-trained assistant was "so impressed that he went back to Germany to the Institute of Spectroscopy. He was actually written up in the spectroscopic journals as having proven the existence of these elements in the Southwestern United States in natural materials."[29] Hudson's land was literally sitting atop one of the richest deposits of platinum group metals in the world, if not *the* richest! And the concentrations, moreover, were anomalously high.

2. Anomalies in Neutron Activation Analysis

But the *real* anomalies were yet to come! Having burned the material far beyond the normal fifteen seconds, Hudson decided to stop the burn at precisely 69 seconds.

---

27 This was ca. 1986.
28 www.asc-alchemy.com/hudson.html, pp. 6–7.
29 Ibid.

> I let the machine cool down and I took a pocket knife and dug that little bead out of the top of the electrode. When you shut off the arc it sort of absorbs down into the carbon and you have to dig down into the carbon to get it out, this little bead of metal.
>
> So I sent this little bead of metal over to Harlow Laboratories in London. They made a precious metals analysis of this bead. I get the report back "no precious element detected." Now this was *one second before the palladium was supposed to start leaving. Yet according to neutron activation, which analyzes the nucleus itself, there were no precious elements detected.*[30]

In other words, Hudson's special spectroscopic analyzer had returned anomalously high concentrations of platinum group metals, yet an English neutron activation analyzer had returned absolutely nothing! Hudson drew an interesting conclusion from this highly contradictory set of data: "Either this material was converted to another element *or it's in a form that we don't understand yet.*"[31] This, as we shall see, was the key to unlock the mystery, not only of his strange material, but perhaps other anomalously behaving matter, including that of the ancient and mediaeval alchemists!

a. White to Red to Rhodium

But Hudson's odyssey of anomalies was still not over. One more final attempt by Hudson to discover the identity of the strange material through yet another expert in metallurgy failed. The metallurgist, after putting the substance through every known test available to him, told Hudson that the material "is not any of the other elements on the periodic table."[32]

Hudson and the metallurgist decided to do one more precipitation experiment, this time with rhodium chloride. And here the alchemical connections were once again in evidence, for the chloride solution produced a now-familiar color:

> The example I use is rhodium because it has a very unique color to the chloride solution. It is a cranberry color almost like the color of grape juice. There is no other element that produces the same color in chloride solution. When my (i.e., Hudson's) rhodium was separated from all the other elements it produced that color of chloride. The last procedure you do to separate the material out is to neutralize the acid solution and it precipitates out of solution as a red brown dioxide.

---

30 www.asc-alchemy.com/hudson.html, p. 7, emphasis added.
31 Ibid., emphasis added.
32 www.asc-alchemy.com/hudson.html, p. 8.

> So we did that...Then we filtered that out. We heated it under oxygen for an hour in the tube furnace then we hydro-reduced it to this gray-white powder: exactly the color rhodium should be as an element. Then we heated it up to 1400 degrees under argon to anneal away the material and it turned snow white. [33]

Oddly enough, in other words, Hudson's subjection of his enigmatic matter to various chemical treatments and stresses produced almost the same type of sequence of colors as that recorded in mediaeval alchemical tests: first a deep cranberry-red color, the reddish-brown, then a "gray-white powder" and finally a snow-white color.

Hudson's metallurgist, whose name was John, was equally mystified. He decided to re-insert the substance in the special furnace, and under an oxygen atmosphere, heat it, cool it down again, purge the atmosphere with inert gases, and then heat it back up under hydrogen again. This process would reduce any oxide residues since any oxygen left in the sample would combine with the hydrogen to create water, thus purifying the sample. Throughout the process, the metallurgist removed samples at each stage of the process and put samples into vials. Once the process was completed, the metallurgist put the sample into vials and sent three samples to analysis to Pacific Spectrochem, "one of the best spectroscopic firms in the U.S."[34]

When the results came back, the mystery only increased. As Hudson states:

> The first analysis comes back. The red-brown dioxide is iron oxide. The next material comes back: silica and aluminum. No iron present.

No iron!? From the *same* material *three* different chemical results!? Was this a manifestation of the triune Stone? As Hudson quips, apparently just the *act* of putting hydrogen

> on the iron oxide has made the iron quit being iron and now it has become silica and aluminum....We had just made the iron turn into silica and aluminum.

Indeed, they had made iron oxide turn into silica and aluminum by standard chemical practices and yet, those practices certainly would not have ever predicted such a result. Hudson continues:

---

33 Ibid.
34 www.asc-alchemy.com/hudson.html, p. 9.

The snow white annealed sample was analyzed as calcium and silica. Where did the aluminum go? John said "Dave, my life was so simple before I met you." He said, "This makes absolutely no sense at all." He said, "what you are working with is going to cause them to re-write physics books, to re-write chemistry books, and come to a complete new understanding."[35]

The metallurgist presented Hudson with his bill — a whopping one hundred and thirty thousand dollars — which Hudson paid. It was then that John the metallurgist told Hudson something else...

b. Platinum in the Arizona Desert

Hudson's metallurgist informed him that he had checked the unusual soil on Hudson's property at least fifty different ways, and the results were nothing less than supremely astonishing, for the metallurgist informed him that his soil contained highly anomalous concentrations of the platinum group metals:

> "...You have," (he told Hudson), "four to six ounces per ton of palladium, twelve to fourteen ounces per ton of platinum, a hundred fifty ounces per ton of osmium, two hundred fifty ounces per ton of ruthenium, six hundred ounces per ton of iridium..." The exact same numbers that the spectroscopist had told me were there.[36]

These concentrations were so unusually high that the metallurgist demanded Hudson take him to the area the samples came from, and allow him to take his own samples, which he did. When these samples were analyzed, they again returned the same highly anomalous concentrations of platinum group metals.

Working on this problem from 1983 until 1989, Hudson had employed "one PhD chemist, three master chemists, (and) two technicians," all working fulltime.[37] Hudson and his associates during this period had "learned how to buy rhodium tri-chloride from Johnson, Mathew and Ingelhardt as the metal and we learned how to break all the metal-metal bonding until it literally was a red solution but no rhodium detectable. And it was nothing but pure rhodium from Johnson, Mathew, and Ingelhardt."[38]

But what had *initiated* Hudson down the path of his modern alchemical quest was not the pure form of these elements, it was the raw soil and ore in his farmland. And in this too there is an alchemical parallel, for as the modern

---

35 Ibid.
36 Ibid.
37 www.asc-alchemy.com/hudson.html, p. 9.
38 Ibid.

day esotericist and scholar of all matters occult and alchemical, A.E. Waite, observed, alchemists

> have always accounted the dissolution of metals as the master key to this art, and have been particular in giving directions concerning it, only keeping their readers in the dark as to the subject, whether ores, or factitious metals, were to be chosen: nay, when they say most to the purpose, then *they make mention of metals rather than the ores, with an intention to perplex those whom they thought unworthy of the art.*[39]

Inadvertently, Hudson had confirmed the alchemical assertions of a man writing fully a century before him!

C. General Electric Sees the Explosions: A New Energy Technology?

During this period, Hudson learned something that would forever change the nature of what had become for him a personal quest, and impress upon him the importance of the phenomenon he had discovered; the American defense contractor General Electric, he learned, "was building fuel cells using rhodium and iridium."[40] Hudson arranged to meet with the General Electric researchers conducting this research at their Waltham, Massachusetts center.

To his surprise, Hudson had learned that the GE engineers had experienced the same unusual "explosions" as well.[41] Hudson arranged to send some of his own mysterious rhodium to the engineers for their analysis, and to see if it would work in their fuel cells. In the meantime, GE sold its fuel cells technologies to United Technologies. As a result of this, GE's fuel cell engineers formed their own company, and it was this company that actually performed the experimentation and analysis of Hudson's mysterious rhodium.[42]

Once again, the material seemed to defy all standard analysis and behavior:

> When our material was sent to them, the rhodium, as received, was analyzed to not have any rhodium in it. Yet when they mounted it on carbon in their fuel cell technology and ran the fuel cell for several weeks, it worked and it did what only rhodium would do...

---

39 A.E. (Arthur Edward) Waite, "On the Philosophers' Stone," from *Collectanea Chemica* (London, 1893), www.levity.com/alchemy/collcgem.html, p. 16, emphasis added.
40 Ibid., p. 10.
41 Ibid.
42

After three weeks they shut the fuel cells down and they take the electrodes out and sent them back to the same place that said there was no rhodium in the original sample, and now there is over 8% rhodium in the rhodium....

So these (former) GE people said "Dave, if you are the first one to discover this, if you are the first one to explain how to make it in this form, if you are the first one to tell the world that it exists, then you can get a patent on this."... Then they told me that if someone else discovered it and patented it, even though I was using it every day, they could stop me from doing it."[43]

And thus was born Hudson's patent for so-called ORME, or "monatomic," elements, which he filed for ORME gold, ORME palladium, ORME iridium, ORME ruthenium, and ORME osmium. The term ORME is an acronym standing for "Orbitally Rearranged Monatomic Elements." The reason for and meaning of this unusual title will be explored in the next chapter. For now, there was one more highly significant anomaly, to be explored with the strange material, the most significant anomaly of them all!

D. The Stunning Mass Anomaly

It was while filing patents for his strange materials that Hudson noticed this new anomaly, the most significant of all the anomalies his strange matter exhibited.

After filing his patents, Hudson was contacted by the patent office demanding more data about the substances prior to issuing its patent. Hudson had noticed that when the substance was reduced to its white powder state, that it had anomalous weight gains — on the order of 20–30% — when brought into the atmosphere. While weight gains for material of this sort in atmosphere are normal, they are not on the order of a fifth to a third of the original weight of the sample! Nonetheless, Hudson had to offer the patent office some explanation for the anomalous gain.

Deciding to use a machine called a thermo-gravimetric analyzer, Hudson began tests. The machine allowed total control of the sample in precisely controlled atmospheres that would permit oxidation, hydro-reduction, annealing, all the while permitting the sample to be heated and cooled in precise amounts, and weighed throughout whatever process it was being subjected to.

What happened next was stunning, and it is best to let Hudson describe it in his own words:

---

43   www.asc-alchemy.com/hudson.html, p. 12.

We heated the material at one point two degrees per minute and cooled it at two degrees per minute. What we found is when you oxidize the material it weighs 102%, when you hydro-reduce it, it weighs 103%. So far so good. No problem. *But when it turns snow white it weighs 56%. Now that's impossible.*

So when you anneal it and it turns white it only weights 56% of the beginning weight. If you put that on a silica test boat and you weigh it, it weighs 56%. *If you heat it to the point that it fuses into the glass, it turns black and all the weight returns. So the material hadn't volatized away. It was still there; it just couldn't be weighed any more.* That's when everybody said this just isn't right; it can't be.

Do you know that when we heated it and cooled it and heated it and cooled it and heated it and cooled it under helium or argon that when we cooled it, *it would weight three to four hundred percent of its beginning weight and when we heated it, it would actually weigh less than nothing. If it wasn't in the pan, the pan would weigh more than the pan weighs when this stuff is in it.* [44]

In other words, Hudson's strange material was not only exhibiting highly anomalous chemical behavior that defied ordinary techniques of analysis, it was now exhibiting highly anomalous *mass* properties as well, to such an extent, that the material seemingly could cause other materials it came into contact with to lose a percentage of *their* mass as well, such as the pan that held the sample! And notably, once again, the various properties the material exhibited seemed to be signified by the *color* of the material — black for mass gain, fine white powder for anomalous mass loss — exactly as the mediaeval and ancient alchemists always insisted!

While Hudson had not yet made the connection to alchemy, he had nonetheless decided to get serious, and find out just exactly what was going on, and just exactly what sort of treasure trove of "gold" he held in his hands. But this *new* story requires a chapter of its own.

---

44 www.asc-alchemy.com/hudson.html, p. 13, emphasis added.

4

Transmutations
Torsion Superdeformities and the New Nuclear Physics

•••

"The kinematic and dynamic moments of inertia of several superdeformed bands are calculated as a function of the rotational frequency..."
Y.R. Shimizu, E. Vigezzi, and R.A. Broglin[1]

A. Getting Serious with Hal Puthoff and the 44% Mass Loss Anomaly

Faced with the strange anomaly of the loss of significant mass of the white powder, and the demands of the patent agency to explain the peculiar mass loss data, Hudson sought out, and found, an explanation for the enigmatic property. The explanation came in the form of one of America's, and indeed the world's, most famous and capable theoretical physicists, Dr. Hal Puthoff. In his search of the physics literature to explain the anomalous behavior of his white powder, Hudson had run across a paper that Puthoff had published in the prestigious peer-reviewed journal *Physical Review*, in March of 1989.[2]

In that paper Puthoff had in fact calculated what happens when superconducting matter reacts in two rather than three dimensions. It loses four-ninths (4/9) of its weight, leaving five-ninths (5/9) of its mass, and

---

[1] Y.R. Shimizu, E. Vigezzi, and R.A. Broglin "Inertias of super-deformed bands", *Physical Review C*, (Volume 41, Number 4, April 1990, 1861–1854), p. 1861, from the abstract.

[2] There is also another interesting paper co-authored by Dr. Puthoff: Bernhard Haisch, Alfonso Rueda, H.E. Puthoff, "Inertia as a Zero-Point-Field Lorentz Force," *Physical Review A,* Volume 49, Number 2, pp. 678–694.

five-ninths is…..*exactly 56% percent!* Hudson had found the one physicist seemingly able to explain why his material in white powder form only weighed 56% of its original mass. Deciding to go meet Puthoff, Hudson took some of his material and his data to Austin, Texas, and presented Puthoff with the actual experimental data that confirmed his mathematical predictions.

During their discussions, Puthoff told Hudson that "when this material only weighs 56% of its true mass, you do realize that this material is actually bending space-time."[3] Such a material, Hudson noted, was what Puthoff "called exotic matter in his papers."[4] Hudson had, in other words, literally stumbled across some of the exotic matter that forms so much of the quest of modern theoretical physics — not to mention mediaeval alchemy — and it was there right beneath his feet in the soil of his farm, and it was not really all that exotic at all. It was ordinary chemical elements, but in some sort of state not hitherto known.

But that wasn't *all* that Puthoff told him.

If the mysterious white powder was indeed losing 44% of its mass, then, said Puthoff, "theoretically it should be withdrawing from these three dimensions…it should not even be in these three dimensions."[5] In other words, the material was behaving exactly as one of Puthoff's "two-dimensional superconductors."

B. Superconductivity

So what was his white powder?

The occurrence of the mention of superconductivity in Puthoff's papers naturally impelled Hudson into an investigation of these strange objects. Hudson begins by noting one of the most standard tests of superconductivity, which may be appreciated by comparing superconducting material to an ordinary metal current carrier, such as a copper wire. Everyone knows that if one passes a magnet close to a wire, running it back and forth, the magnetic field of the magnet will induce a small electrical current in the wire, which, with appropriately sensitive detection equipment, will be detected. But this is not at all what happens around a superconductor.

Hudson states what happens around a superconductor with his typically colorful descriptions:

> If it's a superconductor as you apply a magnetic field it goes negative. It literally eats the magnetic field. It feeds on the magnetic field and takes it

---

[3] www.asc-alchemy.com/hudson.html, p. 16.
[4] Ibid.
[5] Ibid., p. 17.

inside itself. Negative inductance in a positive applied magnetic field is the proof of a superconductor.

In other words, if you had a machine that was a superconductor when it passed by ordinary power lines, it would cancel the voltage potential of the power lines. Or if it passed by a home that had electric appliances it would literally turn them off and cause them to flicker and go off.[6]

But what had all this to do with what Dr. Hal Puthoff told him? As far as Hudson was concerned, the ability of his anomalous material to literally leave these three dimensions meant

> that if you had a machine that would do that, it could literally move in space time, is what Hal (Puthoff) was saying? That it could disappear and reappear in space time....
>
> A superconductor is billions and billions of atoms all acting like one big macro atom. And so literally you make yourself a vessel that you can climb inside of that superconducts and you energize it and you exclude all external magnetic fields including gravity. And you are now in this world but you are not of this world.[7]

Hudson's material was capable, in other words, of the manipulation of the fabric of the physical medium, of "space-time," itself.

While he did not yet know it, David Hudson now held the ancient Philosophers' Stone in his hands.

1. Consciousness and Superconductive Behavior of DNA

How Hudson came to the conclusion that he was actually holding the alchemical Philosophers' Stone is itself one of the more intriguing aspects of his story. It began when he noticed that a connection was being made in the scientific literature between superconductivity and biology, with all the rich implications for the idea of consciousness that this entails.

While researching superconductivity, Hudson ran across a little paper with the revealing title "Evidence from Activation Energies for Superconductive Tunneling in Biological Systems at Physiological Temperatures" by Freeman A. Cope.[8] I must confess that when I read the actual title of this article, I was dumbfounded, for the title clearly implied that superconductivity could and

---

6 www.asc-alchemy.com/hudson.html, p. 18.
7 Ibid.
8 Freeman A. Cope, "Evidence from Activation Energies for Super-conductive Tunneling in Biological Systems at Physiogical Temperatures," *Physiological Chemistry and Physics 3* (1971), pp. 403–410.

did occur in biological systems at ordinary "room temperature" so to speak, whereas everything I knew from the literature implied that superconductivity normally only occurred in extremely cold temperatures, well below those in which life could actually live and thrive.

But an actual glance at some of the statements in the article left me even more stupefied:

> Considerable evidence for semiconduction and for solid-liquid interfacial electron conduction in biological systems has been obtained from kinetic analyses, which have been supported by electron mobility measurements using the microwave Hall effect and pulsed electron beam techniques, and by the finding of a low semiconduction activation energy in the dried enzyme cytochrome oxidase.

That's nice, but what does it *mean?* Hang on, for the article continues with unusual clarity:

> In the present paper, evidence for another class of solid state biological process is given. It is suggested that *single-electron tunneling between superconductive regions may rate-limit various nerve and growth processes. This implies that micro-regions of superconductivity exist in cells at physiological temperatures, which supports theoretical predictions of high temperature organic superconduction.*
>
> Superconduction is the passage of electron current without generation of heat and hence with zero electrical resistance. Such behavior has been observed only in organic materials and only at temperatures below approximately 20°K, although theory predicts that superconduction might occur in organic materials at room temperatures. The conduction of electrons *across interfaces* between adjacent superconductive layers behaves differently from current across ordinary solid junctions. Electron tunneling currents across interfaces between superconductive layers or regions have been predicted and demonstrated to have a particular form of temperature dependence...
>
> ...*Little...has suggested DNA as the sort of biological molecule along which electrons might superconduct...* [9]

In other words, at ordinary temperatures for living organisms, and under certain temperature conditions, their DNA molecules could actually superconduct electricity! The article even went on to stress the fact that this phenomenon

---

9   Freeman A. Cope, "Evidence from Activation Energies for Super-conductive Tunneling in Biological Systems at Physiological Temperatures," *Physiological Chemistry and Physics 3* (1971), 403–410, pp. 403–404, emphasis added.

appeared to be tied most closely with nerves. Perhaps this had something to do with the alchemical claims that the Great Elixir, the Philosophers' Stone, also conferred healing and longevity properties on those who consumed it.

But what exactly in the DNA could account for this behavior, and what in DNA could possibly link it to Hudson's strange white powder? Hudson set out to answer these questions, and, focusing on the fact that nerve tissue was somehow involved in the phenomenon, procured the brains of cows and pigs, and decided to do an experiment. Hudson immersed the brains in alternating sulfuric acid and water several times to rid the brain matter of all carbons and nitrous compounds. What was left was dry matter, approximately five percent of which was rhodium and iridium — both platinum group metals! — in the high-spin state, the same state as Hudson's fine white powder![10]

In other words, Hudson had found a possible connection, via his white powder high-spin-state platinum group metals, and via the superconductive properties of DNA, to the ancient alchemical insistence that the alchemist himself must in some sense be transformed or purified in order to confect the Philosophers' Stone, for consider, we now have:

1) matter whose mass exists partly in (56%), and partly outside (44%), of this ordinary three dimensional space (or four-dimensional space-time);
2) matter which only exhibits this property after going through intensely stressful chemical processes in which the alchemical "sequence of colors" is more or less confirmed, and which, as indicated in those texts, exists in a fine white powder form;
3) these properties are apparently related to superconductive properties;
4) there is a relationship to DNA's superconductive properties, which might account for ancient alchemical warnings that the alchemist himself must be transformed in order to confect the Philosophers' Stone.

2. Gamma Ray Bursts

Hudson decided on further tests of his mysterious material to find out exactly what the nature of the "explosions" was that he had first observed in his material, the very same explosions that set him off on his quest to understand the substance. Purchasing an arc furnace and pumping out all the air and then filling it with helium, he stirred his material with a tungsten electrode about the size of a human thumb.[11] Supposedly the furnace was

---

10 www.asc-alchemy.com/hudson.html, p. 19.
11 www.asc-alchemy.com/hudson.html, p. 20.

supposed to be able to heat any substance thirty or forty times before the tungsten electrode would burn out.

But this was not what happened:

> We didn't even get a second out of this thing. So we sent it to the manufacturer, got another electrode, put it back in it, put back on, closed it back up, vacuumed out the out, put in the inert gas, struck another arc, bzzp, shut off. Opened it up again and the tungsten electrode is all molten into this powder.
>
> What we found when we analyzed the powder after we did this, it wasn't the same element it was before we did this. And what we also found is that there was an amplification of heat about two thousand times. It was not chemical heat, it was nuclear heat.

How did Hudson know this? He knew because

> all the wiring in the laboratory was beginning to crumble and fall apart. You could go up to copper wires and do that and they would just go to powder.
>
> The glass beaker sitting in the laboratory near the furnace was getting full of little air pockets in the glass and when we would pick them up they would fall apart. And that's radiation damage. There is no other explanation for it.[12]

Consulting nuclear laboratories, Hudson determined that his "explosions" were releasing 25,000 electron volts of *gamma* rays! The "explosions" were gamma ray bursts! In other words, the combination of Hudson's strange high-spin-state powder, when subjected to electromagnetic radiation of the sun — remember his initial observations were formed by drying the substance in sunlight — or of the arc furnace, resulted in the same sort of nuclear transmutations one might associate with normal processes of nuclear fission via neutron bombardment, yet no neutron bombardment was occurring! The stresses on the substance were all electrical and magnetic in nature.

3. *Hudson Discovers the Alchemical Connection*

It was at this point that Hudson discovered, finally, the alchemical connection.

> ...my uncle came up with this book in 1991 called *Secrets of the Alchemists*. I said "I'm not interested in reading about alchemy....I want to know about

---

[12] www.asc-alchemy.com/hudson.html, p. 21.

chemistry and physics. He said "Dave, it talks about a white powder of gold" I said "Really?" And so I began to look into alchemy. And the Philosophers' Stone, the container of the light of life was the white powder gold.

Now I said "is there a chance that this white powder of gold that I have, could it be the white powder of gold they're talking about? Or it is possible that there are two white powders of gold? Now the description says it is the container of the essence of life; it "flows" the light of life. Well, that we had proven. It's a superconductor. It "flows" the light that is in your body. (The alchemists) claimed that it perfects the cells of the body.[13]

The book was, in fact, published by Time-Life books, and stated that it was a belief of the alchemists that partaking of the white powder gold actually helped induce longevity.[14] But Hudson also found, as he investigated alchemical texts, that it all went "back to a man the Hebrews called Enoch, the Egyptians called Thoth," and who in Greece was the now well-known "Hermes Trismegistus," the Thrice-Great Hermes.[15]

And with Hermes-Thoth, we have closed the circle, and returned to the place and the concepts from which the quest began: to Egypt, and to its notions of an underlying physical substrate, a *materia prima,* which was transmutative, a physical substrate, moreover, as I have demonstrated in my previous books, that was believed to have spatial-cellular properties like an organism,[16] or a lattice structure, like a crystal,[17] that could *create* information from an initial condition of non-equilibrium and a quasi-analogical process,[18] and that was a common medium to mind or consciousness, energy, and matter.[19]

C. Superdeformities and the New Nuclear Physics

But how, thought Hudson, could one rationalize all this *scientifically?* How could one account for all the strange and anomalous behavior of this "white powder gold"? Hudson observed that the U.S. Naval Research Facility had known and proven that biological cells communicated with each other via some process involving superconductivity, but that they could not quite determine exactly *what* was superconducting nor how it was doing it. For

---

13 www.asc-alchemy.com/hudson.html, p. 21.
14 Ibid., p. 30.
15 Hudson was able to find the alchemical connections in the Egyptian *Book of the Dead,* q.v. www.asc-alchemy.com/hudson.html, pp. 31-32.
16 See the entire third chapter of my *Giza Death Star Destroyed* (Kempton, Illinois: Adventures Unlimited Press), pp. 38-110.
17 See my *Giza Death Star Destroyed* (Kempton, Illinois: Adventures Unlimited Press), pp. 54-62; 65-68; 185-189; 247-263.
18 Ibid., pp. 99-150; 222-245.
19 Ibid., p. 245.

Hudson, however, it was obvious: the reason nothing unusual was showing up in the navy's chemical analyses was the same reasons he had encountered in his own quest to identify the strange material: they were, as he called them, his high-spin-state "'stealth atoms' at work. No one knows they are there, because they don't identify by normal instrumental analysis."[20]

His explanation for this behavior, however, is even more intriguing, and suggests how perceptively he had absorbed not only the alchemical texts he was now studying, but also their modern scientific analogues:

> What they found, was that the nucleus of these atoms deforms, goes to a high spin state, called a high spin nuclei [sic], and theoretically the high spin nuclei should be superconductors, because high spin nuclei pass energy *from one atom to the next with NO NET LOSS OF ENERGY...* When you understand that a superconductor flows with only a single frequency of light, in fact, that light is a NULL light. *In other words, it consists of two waves that are mirror images of each other. Because of this mirror symmetry, there is NO WAVE, it APPEARS to cancel.* [21]

This requires some rather careful unpacking, and understanding in the context of the previous discussion concerning the work of Hal Puthoff.

Hudson is here maintaining several things:

1) Insofar as the superconductive effects of high-spin-state atomic nuclei are concerned, this may somehow be due to a *resonance* effect existing between each nucleus;
2) Such energy and information transfers are *not* bound by normal laws of thermodynamics in energy transmission, for there is no net loss of energy, in other words, the transference of energy or information may not even be occurring by standard linear transmissions of energy; some *other* unknown mechanism must be at work, a mechanism that is perhaps induced by the high spin state itself;
3) A portion of Hudson's "white powder gold," according to Puthoff, was existing in our normal space-time, and a portion of it was *not,* but *was existing in a kind of lower-dimensional sub-space*;
4) This resonance effect, moreover, may be due to a peculiar zero-summing of ordinary vectors of electromagnetic energy.

---

20   www.asc-alchemy.com/hudson.html, p. 31.
21   www.asc-alchemy.com/hudson.html, p. 31, italicized emphasis added, capitals emphasis in the original.

As will be seen in the final portions of section three of this book, the idea of "sub-spaces" capable of conveying information is part of the geometry of a rather interesting and breathtaking theory of physics. But the idea of physical sub-spaces appears to have been part of the alchemical quest all along. As the famous esotericist Manly P. Hall observed:

> As one of the great alchemists fittingly observed, man's quest for gold is often his undoing, for he mistakes the alchemical processes, believing them to be purely material. He does not realize that *the Philosopher's Gold, the Philosopher's Stone, and the Philosopher's Medicine exist in each of the four worlds and that the consummation of the experiment cannot be realized until it is successfully carried on in four worlds simultaneously according to one formula.* [22]

The phrase "four worlds simultaneously" is suggestive enough of an object in normal space-time also inhabiting higher (or lower) dimensional subspaces, but more importantly, the alchemical quest is carried out "according to one formula," implying that one chemical recipe will grant access to these four worlds simultaneously. Interestingly enough, as we shall see in part three of the present work, even the number of "*four* worlds" is in exact conformity with at least one new theoretical physics model.

Even more intriguing is the fact that some alchemists were quite clear, according to E.J. Holmyard, that the Philosophers' Stone did not occupy space:

> Dastin seems to endow the elixir with a spiritual nature, so that although it must be confined in some kind of matter, for otherwise it could not be manipulated, *it nevertheless occupies no space*; an idea to which we find an interesting parallel in the views of Paracelsus.[23]

For an ancient or mediaeval alchemist, occupying "no space" would be almost the same, if not identical, to a modern physicist's conception of occupying a *different* space simultaneously as it was occupying, in part, the normal space-time of everyday existence. Indeed, something very much like this is implied by Holmyard's remark that the "spiritual nature" of the Philosophers' Stone must be "confined in some kind of matter, for otherwise it could not be manipulated." Nonetheless, it "occupies no space."

For all this, it is, however, the fourth point in the previous summary that must really preoccupy our attention: "This resonance effect is, moreover, due

---

22   Manly P. Hall, *The Secret Teachings of All Ages,* Reader's Edition, p. 508, emphasis added.
23   E.J. Holmyard, *Alchemy,* p. 151, emphasis added.

to a peculiar zero-summing of ordinary vectors of electromagnetic energy." In its high-spin-state, in other words, matter appears to be existing in almost a kind of self-contained space-time "bubble," wherein all the vectors of electromagnetic force it normally exhibits, and which therefore permit normal spectroscopic chemical analysis, cancel out or "zero sum" by being mutually opposed in pairs of bi-directional waves. Under this condition, it is obvious that normal spectroscopic chemical analysis would simply break down and be unable to analyze the material properly, thus accounting for Hudson's early failures to obtain any consistent chemical analysis of his white powder.

But there is something else here, and it should give one pause, for this zero-summing *is precisely the necessary condition for scalar, or torsion, physics, a physics capable of effecting action at great distance with no diminution of energy; a physics capable of potentially great good for energy production, or great evil for creation of hugely destructive weapons.* [24] And clearly, *it is the fourth, zero-summed, high-spin-state condition that makes possible the third condition, of a state of matter capable of a partial existence in normal space-time, and partial existence in a kind of "sub-" or "hyper-space."*

1. Hudson's Physics Papers Sources and a Methodology

But could Hudson in fact find substantiation for these radical views? Indeed he could, and even more to the point, the previously outlined summary is precisely that, a summary of what he found. But what precisely did he find in the physics literature? He found a series of papers dealing with "new radioactivities," "superdeformed nuclei" brought on by "high spin states," and entirely new models of nuclear fission and transmutation *not* brought about by neutron bombardment and fracturing of an ordinary non-high-spin

---

24 For this whole principle and technique of analysis, see my *Giza Death Star Deployed*, pp. 170–193, and my *SS Brotherhood of the Bell*, pp. 200-241. It was the famous British physicist, the brilliant E.T. Whittaker, whose paper "On the partial differential equations of mathematical physics," analyzed the mathematical scalar potential into pairs of bi-directional *longitudinal* waves of stress and rarefaction in the medium itself, moving superluminally. In other words, the normal dimensionless "scalar potential" became an entity possessed of a *discrete and quantized structure*; quite an accomplishment, especially considering that his paper was published before Special Relativity (in 1903 as compared to Einstein's paper of 1905), and this some three decades before quantum mechanical theory began to be fully fleshed out. This remarkable paper is rendered even more remarkable by the fact that Whittaker clearly implies that the physical substrate or aether not only has *dynamic structure* of stress and rarefaction, but that he clearly implies that it is a kind of *sub-space* of normal space, or, alternatively, that it is a kind of "hyper-dimensional" space lying "inside of," or "behind," or "above" normal space. These "hyper-spaces," moreover, Whittaker clearly implies to exist in a kind of harmonic series, i.e., their defining characteristics are in terms of *frequency resonance*. Again, as will be seen in part three, similar conceptions lie behind a breathtaking physical theory now gaining some quiet attention in the corridors of engineering and applied science institutions and agencies.

nucleus. He details the papers to which he refers in his public lectures, and accordingly, our methodology here will be simply to cite what those papers themselves say about these subjects, and occasionally to point out Hudson's own interpretations of them.[25]

2. Formal Definition of Superdeformity

A convenient definition of these new "superdeformed" nucleonic states is given in the following short paragraph:

> ...those states known as "superdeformed" (SD) (are) where the nucleus acquires a very elongated shape that can be approximately represented by an ellipsoid where the ratio of the long to the short axis is considerably larger than that of normal deformation ~1.3:1. Within the framework of the anisotropic harmonic-oscillator model one can expect the existence of favorable shell gaps that appear regularly as a function of deformation and nucleon number.[26]

Reading between the lines a bit, it does not take much to see what is being said or implied: superdeformed atomic nuclei are "flattened out" due to the much higher rate of spin of those nuclei. Rather than tiny little "spheres" of protons and neutrons, they are now elongated dramatically. Moreover, note that a model is even being hinted at for understanding these shapes: "shells," recalling the early Bohr atom with its electron shells or orbits, each representing a stage of the excitement of the electron. A similar structure, in other words, is now being proposed for the atomic nucleus of these superdeformed nuclei themselves!

Moreover, the article also makes the significant observation that these states are the historical product of investigations of nuclear isomers: "The first observation of SD nuclei goes back to the discovery of fission isomers and the identification of the rotational bands built upon them."[27] Isomers are unusual isotopes of elements that have almost all of their energy locked up *precisely in their high spin state, or in the angular momentum of their nuclei*,[28] so that if one could figure out a way to unlock that energy suddenly, one would have a new and efficient source of energy, as well as a frightening potential for a bomb,

---

25 Hudson's sources are cited at www.asc-alchemy.com/hudson.html, pp. 55–59.
26 A.O. Macchiavelli, J. Burde, R.M. Diamond, C.W. Baeusang, M.A. Deleplanque, R.J. McDonald, F.S. Stephens, and J.E. Draper, "Superdeformities in [104, 105] Pd," *The American Physical Society*, August 1988, 1088–1091, p. 1088.
27 A.O. Macchiavelli, J. Burde, R.M. Diamond, C.W. Baeusang, M.A. Deleplanque, R.J. McDonald, F.S. Stephens, and J.E. Draper, "Superdeformities in [104, 105] Pd," p, 1088.
28 See my *SS Brotherhood of the Bell*, pp. 294–296.

for imagine something spinning at an ultra-high rate of revolutions, and then suddenly stopping it, causing it to fly apart in a sudden burst of energy.[29]

3. Properties of Superdeformity
a. Spontaneous Fission and New Models of Fission in Superdeformed States

As noted above, one of the most unusual properties of superdeformed nuclei is to undergo spontaneous nuclear fission without neutron bombardment. This is a *completely new* and hitherto unknown mechanism of fission. The mechanism is explained by the "shell model" of the nucleus being adopted by scientists to explain the superdeformed nucleus and its properties:

> In particular, new shell gaps appear by inducing a quadrupole distortion in the nuclear shape, where the ratio of the major to minor axis is 2:1... Such deformations play an important role in the process of spontaneous fission, where the 2.1 configuration is connected with the second minimum of the fission barrier, as well as in heavy ion collisions, leading to resonant molecular-like behavior.[30]

Note the extremely elongated shape of the high-spin superdeformed nucleus in a 2:1 ratio. This induces "gaps" in the shells of distributions of protons and neutrons in the nucleus, an important concept as we shall see in a moment.

But one can imagine what is taking place by drawing an analogy to an ice-skater, spinning and drawing in his arms. As he does so, he spins faster, as the angular momentum of his contracting arms is transferred closer and closer to the center of rotation. If we can imagine the skater being made of rubber, he would begin to elongate or, depending upon one's point of view, "squish" slightly. A similar phenomenon is occurring in the superdeformed nucleus. Notice also the final comment: the "bonding" between such atoms begins to occur via resonance in the frequency of spin states of the nucleus they share. Ordinary chemical electron bonding is thus no longer the main mechanism of molecular bonding.

An article in the March 1990 edition of *Scientific American* explains the significance of the discovery of the new form of nuclear fission:

---

29   This possibility was behind a DARPA (Defense Advanced Research Projects Agency) project in the 1990s to investigate the possibility of using hafnium 178 isomer to create a bomb of stupendous explosive power. Not all are satisfied, however, that the project was based on sound science; see the book *Imaginary Weapons: A Journey through the Pentagon's Scientific Underworld*, by Sharon Weinberger.

30   E. Vigezzi and R.A. Broglia, "Inertias of superdeformed bands," *Physical Review C, The American Physical Society,* April 1990, 1861–1864, p. 1861.

The discoveries settled a 40-year old quandary in nuclear physics. Until the 1980s it appeared as though the nuclear fragments from radioactive processes came in roughly three sizes: four, 100 or 200 nucleons – a term that refers to both [sic] protons and neutrons. In the four-nucleon range is the alpha particle, or helium nucleus. If an alpha particle emerges from an atom, it leaves behind a nucleus composed of approximately 200 nucleons, in the 100-nucleon range are the fragments from fission, a process in which a heavy nucleus splits roughly in half. The restricted range of sizes raised an intriguing question: Why did a nucleus not emit a fragment composed of other quantities of nucleons — why not 14 or 24?

Today it is known that a nucleus can indeed eject a fragment of this size or any other. These new radioactivities form when a large number of nucleons within the nucleus spontaneously rearrange themselves in certain configurations. Because these large-scale rearrangements occur at random, the emission of a new radioactivity is in general a much rarer event than, say, the emission of an alpha particle. By the end of the 1980s, physicists had succeeded in observing many of these new nuclear ambassadors.[31]

The article goes on to explain how the "nuclear shell model" helps to explain the new process of fission:

The resemblances between nuclear shell structure and atomic shell structure are striking. If the electrons of an atom completely fill one or more shells, as is the case for helium and neon, the atom is stable: it is chemically inert. If the shell of a nucleus are completely filled, as are those of calcium and lead, the nucleus is stable and consequently spherical.

The first nuclear shell can be filled with as many as two protons and two neutrons; the second shell can be filled with up to six of each nucleon; the other shells are also filled with a certain number of protons and neutrons. The result is that one can usually predict the stability of a nucleus just by counting the number of protons and neutrons. Stable nuclei usually consist of a "magic number" of protons or neutrons; that is, they have 2, 8, 20, 28, 40, 50, 82, 126 or 184 protons or neutrons. Nuclei that have double magic numbers are particularly stable — for example, calcium 48 (20 protons and 28 neutrons) or lead 208 (82 protons and 126 neutrons).[32]

But the shell model, as was seen, is not the only geometry at work in superdeformed nuclei; the other is the elongation, or its converse, the

---

31 Walter Greiner and Aurel Sandulescu, "New Radioactivities," *Scientific American,* March 1990, 58–67, p. 58.
32 Ibid.

"squishing" or flattening out that can occur in a *low* spin state, that occurs in the high-spin state:

> ...(The) shell model assumes a somewhat rigid structure, the collective model holds that the outer part of the nucleus can deform when the outer nucleons move with respect to the nucleons of the inner nucleus. This collective motion, or deformation, derives from the liquid-drop model.
>
> Most nuclei are prolate spheroids (cigar-shaped); some are oblate (disk-shaped). These deformations require that the nucleus whose shape changes slowly as energy is added is called a hard nucleus; a nucleus whose shape deforms rapidly from additional energy is referred to as soft.[33]

But while these models explained why unstable nuclei in the shell model could assume elongated high spin, or alternatively disk-like shapes, it did not yet explain why scientists were observing new forms of fission that did not fit in the previously known model.[34]

The answer occurred when scientists realized that, in high-spin situations, nucleons in the outer shells could begin to pull away from the original center of the nucleus, and form a kind of "proto-nucleus" with a second center of a cluster of nuclei within the larger nucleus structure.[35] Thus, while in its high-spin state, a heavy nucleus may be chemically analyzable as a particular element, yet it will contain a *substructure* of two "clusters" of nucleons, that, in some cases, can fission into two completely different elements *asymmetrically*, that is, the fission products do not divide more or less evenly, but into different products that "differ greatly in mass and charge," and these fragments, moreover, may be and usually are "several times larger than an alpha particle."[36] This "two center shell model" not only led to the prediction of new radioactivities, but new elements and isotopes as well.[37]

b. Superconductivity

As was seen previously, yet another property of superdeformed atoms is their ability to superconduct, and indeed they do so in some circumstances in only two dimensions. One of the most interesting, but technical, studies that Hudson consulted was an article that indicated that in some cases the normal processes of superconductivity itself did not hold in situations of

---

33   Walter Greiner and Aurel Sandulescu, "New Radioactivities," *Scientific American,* March 1990, 58–67, p. 61.
34   Ibid.
35   Ibid., p. 62.
36   Ibid.
37   Ibid.

extremely high angular momentum within the atomic and nuclear system of the superconductor.[38] And this leads us to…

4. The *All-Important* Principle of Superdeformity: Rotation and Angular Momentum

By now it should be obvious what the main mechanisms in these superdeformities, new asymmetrical fission products, and new radioactivities are: rotation and angular momentum. Indeed, the abstract of one of the papers consulted by Hudson even put it with rare, non-technical succinctness: "The kinematic and dynamic moments of inertia of several superdeformed bands *are calculated as a function of the rotational frequency.*"[39] As the article's authors, E. Vigezzi and R.A. Broglia, acknowledge, the whole "discovery of superdeformed rotational bands during the past years opens a new chapter in the study of nuclei under conditions of extreme deformations and angular momenta."[40] But the real revelation occurs in a short comment toward the beginning of their article: "It is well established that the spectra of rapidly rotating nuclei reveal two distinct components in the buildup of the total angular momentum, *corresponding to angular moment of orbital angular momentum of individual particles and to collective rotation.*" [41] In other words, the total system of deformation was comprised of the angular momentum of individual particles in the nuclear shells, but to the total rotation of the proto-nucleus "cluster" which they formed. In all cases, the superdeformity "was determined self-consistently as a function of the angular momentum."[42]

And, for those who've been following the hyper-dimensional "tetrahedral physics" model of Richard C. Hoagland over the years, there is an intriguing additional datum. Let us recall what the principle feature of this physics is. A tetrahedron, if circumscribed or embedded in a rotating sphere, with one vertex of the tetrahedron located on a pole of the axis of rotation, will have the other three vertices of the tetrahedron touching at the latitude of 19.5 north or south on the surface of that sphere, depending on which pole of the sphere one orients the vertex on the axis of rotation. Hoagland has observed that within any massive rotating body, such as a planet or a star, there appears to

---

38   See Mohit Randeria, Ji-Min Duan, and Lih-Yor Shieh, "Bound States, Cooper Pairing, and Bose Condensation in Two Dimensions," *Physical Review Letters,* Volume 62, Number 9, 27 February 1989, 981–984, p. 981.
39   E. Vigezzi and R. A. Broglia, "Inertias of superdeformed bands," *Physical Review C, The American Physical Society,* Volume 41, Number 4, April 1990, 1861–1864, p. 1861.
40   Ibid.
41   Ibid., emphasis added.
42   Ibid., p. 1863.

be, as a universal feature of this geometry, upwellings of energy at that latitude in those bodies, upwellings that are, moreover, vorticular in nature.

Interestingly enough, energy seems to gate "into" one of the bands of palladium 104 and 105 superdeformities at a most unusual rate of spin: "*The feeding of the $g_{7/2}$ band seems to take place at a spin of 39/2 and therefore the deexcitation patterns suggests a spin of 43/2 for the first observed level of the new rotational band, with an uncertainty of ± 2 units.*"[43] In other words, energy appears to gate into the system when the spin frequency is 39/2, or *19.5 units,* and appears to deexcite at a spin frequency of 43/2, or *21.5* units. This is an astonishing correlation of Mr. Hoagland's "tetrahedral physics" model, and moreover, is a possible indicator of its scale invariance, for in this case, it is occurring at the nuclear, and not planetary, scale.

5. Mercury

Not surprisingly, these superdeformed nuclei also exhibit very anomalous rates of radioactive decay, a point that will assume some significance in part three. Moreover, in weakly deformed nuclei around the *mercury* region of the periodic table, there are highly anomalous discontinuities of expected energies within nuclei, on the order of about 1 MeV (one million electron volts). Mercury, in other words, appears to be the most peculiar element of all, with nuclei already, in their natural state, weakly deformed.[44] One can only imagine — as we shall do in part three — what happens when mercury is subjected to *high*-spin states, and the superdeformities, fission products, and "new radioactivities" that might result. It is, however, interesting to note that Hudson has seen the implications of the high-spin superdeformed state of palladium for cold fusion, and the nuclear transformations that Pons and Fleischmann and others have observed, for as has been seen, in some cases, these superdeformed nuclei can indeed fission asymmetrically, and produce new elements out of their constituent "proto-nucleus clusters."[45]

At the end, what is one to make of Hudson's quest?

It was Hudson himself who uncovered the resemblance of his discoveries to the claims and practices of alchemy, but even he missed, for all this, perhaps the most significant connection of them all: the strange mass loss anomaly that his material exhibited, for here too, as elsewhere, there is an exact alchemical

---

[43] A.O. Macchiavelli, J. Burde, R.M. Diamond, C.W. Beausand, M.A. Deleplanque, R.J. McDonald, F.S. Stephens, and J.E. Draper, "Superdeformation in $^{104,\,105}$Pd," *Physical Review C, The American Physical Society,* Volume 38, Number 2, August 1988, 1088–1091, p. 1089, emphasis added.

[44] C.S. Lim, R.H. Spear, W.J. Wermeer, and M.P. Fewell, "Possible discontinuity in octupole behavior in the Pt-Hg region," *Physical Review C,* Volume 39, Number 1, March 1989, 1142–1144, pp. 1142, 1144.

[45] See Hudson's comments at www.asc-alchemy.com/hudson.html, p. 59.

foreshadowing of his discoveries.

In fact, one famous mediaeval alchemist and theologian, Roger Bacon, noted that the Stone could exhibit anomalous weight *gain:*

> The second multiplication is an *Augmentum quantitatis* of the stone with its former power, in such a way that it neither loses any of its power, nor gains any, but in such a manner that *its weight increases and keeps on increasing ever more, so that a single ounce grows and increases to many ounces.*[46]

The title of Bacon's treatise in which this quotation occurs has its own significance as well: *Tract on the Tincture and Oil of Antimony*, for as we shall see in the next part of the book, one of the chemical compositions suggested for the mysterious Soviet "Red Mercury" is precisely the compound mercury antimony oxide, and, if the stories about the "Soviet Mercury" are true, Red Mercury was supposed to be an extraordinarily and anomalously dense and heavy substance.

Even more suggestively, Paracelsus, commenting on "the projection to be made by the mystery and arcanum of *antimony*" states that "no precise weight can be assigned in this work of projection.... For instance, that Medicine tinges sometimes thirty, forty, occasionally even sixty, eighty, or a hundred parts of the imperfect metal."[47] While the basic sense of this passage is usually taken to mean that the "tincture of antimony" is able to transform and transmute an anomalous amount of material in proportion to its own weight, there might indeed be another meaning, since "no precise weight can be assigned in this work of projection." In other words, perhaps Paracelsus himself had observed some anomalous weight gain or loss in his alchemical tinctures.

In any case, it is as unlikely as it is true, that the anomalous behavior of Hudson's "white powder gold" exhibited its most anomalous behavior, not in what it was doing in and of itself, but in what it did to Hudson, by leading him to a serious study not only of ancient and mediaeval alchemy but also of modern nuclear theory. In this, perhaps, his material had affected a truly alchemical transmutation, for Hudson, like all other alchemists, had ended his quest with his thoughts and outlook utterly transformed.

---

46   Roger Bacon, *Tract on the Tincture of Oil and Antimony,* www.levity.com/alchemy/rbacon2.html, p. 12, emphasis added.
47   Paracelsus, *The Aurora of the Philosophers,* p. 20.

Conclusions to Part Two

∴

"As we progress, we shall see that the Stone of Paradise (which is heavier than gold, but lighter than a feather) is no myth of the distant past. It now holds a primary position in the world of modern physics, with its baffling weight ratios fully explained as a scientific fact."
Sir Laurence Gardner[1]

Hudson's unique quest to understand the anomalous white powder he had extracted from his soil samples permits us now to draw certain tentative conclusions about the relationship between ancient and mediaeval alchemy and modern physics:

1) The basic principles and assertions of certain alchemical texts are verified, namely:
   a) *with respect to the color sequence indicating successful confection of the Philosophers' Stone:* Hudson, via known and standard chemical and physical techniques, was able to replicate the overall color sequence of the derivation of the Philosophers' Stone. As was seen, two of these colors, a cranberry-red color, almost the color of grape juice, signified a stage in the process which ended with a fine "white powder of gold";
   b) *with respect to the composition of the Philosophers' Stone:* As was seen from chapter one, many alchemical texts stressed the *powder form* of the Philosophers' Stone, a form amply demonstrated by Hudson's material;

---

1 Sir Laurence Gardner, *Lost Secrets of the Sacred Ark,* p. 22.

c) *with respect to the Philosophers' Stone "occupying no space yet being confined in matter":* As was seen, one of the most significant anomalies exhibited by Hudson's white powder was its unusual mass loss anomaly, an anomaly that physicist Hal Puthoff explained by maintaining that some of the material was actually existing in a wholly different space and time, a "sub-" or "hyper-space." Thus, alchemical texts that indicate the existence of the Philosophers' Stone in different "worlds" would appear to be capable of interpretation along hyper-dimensional physical models.

2) In turn, these anomalous properties appear to be based upon:
   a) Extraordinary or non-ordinary geometries and shapes of atomic nuclei in elements within the platinum group to mercury range in the periodic table of the elements; these are in turn the result of,
   b) extremely high-spin states of the nuclei; furthermore,
      i) in *some* cases these states are excited at a spin frequency of 19.5, and deexcited at a spin frequency of 21.5, in apparent correlation of the "tetrahedral physics" model of popular Mars anomalies researcher Richard C. Hoagland; that is to say, energy appears to gate, in some elements, into the system at 19.5, and to exit or deexcite at 21.5, indicating the possibility that Hoagland's model might be scale invariant;
   c) the total angular momentum of the system, which is composed of two sub-systems:
      i) the individual particles in the nuclear shells; and,
      ii) the total angular momentum in the "proto-nuclear cluster" of the superdeformed nucleus;
3) These superdeformed atoms apparently bond in "quasi-molecular fashion" via a resonance phenomenon related to the spin frequencies and angular momentum of their various nucleus shells;
4) These superdeformed atoms likewise can undergo spontaneous asymmetrical fission *without* neutron bombardment, and often can yield extremely high bursts of gamma radiation when deexciting from their high-spin state. The weaponization potential and implications of this phenomenon will be explored in the next two parts of the book. Similarly, superdeformed nuclear isotopes in a high-spin state do not decay at standard rates of radioactive decay.

The significant and crucial role of angular momentum and rotation throughout Hudson's alchemical quest and in the physics of nuclear isomers and superdeformed nuclei, as exhibited by the existence of a portion of Hudson's white powder in an entirely different sub-space or hyper-space, points to a deeper underlying physics, to that of the folding and pleating of space-time that results from the phenomenon known as *torsion*. And to understand the effect of torsion on matter and the formation of exotic states of matter, we must cross the ocean, to Russia, and to Germany, and visit the groundbreaking, and breathtaking, work of Russian theoretical and experimental physicist Nikolai Kozyrev, and the even more astounding work of Nazi Germany, and a postwar German physicist all but unknown to American physics, who, however, has come to the definite attention of NASA...

Part Three
THE SOVIET "MERCURY"

∴

"With beating heart, Flamel watched for the revealing colours. They came, and in the correct sequence: from grey to black, 'the crow's head,' then from black to white, the white first appearing like a halo round the edge of the black, and the halo then shooting out white filaments towards the centre, until the whole mass was of a perfect white. This was the white elixir.... Sure at last that he had achieved mastery of the Art, he replaced the rest of the elixir in the flask and continued the heating. Now the rest of the colours appeared one after the other: the white turned to the iridescence of the peacock's tail, this to yellow, the yellow to orange, the orange to purple, and finally the purple to red — the red of the Great Elixir."

E.J. Holmyard,
Alchemy, p. 245.

"Red" Mercury
Hoax, Code Name, Intelligence Operation, or Genuine Article?

∴

> *"And it's also strange that, together with the mercury, they used ...antimony. The alchemist of the XII century Rtefio talks about a special tincture containing antimony and sublimate mercury which would have had spectacular effects. Scientists admitted only in 1968 that a compound called mercury ...antimony oxide could be realized...."*
> The Secret Book of Artrephius [1]

The Russian part of this alchemical quest begins, oddly enough, at almost the same time as Hudson's was ending, with the "Red Mercury" nuclear materials trafficking scare of the 1990s. During that decade, stories began to appear in the Western European and American media that agents of various "nations of concern" — Libya, Iraq, Iran, North Korea, and so on — were caught red-handed trying to buy a substance called "Red Mercury" that had been smuggled out of the old Soviet Union's secret laboratories and nuclear stockpiles. One would expect, of course, that such nations would be more interested in acquiring plutonium-239 or uranium-235, or the reactor and enrichment technologies to make them. But no, their agents were seized after trying to purchase at enormous sums a cherry-red liquid goo called "Red Mercury." And in that, there lies yet another difficult tale.

---

[1] "The Secrets of Mercury," mmmgroup.altervista.org/e-mercur.html, p. 5, citing *The Secret Book of Artrephius*, p. 6.

Part of the difficulty concerning the Red Mercury story lies in the completely ambiguous and often contradictory explanations that were offered for the whole episode, not to mention the dubious explanations put forward for the substance itself. The difficulty is further compounded by the fact that a broad range of magical and indeed "alchemical" capabilities were alleged for the substance, and that it, like the alchemical Philosophers' Stone itself, was confected by a rather arcane "recipe" whose constituent chemical elements were only vaguely known or guessed at. And like the Philosophers' Stone itself, its alleged uses seemed to change according to the needs of the buyer. In this, it has all the hallmarks of a modern retelling of the quest of mediaeval alchemy. Many, who noticed this odd resemblance, were therefore inclined to write off the whole strange episode as a deliberately concocted hoax. The problem with that explanation, as we shall see, is that it does not really do justice to the whole picture. Indeed, if it were a hoax, why then did nations such as Libya, Syria, Iraq, Iran, and North Korea fall for it? There had to have been enough truth in the hoax to hook such buyers, who certainly had scientists sophisticated enough to smell a fraud, if indeed that is what it was.

So, what exactly *is* "Red Mercury" and what has it to do with the modern alchemical quest for the Philosophers' Stone?

A. Various Explanations of the Red Mercury Scare
1. *The "Simple Hoax" Explanation*

The explanations for the Red Mercury scare are no less diffuse and ambiguous as the claims for the substance themselves. Some speculate that the scam was nothing more or less than a complete and simple hoax, perpetrated on an unsuspecting — and by implication, somewhat dimwitted — nuclear terrorism community by the psychologically much more sophisticated Russian intelligence services, which sold nothing more than depleted uranium, spent reactor fuel, cobalt or cesium, or other such materials to their "marks" for enormous sums of hard cash.[2] But how does one account for the fact that the Russians were so successful in these efforts? Surely the purchasers — which included scientifically sophisticated nations such as South Africa, Iraq, and Iran — would have had the scientific prowess to detect a fraud. Indeed, this highlights the strong possibility that the whole Red Mercury episode "seems to have been so widespread and common is likely related to the fact that there is some truth in the claims made by the con artists."[3] The question

---

2   See for example the explanations of Kenley Butler and Akaki Dvali in "Nuclear Trafficking Hoaxes: A Short History of Scams Involving Red Mercury and Osmium-187" at www.nti.org/e_research/e3_42a.html, p. 1.

3   Kenley Butler and Akaki Dvali, "Nuclear trafficking Hoaxes: A Short History of Scams Involv-

intensifies when one considers the additional, and contradictory, fact that "much-publicized statements from British, Russian, and U.S. government officials" exist, statements that in no uncertain terms assert that "no material matching the properties of red mercury exists, and no such material is used in the construction of nuclear weapons. How, then, did red mercury become the commodity of choice for con artists and unwitting buyers?"[4] This problem becomes very acute, since long after such official denials were made public, the United Kingdom nevertheless brought two men to trial for conspiracy in an alleged plot involving nuclear terrorism and Red Mercury! So, if it was a complete hoax, why the trial? And if not a complete hoax, why the official denials? One explanation for the denials may lie in those nations' concern that a real threat was involved, and their wishes to reassure their populations, and curtail the black market trade in nuclear weapons materials.

But convenient as that explanation may be, it still does not dissolve the problem of the various explanations. Indeed, as one observer of the controversy pointed out, the term "Red Mercury" may itself be nothing more than a code name, with the word "Red" denoting its origin in Communist Russia, and "Mercury" simply being a code word for the high heats and pressures involved with thermonuclear fusion. The substance, whatever it was, may not therefore be a compound of mercury at all.[5] As will be seen, these speculations are in line with the alleged "recipe" of Red Mercury itself.

2. *The Anti-Terrorism Counterintelligence Hoax*

A somewhat more sophisticated version of the "hoax" explanation is that the whole Red Mercury episode was a cleverly conceived anti-terrorism counterintelligence operation launched by the Russian government, with the probable collusion of Germany, the United Kingdom, and other Western governments. The ostensible goal of this alleged operation was to infiltrate, expose, and implicate terrorist cells attempting to procure nuclear weapons materials on the black market that emerged in the wake of the Soviet Union's collapse.

Supporters of this explanation point out the fact that the post-Communist Yeltsin government of Russia began to run articles about Red Mercury in the Russian media, articles which were soon picked up and circulated further by Western media agencies. With the appearance of the stories, a black market for the substance was created, and Russia's vast intelligence capabilities swung into action in concert with Germany's *Bundesnachrichtendienst,* Britain's MI-6,

---

ing Red Mercury and Osmium-187," www.nit.org/e_research/e3_42a.html, p. 1.
    4 Ibid., p. 2.
    5 See the comments of William Yerkes posted at chemistry.about.com/od/chemistryarticles/a/aa100404a.htm, p. 1.

and America's CIA, shutting down potential terrorist threats. The stories made the substance almost impossible to resist, for it was alleged that Red Mercury could actually detonate an H-bomb *without the need of an atomic bomb as the fuse to set it off.*

For the would-be nuclear terrorist, this was a literal Allah-send, since hydrogen bombs are vastly more powerful than atomic bombs, much "cleaner," and, if no A-bomb is involved, would be much easier to engineer and deploy.

Red Mercury thus afforded a short cut to a more powerful weapon without the need for nuclear reactors and uranium enrichment facilities, all of which are large, stationary, and costly facilities easily targetable by their potential enemies. Indeed, since on the standard physics model an A-bomb is needed as the "fuse" to detonate an H-bomb, nuclear non-proliferation efforts have always focused on the need to limit and closely monitor fissile materials and fuel enrichment technologies, since these are the technologies that require such comparatively large and stationary facilities. The Red Mercury legend thus did an end run around all of that.

While there is much to commend this interpretation, and in fact, I believe something very much like this is involved in the origins of the Red Mercury scare, the question that hung over the "simple hoax" explanation also hangs over this one as well: how could scientifically sophisticated nations fall for it, unless there was some element of truth mixed into the legend?[6]

3. *The BBC's Problematical Statements of 2006*

Perhaps the most baffling and problematical part of the Red Mercury story is the fact that it is not a story that died in the 1990s, as the conventional views of it maintain. It seems to have a contemporary life of its own. On Tuesday, July 25, 2006, BBC reporter Chris Summers authored a story entitled "What is Red Mercury?" concerning the conclusion of a trial of two alleged nuclear terrorists who supposedly tried to buy the mysterious compound. Summers notes that during the trial the substance was exposed as being "something of an urban myth, a substance which was either radioactive or toxic or neither, depending on who you spoke to."[7]

---

[6] One source actually suggested that Saddam Hussein's attempt to procure Red Mercury for Iraq's nuclear program in the early 1990s was one very hidden factor leading to the First Gulf War. See Kirt R. Poovey, "The Red Mercury Nightmare?" at www.prisonplanet.com/analysis_poovey_122602_redmerc.html, p. 1. Poovey states that "the U.S. maintains that the red mercury they have analyzed is fake. However as (Jeff) Nyquist reports, it would appear that 'American intelligence has evidence of something dire, something they don't want to tell us,'" in other words, that there may be a deeper kernel of truth to the whole Red Mercury "pure H-bomb" possibility than meets the eye.

[7] Chris Summers, "What is Red Mercury?", news.bbc.co.uk/1/hi/uk/ 5176382.stm, p. 1.

Were this all that the BBC article said, however, we would have nothing more than a restatement of the "simple hoax" or "terrorist intelligence operation hoax" explanations. While the alleged conspirators were eventually acquitted, what is of interest is not the trial itself, but the BBC's reporting of it. Here is how Summers summed up the various theories then circulating on the internet and in the press, and how he ended his BBC article:

The five main theories are:

- That red mercury is a reference to cinnabar, a naturally-occurring mercuric sulphide. The red pigment derived from cinnabar is known as vermillion.
- That it is a reference to the alpha crystalline form of mercury iodide, *which changes to a yellow colour at very high temperatures.*

(This is sounding familiar!)

- That it is simply referring to any mercury compound originating from the former Soviet Union. The 'red' tag would simply be a legacy of the Cold War era.
- That it is a ballotechnic mercury compound which just happens to be red in colour. Ballotechnics are substances which react very energetically *when subjected to shock compression at high pressure.* They include mercury antimony oxide which, according to some reports, is a cherry-red semi-liquid produced in Russian nuclear reactors. This theory contends that *it is so explosive that a fusion reaction — a nuclear explosion — can be triggered even without fissionable material such as uranium.*
- That it is a military code word for a new nuclear material, probably manufactured in Russia.

Putting all of this into context, Summers observes that

> In the early 1990s, in the wake of the collapse of the Soviet Union, several articles were published claiming that a pure fusion device had been invented.
> It reportedly weighed around ten pounds and was no bigger than a baseball.
> If such a device existed, and was capable of triggering a nuclear explosion, the threat to the world — especially the Western world — would be catastrophic.

THE PHILOSOPHERS' STONE | 127

But no such bomb has been discovered and nobody — not even Osama bin Laden from his mountain base in Afghanistan or Pakistan — has even threatened to use one.

So is red mercury just a hoax?

Let us hope so. [8]

And on that ambiguous and disturbing note, Summers' article ends and the questions begin once again.

The first of these questions concerns both the content and timing of Summers' article, for it was published *long after* the "threat" had long since been denounced as a complete hoax not only by the Western powers but by Russia herself, yet, Summers' article reproduces the main alleged capabilities of Red Mercury — its ability to trigger an H-bomb — and the proliferation nightmares it invokes. Moreover, unlike the earlier official and unambiguous denunciations of those governments, the BBC article ends, not with a reassuring restatement of those denials, but with a disturbing question and a less-than-confident statement of "hope." Since the BBC is well-known as a media organ of the British government, it would thus appear that earlier official denunciations have retreated once again into questions and vague hopes that the Red Mercury legend is not true.

There are two equally disturbing ways of interpreting this most recent official retreat from the variations of the "hoax" explanation. First, one might see in the BBC's cautious article an attempt to prepare the climate of Western public opinion for further assaults within its territory, of a nuclear or, even worse, a thermonuclear nature, from whatever source. The corollary to this, of course, is to prepare the climate of opinion in the West for further curtailments of personal sovereignty and liberty and for an increase in the police state measures and culture beginning to emerge in Western nations in the wake of the 9/11 attacks.

But the second way of interpreting the BBC's mystifying "about-face" is that the Western intelligence agencies themselves suspect that there was, after all, some kernel of truth in the whole Red Mercury story. Thus, while the story may indeed have been concocted as an elaborate anti-terrorism operation in collusion with Russia, the Russians themselves may have been running a psychological or disinformation operation *against* the West *within* an operation ostensibly being run in *cooperation* with it. As such, there had to be, at a multitude of levels, *a buried kernel of truth in order for the whole thing to work*, not only against its terrorist targets, but the secondary target, the West, itself.

---

8   Chris Summers, "What is Red Mercury?" news.bbc.co.uk/1/hi/uk/5176382.stm, pp. 2–3, emphasis added.

If so, then what *is* that hidden kernel of truth? What could it be? To answer these questions, we must turn to the alleged uses and purported recipe of Red Mercury, with a view to a possible scientific rationalization of them. It is the science of the substance itself that constitutes both the mystery and the possible solution to the Red Mercury riddle.

B. Its Alleged Uses

In the legend of Red Mercury, the substance has such a variety of alleged uses and properties that it might with some justification be described as magical or alchemical. It is described as being a powerful "ballotechnic" explosive in its own right such that an amount no larger than a hand grenade could blow an entire ocean liner out of the water. It has also been noted that one consistent element in the legend is that it is also capable of functioning as a detonator in an H-bomb without the need for an atomic bomb as the fuse. But there are obscurer reports dating from the earliest period of the emerging legend that also describe it as a kind of "radar stealth paint," and even more curiously, as a "stimulated gamma ray emitter." Can any of these alleged properties be scientifically rationalized, much less all of them *together*? The answer requires a closer look at each of them.

1. *"Stealth Paint"*

An article in the magazine *New Scientist* appeared in 1996 that summed up the dilemma of the wide variety of uses alleged for Red Mercury. Outlining a report prepared on the mysterious substances by Los Alamos National Laboratory, the article noted that

> The supposedly top secret nuclear material was 'red' because it came from Russia. When it resurfaced last year in the formerly communist [sic] states of Eastern Europe it had unaccountably acquired a red colour. But then, as a report from the US Department of Energy reveals, mysterious transformations are red mercury's stock in trade.
>
> The report, compiled by researchers at Los Alamos National Laboratory, shows that in the hands of hoaxers and conmen, red mercury can do almost anything the aspiring Third World demagogue wants it to. You want a short cut to making an atom bomb? You want the key to Soviet ballistic missile guidance systems? *Or perhaps you want the Russian alternative to the anti-radar paint on the stealth bomber?* What you need is red mercury.[9]

---

9 Cited in Wikipedia, "Red Mercury," www.en.wikipedia.org/wiki/Red_mercury, p. 1.

Stop and consider this list carefully, for it contains significant clues, as will be seen.

First one has the now standard and persistent *core* of the Red Mercury legend, namely, that

1) It has something to do with a powerful explosive, able to simplify atom bomb construction, or, alternatively as has been seen, able to detonate the far more powerful H-bomb *without* an A-bomb fuse;
2) It is thus a powerful conventional explosive, and therefore valuable as such, in its own right, for in order to compress fusion fuel to the necessary heat and pressures to achieve nuclear fusion, it would have to be extraordinarily powerful.

But now we encounter two new alleged uses, seemingly quite unrelated to the above uses:

3) It has something to do with ballistic missile guidance; and,
4) It has something to do with radar stealth or camouflage, i.e., it has electromagnetic shielding or screening properties.

Bear these four points in mind, for as will be seen, there *is* a unified way to speculatively rationalize all of them scientifically, and as such, there may not only be a kernel of truth to the whole Red Mercury story, but that kernel of truth in turn implies that Russia's and the West's own subsequent attempts to debunk the whole story may *itself* be a component in disinformation intelligence operations.

2. "Ballotechnic Explosive"

The core of the Red Mercury story, as noted above, is that the substance was a powerful new type of "conventional" explosive known as a "ballotechnic." To understand this part of the story with the full implications of this component of the Red Mercury legend, it is necessary to go back to what I wrote about it in my book *The SS Brotherhood of the Bell*. I will cite the passage without the customary indented block quotation, but with its own block quotations, in its entirety:

"Red mercury, or mercury antimony oxide — chemical symbol $Hg_2Sb_sO_7$ — enjoyed a short, if notorious, career as the nuclear threat of the nineteen nineties. The story broke more or less simultaneously in various parts of the world, as the mysterious substance appeared to be behind a series of murders in the black market arms trade in post-apartheid South Africa, blocked smuggling attempts in the then recently-reunified Germany, and according to some stories, was even being sought by such 'nations of concern' as Libya and Iraq as a basis for their own nuclear weapons programs. Then, almost as soon as the mysterious compound appeared, denunciations of the whole substance and subject as a 'hoax' were issued by the United States Atomic Energy Commission and various other national and international nuclear regulatory agencies.

"But one physicist who did *not* dismiss the story as a pure hoax was the American inventor of the neutron bomb, Dr. Sam Cohen. For Cohen, the possibility of 'pure fusion' bombs — that is, hydrogen bombs that do *not* require an atom bomb as their trigger — was brought home to him while he was on a visit to the Lawrence Livermore Laboratories during a visit he made there in the spring of 1958. During this visit, Cohen was briefed on a pure fusion bomb project.

> This device, code-named DOVE, fascinated me. It contained no fissile material; rather, its explosive power derived from heavy hydrogen — deuterium and tritium. Because of its extremely low nuclear cost and its high yield – comparable to that of a very large conventional bomb — it would in a military application, represent a revolutionary new class of weapons. A device of this nature, having the yield the equivalent of 10 tons of TNT, could kill enemy troops out to hundreds of yards, with no significant urban destruction and contamination.[10]

The interest such a device held for the American military was more than just theoretical, for such devices would cost 'roughly one-hundredth that of a battlefield fission weapon, meaning that these things could be turned out by the hundreds.'

"The theory behind such a device was simple.

> The most promising approach was to use a large spherical high-explosive charge to concentrate the explosive energy in a very small capsule containing deuterium and tritium. In theory, this would cause the desired thermonuclear reaction. *The program proceeded for some years and finally was terminated for lack of progress. Later, the Los Alamos laboratory had a go at it. But to my disappointment, and theirs, the problem remained intractable. The program was ultimately ended.*[11]

---

10   Joseph P Farrell, *The SS Brotherhood of the Bell* (Kempton, Illinois: Adventures Unlimited Press, 2006), p. 279, citing Sam Cohen, "The Dove of War," *National Review* (December 17, 1995) 56–58, 76), p. 56.

11   Joseph P Farrell, *The SS Brotherhood of the Bell* (Kempton, Illinois: Adventures Unlimited

That is, theoretically, it should be possible to take the implosion detonator for a conventional atomic bomb, and instead of using it to compress a critical mass of plutonium to initiate fission, one could replace the plutonium with deuterium or tritium, compressing it sufficiently to increase the energy and density, and hence the statistical probability of collision (fusion) of heavy hydrogen atoms, and voila! One would have a 'small' hydrogen bomb without the need of an atom bomb to detonate it.

"But the effort failed, and it should be obvious why: no conventional explosive possessed sufficient brisance to compress the heavy hydrogen to pressures sufficient to initiate fusion reactions. This affords a clue to what Dr. Cohen is *not* telling in his article: the United States was searching for a *conventional* explosive of sufficient bursting power that, when used in an implosion detonator, would compress heavy hydrogen to fusion energies and pressures. If such could be found, then the atom bomb would become as extinct as the dodo bird, for two obvious reasons. First, if such a conventional explosive *could* be found, then it could be used as a powerful explosive in its own right, replacing the need for small yield strategic and tactical fission weapons, since it would be far smaller than a fuel air bomb of similar yield, and far less costly than its fission counterparts. Secondly, if such an explosive could be found, then, as Cohen intimates, it could be used as the detonator for a very small, 'clean,' neutron-emitting hydrogen bomb, or, as Cohen does *not* intimate, as the detonator for the city-and-county-cracking blockbuster strategic hydrogen bomb. In either case, the cost would be far less than a conventional thermonuclear bomb.

"However, as Cohen relates, there the story ended, *until* the crack-up of the Soviet Union and the new Russian Republic's willingness to be more open about its nuclear weapons research.

> Several years before Livermore began DOVE, the Soviets had started their own "pure-fusion" development. Unlike the U.S. they were quite open about it, claiming it was directed solely for peaceful applications. In 1957, Soviet nuclear-weapon designer I.A. Astsimovich presented a paper in Geneva describing experiments done in 1952, based on the same high-explosive implosion technology used in DOVE. He claimed progress had been made. Shortly thereafter, however, Soviet researchers stopped all public mention of the project.
>
> On the other hand, the Soviet military had no hesitation in writing about such devices in their open military literature. In 1961, Colonel M. Pavlov, writing in *Red Star*, discussed almost precisely what I had briefed Paul Nitze

Press, 2006), p. 279, citing Sam Cohen, "The Dove of War," *National Review* (December 17, 1995) 56–58, 76), p. 56, emphasis added.

on. Pavlov's calculations of weapons effectiveness were almost identical to mine, which were classified. This indicated to me that although the Soviets were not talking about research on DOVE, they were doing it.[12]

Then, in 1992, a Russian nuclear weapons expert revealed details about what the Russians called 'third generation nuclear weapons,' weapons that could 'double the yield' with a 'hundredfold reduction of weight compared to existing weapons.'[13] Cohen cites another Russian authority on the subject as stating 'You can drop a couple of hundred little bombs on foreign territory, the enemy is devastated, but for the aggressor there are no consequences,'[14] for with such weapons there is none of the deadly, long-lasting radioactive fallout. Again, one is reminded of the Nazis' use of fuel-air bombs in their rocket batteries on the eastern front, only in this case, it really *is* a combination of the phrases 'tactical nukes' and 'carpet bombing.'

"As Cohen explains, the 'doubling of yield' with a 'hundredfold reduction in weight' clearly indicates that the Russians were *not* talking about standard battlefield tactical nuclear weapons, since even the most pure plutonium still had to be at least a few hundreds of grams simply to have enough material to generate spontaneous fission. Below a certain threshold of weight, fission was impossible, and a one hundredfold reduction would make a fission weapon inconceivable. The Russians therefore had to have been talking about a pure fusion weapon, about 'some version of DOVE, *based on a detonation technology that doesn't exist in the United States.*'[15] In other words, there were only two options for interpreting the Russians' remarks: either they were lying, or they had discovered the holy grail of thermonuclear bomb engineering, a conventional explosive with enough brisance to compress heavy hydrogen to fusion pressures and energies.

"And with this Cohen comments, albeit only briefly, on a whole new type of conventional explosive, of which red mercury is but one substance:

> In recent years unclassified research has been conducted on a new class of materials (including red mercury), referred to as ballotechnics. *These materials use a number of elements in low density powder form.* When they are subjected to high-pressure shock compression, chemical reactions take place which under

---

12 Joseph P Farrell, *The SS Brotherhood of the Bell* (Kempton, Illinois: Adventures Unlimited Press, 2006), p. 281, citing Sam Cohen, "The Dove of War," *National Review* (December 17, 1995) 56–58, 76), p. 56.
13 Ibid.
14 Ibid.
15 Joseph P. Farrell, *The SS Brotherhood of the Bell* (Kempton, Illinois: Adventures Unlimited Press, 2006), p. 282, citing Sam Cohen, "The Dove of War," *National Review* (December 17, 1995) 56–58, 76), p. 57, emphasis added.

certain conditions can produce energy concentrations considerably in excess of those from high explosives. Ballotechnics therefore offer a significantly greater prospect for success in attaining a very low yield pure-fusion weapon than the high explosive techniques we and other nations have explored.[16]

Cohen also notes that red mercury was allegedly developed in the former Soviet Union precisely as a detonator for nuclear warheads. Indeed, its efficiency as a detonator was so great that a bomb the size of a hand grenade would be sufficient to blow a large ship out of the ocean.[17] After taking note of the fact that the CIA and various other American agencies dismissed the 'red mercury' story as a hoax which they nevertheless were taking seriously — 'whatever that means,' Cohen quips[18] — he then says nothing more about it."[19]

And there my survey of the initial stages of the Red Mercury legend in *The SS Brotherhood of the Bell* ended. A number of important points must be summarized before we proceed:

1) The United States, and the Soviet Union, beginning in the 1950s, were searching for a means of detonation of hydrogen bombs *without* the need of an atom bomb as the "fuse." They were thus searching for a relatively "fallout-free" form of thermonuclear bomb;
2) This in turn implies that both nations were searching for a *powerful* and wholly *new* type of conventional explosive able to compress heavy hydrogen to fusion pressures, heat, and energies, which would be a powerful conventional explosive in its own right;
3) The American program, at least, is a matter of public *record,* and thus, *the kernel of the Red Mercury legend — that dealing with the allegation of the creation of a whole new class of very powerful "conventional" explosive — is far from being a hoax, but is true.* Small wonder, then, that Western intelligence agencies were quick to dismiss the whole Red Mercury scare as a hoax, for if the story were really appreciated for what it implied — a quick, atom-bomb-less route to the hydrogen bomb — nuclear proliferation became a nightmare, as thermonuclear wars, and terrorism, became a venture free of the fallout consequences to an aggressor state or terrorist group employing it;

---

16   Ibid., emphasis added in this book, absent in *The SS Brotherhood of the Bell.*
17   Joseph P. Farrell, *The SS Brotherhood of the Bell* (Kempton, Illinois: Adventures Unlimited Press, 2006), p. 282, citing Sam Cohen, "The Dove of War," *National Review* (December 17, 1995) 56–58, 76), p. 97.
18   Ibid.
19   *The SS Brotherhood of the Bell,* pp. 278–282.

4) Moreover, as Cohen rightly notes, Soviet sources and literature were not only quite *open* about their quest for the pure fusion bomb, they were also involved in the quest fully three *decades* before the emergence of the Red Mercury story in the 1990s! The Soviets were, in fact, researching the possibility in more or less the same time frame that the United States was. One has only to ponder, then, what might actually have been accomplished in those three decades after the early 1960s, when those Soviet sources suddenly became very quiet about the nature of that research.

5) Finally, the *timing* of the sudden Soviet silence on their research, after such a period of openness, is quite important; according to Cohen, the first mention of Soviet research occurred in 1957, and referred to experiments begun in 1952, and the last more or less open references appeared in 1961, close to the timing of Russia's detonation of its massive fifty-seven-megaton "Tsar" hydrogen bomb, and a few months later, the Cuban missile crisis.

Note those dates, for the *terminus ante quem* would place the beginnings of the research a scant seven years after the end of the Second World War, and the *terminus post quem* would not only place it close to the dangerous nuclear confrontation of the Cuban Missile Crisis, but, as we shall see in the next chapter, to a period of the most highly classified secret Russian research of a seemingly entirely different matter altogether. Finally, these dates would coincide with the period of open atmospheric tests of hydrogen bombs by both the United States and the Soviet Union. These three things — the close proximity to World War II, the period of atmospheric hydrogen bomb testing, and the sudden silence of Soviet sources after the early 1960s — will be significant clues to bear in mind throughout the rest of this book.

But what of "ballotechnic" explosives themselves? How would they really work? One method — missed by most commentators — is that ballotechnics may not even be explosives in the conventional sense at all. A conventional explosive releases its energies in the form of rapidly expanding gases. But a ballotechnic explosive might actually accomplish its peculiar and unique detonating qualities by an entirely different mechanism: the sudden release of ultra-high heat and electromagnetic radiation. Instead of generating a compression shock wave of *pressure* from detonation of an implosion core around an atom bomb, a ballotechnic might release such intense heat in the form of electromagnetic radiation that fusion would be initiated.[20]

---

20 See the article "Ballotechnic nuclear bomb," www.everything2.com/index.pl?node= ballotechnic%nuclear%20bomb, p. 1.

3. A "Stimulated Gamma Ray Emitter"

In this regard, it is interesting to note that one alleged property of Red Mercury was speculated to be its capacity as a high emitter of gamma rays under certain conditions of stimulation. One commentator on the internet, William Yerkes, posted the following comment to Anne Marie Helmenstine, Ph.D.:

> I don't think it's a mercury derivative at all — I think the label "mercury" is a sort of metaphor, suggested by the proximity of the planet Mercury to the Sun and also, perhaps, by the association in the mind between temperature measurement and mercury.... "Mercury" seems to be an obscure and oblique reference to high temperatures and, therefore, perhaps, to fusion. And the name may well have been a project name, later adapted to the product itself. As to "red," this is the color usually used to signify danger, and often to signify "heat" as well. Also apropos, if we assume the stuff exists. Assuming this, I'll call it RM.
>
> The story is that RM is shock-sensitive ballotechnic. I suspect that this too is related only in an obscure somewhat metaphorical way to the material. *I suspect that RM is a stimulated gamma emitter....*It is possibly hafnium 178m2 or another substance that, similarly, can be pumped to a high state and collapsed nicely with the attendant emission of high energy photon(s). I speculate that stimulated by the input "shock" of a burst of gamma or possibly ionizing radiation from an electronic or radioactive "trigger," the nuclei of the "pumped" RM atoms, if they exist, become extremely unstable and rapidly (<1 nano-second perhaps) collapse to a lower energy level, releasing high energy photons in the form of gamma with, presumably, an energy level >5000 (million electron volts). Open sources cite gamma radiation in excess of this level.
>
> The ignition threshold for deuterium-tritium fusion (D+T) is, as I understand things, 5000 (million electron volts).[21]

The mention of the hafnium metastable 178 spin 2 isomer is significant, for this isomer was itself the subject of a recent U.S. project, sponsored by the Defense Advanced Research Projects Agency (DARPA), to achieve a "hafnium isomer bomb," an extraordinarily powerful weapon.

As mentioned in my book *The SS Brotherhood of the Bell,* a nuclear isomer has most of its energy locked up in its *high-spin state,* such that, if a method could be found to de-excite such atoms — to suddenly shock and slow them

---

21 William Yerkes, "Thoughts and Speculations," www.chemistry.about.com/od/chemistryarticles/a/aa100404a_2.htm, p. 1.

down, so to speak — enormous amounts of energy would be released for a small investment of material, as the energy loss of angular momentum in such isomers would be suddenly released as photons, i.e., as a burst of high frequency electromagnetic radiation: gamma rays. Indeed, as I also observed in that book, one little-known physical effect, the Mossbauer Effect, was observed when certain nuclear materials, under conditions of acoustic stress, would release *cohered* bursts of gamma rays, i.e., would become the "optical cavity" for a gamma ray laser, itself an extraordinarily powerful weapon.[22] In other words, Yerkes speculates that such cohered gamma ray bursts from a de-excited nuclear isomer might possess high enough energy potentials to initiate a fully fledged thermonuclear fusion reaction without an atom bomb.

There is a highly significant clue here, and the reader will have noticed it at once: *the basic mechanism in use by ballotechnic explosives might very well be their high-spin state, and the sudden "shock" or "slowing down" of that state.* The careful reader will also have noted that one feature of such ballotechnic explosives, as noted by neutron bomb inventor Dr. Sam Cohen himself, was that they existed in a *powder* state. Shades of David Hudson, for not only did *his* material "explode" in sudden bursts of gamma rays, but the testimony of alchemy itself has been that the Stone of the Philosophers radiates "light" and is itself indestructible! Adding all this together, then, it is looking increasingly unlikely that the Red Mercury story was or is a mere hoax; there are too many coincidences piling up, and more importantly, there is a way of rationalizing a scientific basis for its alleged properties.

4. *The Strange Contradiction of DuPont's Mercury Antimony Oxide*

If one examines the record closely, it is seen to contain a peculiar contradiction. As the excerpt from my book *The SS Brotherhood of the Bell* made clear, one possible chemical compound candidate for Red Mercury is the substance mercury antimony oxide. This compound was first synthesized and registered in the international chemical register by the American defense contractor DuPont in 1968, with the registry number 20720-76-7.[23] What is so unusual about this entry is that "no documentation exists to explain its possible uses and the company was not prepared to comment."[24] However, another source implies that it is a liquid explosive of high density![25]

---

22 See the discussion in my *SS Brotherhood of the Bell,* pp. 242–248; 294–296.
23 Eric Singley, "What is Red Mercury?" www.groups.google.com/group /sci.chem/msg/69a33 ee6f25c5073?q=group:sci.chem+ins, p. 4.
24 Ibid.
25 Anne Marie Helmenstine, Ph.D., "What is Red Mercury?" www.chemistry.about.com/cs/chemicalweapons/f/blredmercury.htm, p. 1.

Consequently, if one takes what has been said thus far and combines it together, and if one moreover accepts for the sake of argued speculation that what has been said is true, then one is dealing with the following elements:

1) A liquid explosive
2) of high density
3) whose ballotechnic properties are the result of de-excitation of its high-spin, and perhaps isomeric or superdeformed, states, resulting in an "explosion" or extreme burst of high-frequency gamma rays.

Note also the new peculiar resemblance to the Philosophers' Stone, for in Cohen's version, it is a *powder* resembling David Hudson's monatomic platinum group elements, and in the DuPont mercury antimony oxide version it is a high-density *liquid*. And lest it be forgotten, one of the primary candidates for the alchemical tincture was precisely a tincture of antimony; moreover, mercury antimony oxide would indeed be red in color.

5. The Platinum Metals Group Again: Hudson and the Russian Involvement Reconsidered

The mention of Hudson's platinum group monatomic high-spin-state elements pries open yet another door to the Red Mercury mystery, that of the possible role of platinum group metals — or to be more precise, *a* platinum group metal — in its chemical recipe.

In their article "Nuclear Trafficking Hoaxes: A Short History of Scams Involving Red Mercury and Osmium-187," researchers Kenley Butler and Akaki Dvali note that one little known substance involved in the 1990s Red Mercury nuclear trafficking scare was the platinum group metal isotope osmium-187:

> Two non-fissile substances that frequently have been used by con artists as substitutes for nuclear materials are so-called *red mercury* and osmium-187. Hoaxes involving both substances have become legendary after being the subject of widely reported trafficking attempts throughout the 1990s. *A major reason these scams have been so widespread and common is likely the fact that there is some truth in the claims made by the con artists.* Red mercury is the name given to an alleged nuclear weapons ingredient that does not exist in the form ($Hg_2Sb_2O_7$) and with the characteristics described by nuclear scam artists. Some experts have suggested, however, that red mercury is in fact another name for lithium-6, a substance that *can* be used in the production of compact and highly efficient thermonuclear devices. Osmium-187

is a bona fide nonradioactive material not used for weapon construction, but because it is indeed an expensive commodity and one that is produced through a process similar to uranium enrichment, nuclear traffickers seized on it as a marketable product.[26]

While the authors' attitude toward the Red Mercury story is obvious from the title of their article and their initial comments, the only thing further that they offer about osmium-187 is that "No sooner had red mercury begun to disappear from media reports than nuclear traffickers began touting a new commodity — osmium-187 — as a vital substance for the creation of nuclear weapons. Osmium is a metal of the platinum group used to produce very hard alloys for fountain tip pens, instrument pivots, phonograph needles, and electrical contacts."[27] They also note that the naturally occurring isotope osmium-187 is not included on any U.S. government agency list of prohibited or regulated materials. Its only other possible use, they note, might be construed by con artists to be as a tamper for a nuclear weapon because of its extreme density, a role however for which the much less costly beryllium, with its high neutron reflective properties, is much better suited.[28]

But as we saw in part two, osmium can also exist in the rare high-spin state, a state leading to "superdeformities" of the ordinary atomic nucleus, able to spontaneously fission and capable of "new radioactivities." About *this* possibility, which so mystified David Hudson, they say nothing. Nevertheless, it *exists in the peer-reviewed physics literature.*

There is something else to be noticed here, and it is quite important to the building case. Recall that in the previous section of the book, Hudson actually consulted Russian expertise in order to understand the mysterious properties of his platinum group metals. In fact, *without* that expertise, he could not have proceeded with the analysis of his material.[29] So once again, the Russians appear to have been doing *some* sort of sophisticated analysis of platinum group metals *long before the Red Mercury story broke.* The question is, *why?* Before going into the detailed analysis of the next chapter, we are permitted to speculate here that they had perhaps themselves, as early as the 1960s, already discovered some of the high-spin superdeformed states of this group of metals. If so, then the Red Mercury legend takes on yet another aspect of authenticity, for why else would they have developed the sophisticated analytical techniques they did if they were only analyzing *ordinary* platinum

---

26 Kenley Butler and Akaki Dvali, "Nuclear Trafficking Hoaxes: A Short History of Scams Involving Red Mercury and Osmium-187," www.nti. org/e_research/e3_42a.html, p. 2.
27 Ibid., p. 3.
28 Kenley Butler and Akaki Dvali, "Nuclear Trafficking Hoaxes: A Short History of Scams Involving Red Mercury and Osmium-187," www.nti. org/e_research/e3_42a.html, p. 3.
29 Q.v. pp. 48–50.

group metals, which would be analyzable by other less costly and less involved chemical means?

And there is another connection to Hudson as well that is worth recalling in this context. As noted above, one speculated property for a ballotechnic explosive substance is not only its high-spin state, but as an emitter of *gamma rays*. As we saw in part one, Hudson's mysterious platinum group compound *exploded* when dried in the sun, and, as we also saw, Hudson eventually learned that these "explosions" were due to sudden bursts of gamma ray emissions![30]

C. Its Strangely Alchemical Recipe and the Strange Behavior it Conjured

Beyond all of Red Mercury's strangely alchemical associations — its form as either a powder or liquid, its high density and spin state, its emissions of high bursts of light and gamma rays, and its obvious associations with two of the colors in the sequence of colors denoting a successful confection of the Philosophers' Stone, yellow and red, all of these things being in turn the now familiar properties of the Philosophers' Stone itself — there is also the matter of its strange recipe or, as the case may be, recipes. And this, as they say, is where it gets *really* interesting.

1. The Alleged Classified Russian Report on Red Mercury

An entry into this most interesting aspect of the story is afforded by a strange comment found on one lone internet site, which referred to an alleged classified Russian report on Red Mercury:

> According to a report by Yevgeny Primakov, chief of Russia's external intelligence service, to (then) Secretary of State Gennady Burbulis of March 24, 1992, red mercury is used in the production of high-precision fuses for conventional bombs and fuses for nuclear bombs, in the starting of nuclear reactors, in the production of anti-radar coatings for military hardware and in the manufacture of self-targeting warheads for high-precision missiles.[31]

While there is little new here regarding the alleged uses of Red Mercury, what is unusual is the supposedly high level from which the whole catalogue of alleged applications originates: at the pinnacle of Russia's intelligence and foreign services.

---

30 Q.v. pp. 44–45, 61–62.
31 www.groups.google.com/group/sci.chem/msg/69a33ee6fc5073?q, p. 1.

The site then follows with a clear indication that, in the opinion of the compiler of the information, this too is a part of the hoax!

Certainly this makes a great deal of sense in the context that the whole 1990s nuclear trafficking scare was a deliberate intelligence operation by the Russians in collusion with Western powers against potential terrorist threats. Likewise it makes a great deal of sense if, as has been suggested, the Russians mounted a further disinformation operation against the West wrapped *inside* of their collaboration with the West, for what better way to stamp both operations with the air of authenticity than by the deliberate leak of a contrived "top secret" report? Certainly the Russian intelligence agencies were very skilled, and practiced, in such operations.

2. The Press Campaign and Subsequent Russian Debunking

With this in mind, the subsequent "press campaign" concerning Red Mercury, both inside of Russia and in the West itself, makes a little more sense. In their article mentioned previously, Kenley Butler and Akaki Dvali summarize this initial press campaign, in which even *more* alleged uses for the substance were circulated:

> References to red mercury began to appear in major Russian and Western media sources in the late 1980s and early 1990s. The articles were never specific as to what exactly red mercury was, but the accounts claimed that the substance was a valuable strategic commodity and a necessary component in a nuclear bomb and/or that it was important in the production of boosted nuclear weapons. Supposedly citing a leaked Russian government memorandum, an April 1993 article in the widely-read Russian daily *Pravda* reported that red mercury is "a *super-conductive material used for producing high-precision conventional and nuclear bomb explosives, 'stealth' surfaces and self-guided warheads.* Primary end-users are major aerospace and nuclear-industry companies in the United States and France along with nations aspiring to join the nuclear club, such as South Africa, Israel, Iran, Iraq, and Libya." Red mercury was peddled throughout Europe and the Middle East by Russian businessmen, who made fortunes in the process.... Described as a *brownish powder or a red liquid,* red mercury was said to originate from various locations in the USSR, namely Ust-Kamenogorsk, Kazakhstan, and Krasnoyarsk, Novosibirsk, and Sverdlovsk in Russia.
>
> Western media also carried accounts of red mercury and its nuclear applications. According to a July 1993 article in *Nucleonics Week,* red mercury was a code word used in the USSR nuclear weapons program since the 1950s to describe enriched lithium-6, which, according to the article, can be used

to produce tritium, which, when fused with deuterium, can be used in the fusion stage of a thermonuclear weapon. Lithium-6 received its code name because of the red-hued impurities in the mercury used to produce lithium-6. According to the article, the USSR built a large complex in the early days of its nuclear weapon program to produce and stockpile lithium-6.

The *Nucleonics Week* article was followed by two television programs on red mercury produced by the British Broadcasting Corporation as part of its *Dispatches* series. *Trail of Red Mercury* (1993) and *Pocket Neutron* (1994) presented "startling new evidence" that Russian scientists had designed a simple, cheap, pure fusion weapon or neutron bomb, the size of a tennis ball, using a "mysterious compound" called red mercury. A June 1994 article in the venerable *International Defense Review* quoted Western and Russian nuclear physicists as confirming the existence and destructive capabilities of red mercury. One of those quoted, U.S. nuclear physicist Sam Cohen, to this day continues to write passionately about the nuclear applications of red mercury, which he describes as a "ballotechnic" explosive that, "when ignited, does not actually explode but stays intact long enough to produce the enormous temperatures and pressures sufficient to enable deuterium-tritium fusion."[32]

If the intention of the Russian and Western intelligence services was to concoct a disinformation campaign designed to entrap terrorist cells trading in the black market for nuclear arms and materials, then they could not have done a better job of placing the stories than in these prestigious professional and media journals and outlets. Once again, however, note the reference to high spin in the form of super-conductivity and its peculiarly alchemical-sounding powdered and liquid form.

3. The Strangely Contradictory Statements of British Physicist Frank Barnaby

The subject of Red Mercury's recipe — which would at least provide a measure of corroboration or rebuttal of the idea that the substance was genuine — becomes even more peculiar with the strange behavior exhibited by those directly involved with the story. On February 21, 1992, for example, Russia's then President Boris Yeltsin signed Decree No. 75-RPS entitled "On the Promekologiya Concern." This decree granted the company Promekologiya Company, based in Evkaterinenburg, exclusive rights to "produce, store, transport, and sell 84 tons of red mercury for $24.2 billion

---

32 Kenley Butler and Akaki Dvali, "Nuclear Trafficking Hoaxes: A Short History of Scams Involving Red Mercury and Osmium-187," www.nti.org/e_research/e3_42a.html, p. 2, emphasis added.

over a three year period to a Van Nuys California company called Automated Products International."[33] The decree was later rescinded on March 20, 1993, amid confusion of whether or not any of the supposed material had ever actually changed hands.[34] Following this bizarre turn of events, the accounts of Red Mercury trafficking in Russia dried up and withered away, and a media campaign was apparently mounted to debunk the whole idea of Red Mercury, as government authorities began to deny "its very existence."[35]

As these sources indicate, at the height of the Red Mercury scare, the story went all the way to the top of the Russian government, to the office of the Russian president himself. And as they also indicate, at one point the story involved an American corporation. So whatever was going on — hoax, anti-terrorist disinformation campaign, or genuine article — it came from a very high level within Russia, and implied a measure of non-Russian involvement as well.

This strangely contradictory behavior prompted some strangely contradictory behavior of its own from British physicist Frank Barnaby. At first on record as to his belief that the substance and at least some of the claims for it were genuine, Barnaby noted that "the thing that has to give you pause is the seniority of the Russians who are claiming it exists and has these applications, and it's very hard to see these people involved in a hoax."[36] But then, later on, Barnaby apparently reversed himself, for according to an article in the British newspaper, *The Guardian:* "Nobody would dream of getting that stuff for a dirty bomb," the article reported, noting that Barnaby was a nuclear physicist at the United Kingdom's Atomic Weapons Establishment at Aldermaston during the 1950s. According to the article, Barnaby stated that "For a terrorist it would offer no significant advantages over an ordinary high explosive or, if they wanted a dirty bomb, a radioactive source. To go to the trouble of spending huge amounts of money on red mercury makes no sense at all."[37] It makes no sense, unless of course the substance's alleged property of being able to trigger thermonuclear explosions without an A-bomb as the "fuse" possesses some basis in reality.

---

33   Kenley Butler and Akaki Dvali, "Nuclear Trafficking Hoaxes: A Short History of Scams Involving Red Mercury and Osmium-187," p. 3. As an aside, it is worth recalling that Evkaterinenburg, known as Sverdlovsk during the Soviet era, was the place where the Tsar and his family were murdered by the Communists. It was also the place U-2 pilot Francis Gary Powers was to photograph on his fly-over of the Soviet Union when he was shot down.

34   Kenley Butler and Akaki Dvali, "Involving Red Mercury and Osmium-187," www.strata-sphere.com/blog/index.php/archives/3152, p. 5.

35   Butler and Dvali, "Nuclear Trafficking Hoaxes: A Short History of Scams Involving Red Mercury and Osmium-187," p. 3.

36   "What is Red Mercury?" www.groups.google.com/group/sci.chem /msg/69a33ee6f25c7073 ?q=group:sci.chem+ins., p. 3.

37   David Adam, "What is red mercury?" *The Guardian,* September 30, 2004, www.guardian.co.uk/science/2004/sep/30/thisweekssciencequestions1, p. 1.

What is one to make of all this confusion? Was it a hoax? A deliberate disinformation operation of the Russian government, wrapped within yet another intelligence anti-terrorism operation being run jointly with the West? Or was it the genuine article, and were the subsequent efforts of the Russian government to brand it as a hoax due to the fact that it was doing hasty "damage" control in the wake of a disastrous policy of "openness" on the part of the Yeltsin government?

Clearly, no examination of the record of Red Mercury stories by themselves will yield an answer, for that record is ambiguous at best and contradictory at worst. One must turn to actual statements of its alleged recipe in order to find a possible scientific rationalization that could shed light on the question of whether or not it was a hoax, or the genuine article.

4. Various Recipes and Stable Features

Not surprisingly, one of the first hints at the recipe of Red Mercury came from American physicist and neutron bomb inventor Sam Cohen, who suggested it was manufactured under "ultra high pressure. You knock all these electrons out so it's not the same atom. It pulls a lot more energy per gram than any other explosive that I've ever heard of."[38] When pressed for more information, Cohen also observed that "I think it's very difficult to tell how it was made from examining the substance itself."[39] Ultra-high pressure manufacture, as will be seen, is indeed a strong possibility for being one component in the recipe of Red Mercury.

But the actual *chemical* recipe remains a mystery. Candidates include mercury sulfide — the alchemists' cinnabar — mercury iodide, and of course mercury antimony oxide, which, as already has been seen, has its *own* dubious ambiguity in the record, with some sources claiming it has powerful explosive properties, and with another source — DuPont, its actual chemical inventor and fabricator — declining to talk about it at all! Moreover, one does not have to look far to discover alchemical references to antimony but these do not seem to imply any explosive use! Nonetheless, antimony was a substance of some interest to alchemists in their quest to confect the Philosophers' Stone.

The mediaeval alchemist Roger Bacon, for example, devoted a whole treatise to the subject called *Tract on the Tincture and Oil of Antimony*. Notably, Bacon twice refers to this tincture of antimony as having a red color.[40] Then,

---

[38] "Micro-Nukes — Can and Do They Exist?" www.wtcnuke.com/micronukes.php, p. 4. Cohen's remarks were first reported during the Delmart Vreeland affair and his allegations of government prior knowledge of 9/11.

[39] Ibid.

[40] Roger Bacon, *Tract on the Tincture and Oil of Antimony*, www.levity.com'alchemy/rbacon2.html, pp. 4, 7.

toward the end of the treatise, comes this astonishing statement made in the context of the Philosophers' Stone and its transmutative powers:

> The second multiplication is an Augmentum quantitatis of the stone with its former power, in such a way that it neither loses any of its power, nor gains any, *but in such a manner that its weight increases and keeps on increasing ever more, so that a single ounce grows and increases to many ounces.*[41]

Bacon's comment recalls that of another famous alchemist, Paracelsus, whom we have previously mentioned in connection to this mass anomaly, and who apparently observed more or less the same thing much later:

> No precise weight can be assigned in this work of projection, though the tincture itself may be extracted from a certain subject, in a defined proportion, and with fitting appliances. For instance, that Medicine tinges sometimes thirty, forty, occasionally even sixty, eighty, or a hundred parts of the imperfect metal.[42]

While not saying it in so many words, Bacon and Paracelsus appear to be suggesting the same sort of high mass and density anomalies as were alleged for Red Mercury itself, and doing so centuries earlier to boot!

And even though Bacon does not record for his "tincture of antimony" anything remotely resembling the uses alleged for Red Mercury — much less its use as a detonator in a fusion bomb — Bacon *does* make a very suggestive comment concerning the association of antimony to celestial alignments, and to one particular constellation:

> The Arabs in their language, have called it Asinat vel Azinat, the alchemists retain the name Antimonium. *It will however lead to the consideration of high Secrets* [sic], *if we seek and recognize the nature in which the Sun is exalted, as the Magi found that this mineral was attributed by God to the Constellation Aries.*[43]

Antimony, in other words, was in the alchemical lore deliberately associated not only with celestial geometries in general, but to one constellation in particular: Aries, *Mars*, the god of war. A suggestive connection if ever there was one!

---

41 Roger Bacon, *Tract on the Tincture and Oil of Antimony*, www.levity. com.alchemy/rbacon2.html, p. 12, emphasis added.
42 Philippus Theophrastus Areolus Bombastus von Hohenheim, a.k.a. Paracelsus, "Concerning the Projection to be Made by the Mystery and arcanum of Antimony," *The Aurora of the Philosophers*, www.levity.com/alchemy/paracel3.html, p 20.
43 Bacon, op cit., p. 2, emphasis added.

As is also evident, the alchemists were familiar with some substance of abnormally high density and weight.

In any case, notwithstanding this further ambiguity in the official explanations surrounding mercury antimony oxide and DuPont's reluctance to discuss the compound, in an article in *New Scientist* magazine, reporter Jeff Nyquist suggested that the mysterious Red Mercury even played a role in the recent American allegations of Iraq possessing "weapons of mass destruction," allegations that formed a crucial justification for its invasion and military occupation of that country.[44] *If*, as I have previously suggested, the Anglo-American invasion of Iraq was really concerned about weapons of mass destruction of a far more ancient sort — the fabled Sumerian "Tablets of Destinies" — and *if* as I have also suggested elsewhere that these tablets were crystalline in nature, then indeed there may be an "alchemical" connection, given alchemy's own preoccupation with metals and gems, i.e., with substances having lattice structures, then it may be, as Nyquist suggested, that "American intelligence has evidence of something dire, something they don't want to tell us."[45]

In this respect, some suggested compounds for Red Mercury involve elements with very unique lattice structures, structures that make them efficient neutron emitters. One such suggestion was that the substance was a compound of ordinary mercury and plutonium-239, the fuel for a plutonium bomb. Other mixtures were mercury and californium, or mercury and polonium. In other words, *one stable feature of the alleged recipe seems to be some compound of mercury and an efficient neutron-emitting radioactive isotope.* Such characteristics gave Red Mercury a definite "shelf life," depending on what particular isotope was alloyed with the mercury itself. Indeed, this aspect of its recipe seems to involve the previous one, manufacture under conditions of ultra-high pressure, *for yet another stable feature of its alleged recipe is that it was manufactured in a reactor, under high pressure and neutron bombardment.*

Gathering together all that has been learned about the substance, its alleged uses, properties, and manufacture, we have the following stable elements:

1) It could allegedly be used as a radar-absorbent paint, and possessed radar stealth properties;
2) It possessed super-conductive properties, suggestive of a high-spin state;
3) It was manufactured under ultra-high pressure and in a reactor under neutron bombardment;

---

44   Hirt R. Poovey, "The Red Mercury Nightmare?" www.prisonplanet.com/analysis_poovey_122602_redmerc.html, p. 1.
45   Ibid.

4) The substance was a compound of mercury and other elements, most likely radioactive;
5) It had unique lattice properties that allowed for anomalous and great density and made it an efficient emitter of neutrons and in some versions of gamma rays;
6) In one instance, that of mercury antimony oxide, there appears to be an alchemical connection, suggesting that perhaps alchemical texts themselves were the ultimate instance for the substance;
7) And finally, of course, there is the whole legend that the compound was a special kind of "ballotechnic" explosive, able to trigger a hydrogen bomb without the need of an atom bomb for the detonator.

So once again we are confronted with a substance that would appear to involve at least *one* verifiable property — a high-spin state — which as we saw in part two in our examination of David Hudson's anomalous material, could emit sudden "explosive" bursts of radiation in the form of gamma rays. Moreover, like the mysterious Red Mercury, Hudson's substance exhibited anomalous weight properties. In Hudson's case, the anomaly was in the weight *loss*, whereas in Red Mercury's case, the anomaly is in its unusually high density, its *mass,* its "weight" *gain*. But what was the reason for manufacture under high pressure?

D. Conclusions and Connections

The confection of Red Mercury under high pressure by neutron capture in a reactor suggests the possibility that the Russians, like the Americans, had noticed anomalous yields exceeding calculated values in their H-bomb tests of the late 1950s and early 1960s. And like their American counterparts, they had probably also noticed that these yields *varied with time and therefore with local celestial geometry.* By confecting the substance under these conditions — high pressure and neutron capture in a reactor — they were *perhaps* trying to approximate the conditions inside the fusion capsule at the center of the implosion that set off a thermonuclear bomb, which, again, is one of "high pressure" (and therefore heat) and neutron bombardment. If this line of speculation be true, then the Russians were engaged in materials engineering of a substance that could transduce that very same anomalous energy, the very energy of those local geometries of time itself into this world, the very energies that were being evidenced in thermonuclear explosions. They were indeed perhaps after their own version of the Philosophers' Stone. But in order to understand what the process of their scientific rationalization may have been in doing so, indeed, in order to have a real scientific foundation to the Red Mercury legend, we must turn to the breathtaking and brilliant work of one of Russia's most gifted and virtually unknown physicists, Dr. Nikolai A. Kozyrev, and to the possible reasons his work was so highly classified by the former Soviet Union. And as will be seen in the next chapter, with Kozyrev we are, once again, paradoxically confronted with the persistence of the main tenets of alchemy and its claims for the Philosophers' Stone.

Time is Not a Scalar
Nikolai Kozyrev's Causal Mechanics

•••

"Since, then, our matter is our root for the white and the red, necessarily our vessel must be so fashioned that the matter in it may be governed by the heavenly bodies. For invisible celestial influences and the impressions of the stars are in the very first degree necessary for the work...."
Philippus Theophrastus Areolus Bombastus von Hohenheim,
a.k.a. Paracelsus[1]

Without any possibility of doubt, Dr. Nikolai Kozyrev is one of the unsung giants of twentieth-century theoretical and experimental physics. In him was combined that rare Einsteinian mixture of the profound thinker and the careful observer and experimenter. Even with the advent of the internet, however, this intellectual giant's work remains largely unknown in the West, and even, to a certain extent, in his native Russia.

The reason for his relative lack of fame is quite simple. As researcher David Wilcock, who first brought Kozyrev's work to the attention of the modern Western audience, states, "The awesome implications of his work, and of all those who followed him, were almost entirely concealed by the former Soviet Union...."[2] In other words, Kozyrev's work was so awesome and extraordinary

---
1 Paracelsus, "The Aurora of the Philosophers," *Paracelsus and His Aurora & Treasure of the Philosophers, As also The Water-Stone of the Wise Men: Describing the matter of, and manner how to attain the universal Tincture. Faithfully Englished. And Published by J.H. Owen.* (London: Giles Galvert. 1659). Text may be found at www.levity.com/alchemy/paracel3.html, p. 27.
2 David Wilcock, by Sepp Hasslberger, "Aether, Time, and Torsion," www.blog.hasslberger.com/

in its implications, not only for the development of the foundations of theoretical physics, but also for its dangerous potential applications, that the Soviet leadership wisely classified it at the very highest levels. Indeed, it was Kozyrev, in fact, who laid the experimental basis and outlined the theory of Soviet research into that area of physics often called "scalar" physics, but it might equally, and probably with more justification, be called "torsion" physics. But before one can have an appreciation of the enormous scope and implications of his work, one must have an insight into the man and his extraordinary life.

A. A Brief Curriculum Vitae of Dr. Nikolai A. Kozyrev

Dr. Nikolai A. Kozyrev lived from 1908 to 1983,[3] a time frame that makes him a contemporary of several famous scientists: Einstein, Planck, Heisenberg, Schrödinger, Jordan, Dirac…on and on the list could go. And of course he is a contemporary of yet another unsung physics genius, and a namesake: Nikola Tesla. But there is a deeper connection between the two men, as David Wilcock observes.

Tesla, Wilcock notes, stated in 1891 that the physical medium "behaves as a fluid to solid bodies, and as a solid to light and heat." Moreover, with "sufficiently high voltage and frequency" the medium itself could be accessed. This was, as Wilcock correctly notes, Tesla's "hint that free energy and anti-gravity technologies were possible."[4] It is Tesla's assertion that the medium has fluidlike properties that ties his work directly with the work and thought of Dr. Kozyrev.[5]

Publishing his first scientific paper at the "tender age of seventeen,"[6] Kozyrev graduated from the University of Leningrad when he was twenty years old with a degree in physics and mathematics, and "by age twenty-eight was widely known as a distinguished astronomer who had taught at several colleges."[7] In 1936, however, Dr. Kozyrev ran afoul of the repressive regime of Stalin, during the "purge" crazes of the time, and in 1937 he began eleven long years in the Soviet gulag system until his release in 1946.[8] It was during

---

mt/mt-view.cgi/1/entry/66, p. 1.

3 David Wilcock, "The Breakthroughs of Dr. Nikolai A. Kozyrev," *The Divine Cosmos*, www.divinecosmos.com, p. 1.

4 David Wilcock, "The Breakthroughs of Dr. Nikolai A. Kozyrev," *The Divine Cosmos*, www.divinecosmos.com, p. 2.

5 Ibid.

6 Ibid., p. 3. A Russian source, however, states that Kozyrev published his first paper when around 15-16, which I am more inclined to believe: See Dr. Lavrenty S. Shikhobalov, "N.A. Kozyrev's Ideas Today," p. 302.

7 Ibid.

8 Ibid.

Dr. Nikolai A. Kozyrev

this period of harsh servitude in the Gulag that Kozyrev noticed, and reflected deeply upon, the phenomenon that would guide his subsequent thought and experimentation upon his release. This was the fact that so many different forms of life exhibited asymmetrical properties and a spiraling growth pattern.[9] In this, Kozyrev's approach is similar to that of the Austrian naturalist Viktor Schauberger, who noticed the same spiraling growth patterns, and who, like Kozyrev, would ponder and investigate the underlying physics of this pattern, and come to very similar conclusions.[10] Unlike Schauberger, who was not a trained physicist or mathematician but a naturalist and forester, Kozyrev *was* formally and academically trained in these disciplines. Thus, Kozyrev, like Schauberger, would pursue his observations to the overturning of many of theoretical physics' most prized dogmas, but unlike Schauberger, he would frame his philosophical objections with the formal mathematical precision a physicist could understand and accept.

Wilcock notes that it was during the winter of 1951–52 that Dr. Kozyrev began the first of a series of experiments that would last fully thirty-three years, experiments that were, and are, by any lights, intriguing, and to say the least, controversial.[11] What made them controversial was that Kozyrev viewed the spiraling patterns of nature and of life itself as a manifestation of *time*,[12] and that as a consequence of this view, time itself was not a dimensionless "coordinate point" or a "scalar" as scientists and mathematicians would call it, but that time was *itself* a kind of physical *force*, and a very subtle one at that.

---

[9] Ibid.

[10] Q.v. my book *Reich of the Black Sun: Nazi Secret Weapons and the Cold War Allied Legend* (Kempton, Illinois: Adventures Unlimited Press, 2004), pp. 206–221.

[11] David Wilcock, "The Breakthroughs of Dr. Nikolai A. Kozyrev," *The Divine Cosmos*, www.divinecosmos.com, p. 5.

[12] Ibid.

Dr. Kozyrev in Later Life

But what, *really*, had brought this quiet thinker to such an unusual conclusion?

As mentioned previously, Kozyrev was an astronomer, and in fact he himself "considered the determination of the nature of stellar energy to be the main goal of his scientific activity."[13] In fact, Kozyrev's 1947 doctoral thesis was on this very topic and was entitled "The Theory of the Inner Structure of Stars as the Basis for Research of the Nature of Stellar Energy."[14] As such, Kozyrev is considered to be one of the "pioneers of Russian theoretical astrophysics,"[15] developing a theory of the distant photosphere of stars in 1934. This work, subsequently extended and generalized by the Indian physicist Chandrasekar, became known as the Kozyrev-Chandrasekar theory.[16] Kozyrev is also responsible for a number of other discoveries related to the planets of the solar system. But it is not for this public and known work that Kozyrev is so significant; it is for the classified work that he did for the Soviet Union, work that was, moreover, inspired by his publicly known astronomical discoveries and thought.

In the previous chapter I mentioned the fact that early atmospheric hydrogen bomb testing returned some very anomalous results. For example, when America detonated the first hydrogen bomb — code-named "Mike" — in 1952, its calculated yield was in the neighborhood of six megatons. But the bomb, once fired off, ran away to a yield of about *ten* megatons, a rather significant difference. Moreover, I have speculated, since my first book

---

13   Dr. Lavrenty S. Shikhobalov, "N.A. Kozyrev's Ideas Today," p. 302. In utilizing this source, I have taken the liberty of smoothing out its sometimes uneven, but always clear, translation. Dr. Shikhobalov's article contains an excellent *curriculum vitae* of Dr. Kozyrev on this page.
14   Ibid.
15   Ibid.
16   Ibid.

The Giza Death Star, that such testing revealed anomalous gains (or decreases) in yields *depending upon the location of the test or detonation and the local celestial geometry and time it was conducted.*[17] In other words, there was *another* source of energy being transduced *by* the geometries of the detonation, gating significant amounts of energy into the explosion that could *not* be accounted for by standard calculations of neutron cross-sections, statistical analysis of fission chain reactions, heat and pressure gradients of a fusion reaction, and so on. Something *else* was going on, and this "something else" was the *real* reason for continued nuclear testing by the then two superpowers, and it is the real reason for more recent additions to the nuclear (and thermonuclear) club to continue their nuclear tests; consider only France and China. They are trying to learn those laws.

Dr. Kozyrev's work is directly related to this question, though he approached the problem via a slightly different, though highly relevant, route. In short, when he compared "the observed data about luminance, masses, and sizes of stars,"[18] the observed luminance and radioactivity could *not* be adequately accounted for by the theory that stars are nothing but gigantic hydrogen bombs in a state of perpetual detonation; the theory of thermonuclear fusion alone was inadequate to account for the phenomenon of stars. Indeed, Kozyrev's analysis "brought him to a conclusion that the processes of thermonuclear synthesis cannot serve as a main source of stellar energy."[19] In other words, the fusion-gravity geometry model of standard stellar processes — a geometric model inspired in large part by Einstein's General Relativity and extrapolations from it performed by other scientists — was simply not able to account for the enormous energy pouring out of stars. Some other mechanism altogether was at work.

What mechanism did Dr. Kozyrev tentatively conclude was the source of this anomalous energy output? The Russian physicist Dr. Lavrenty Shikhobalov puts the matter with a succinctness and clarity that belies the significance of what Kozyrev had concluded: "The scientist (Kozyrev) made a hypothesis that Time is a source of stellar energy."[20] In other words, Kozyrev had concluded that the *geometry of local celestial space is a determinant in the energy output of fusion reactions, and that the latter, depending upon that geometry, will "gate" now more, now less, energy into the reaction itself as a function of that geometry.* Kozyrev had, in short, surmised why the Russians — who had no doubt encountered similar anomalous energy yields in their own hydrogen bomb tests — were getting such strange results, results that could *not* be explained on the standard theory and its methods of calculations of yields.

---

17 See my *Giza Death Star*, p. 145.
18 Dr. Lavrenty S. Shikhobalov, "N.A. Kozyrev's Ideas Today," p. 291.
19 Ibid.
20 Ibid.

THE PHILOSOPHERS' STONE | 155

B. An Alchemical Aside: Paracelsus on the Incorporation of Celestial Geometries into Alchemical Apparatus

Before continuing with the examination of Dr. Kozyrev's work, it is worth pausing to reconsider once again the similarity to the assertions of alchemy. As was seen throughout part one of this book, alchemical texts over and over again refer to the importance of the moon, the planets, and more importantly, *the Sun*, to the proper and successful operation and confection of the Philosophers' Stone.

In the epigraph which began this chapter, the most famous (or perhaps, notorious) alchemist, Paracelsus, puts the matter in no uncertain terms:

> Since, then, our matter is our root for the white and the red, *necessarily our vessel must be so fashioned that the matter in it may be governed by the heavenly bodies*. For invisible celestial influences and the impressions of the stars are in the very first degree necessary for the work....[21]

However, note that Paracelsus says *another* significant thing in this quotation, and attention is drawn to it lest it be overlooked: not only are the celestial geometries essential to the confection of the Stone, *but the apparatus, the technology **itself**, must in some fashion embody that influence or allow it to work directly upon the material being made in it.* This, as we shall see, is *precisely* the principle in evidence in one of Dr. Kozyrev's most astonishing pieces of experimental equipment.[22]

C. Tensors, Time, and Torsion

As we shall come to see in the next chapters, torsion forms an essential, and indeed the central, component of Dr. Kozyrev's experiments. Accordingly, it is necessary to have a basic idea of what torsion is, and what it does.

Torsion may be defined as a spiraling motion within the fabric of space and time that folds and pleats that fabric by twisting it. The simple analogy of emptying a soda pop can, and then wringing it like a dishrag, illustrates what torsion does. The spirals in the can literally fold and pleat, and the can's length contracts. This, essentially, is what torsion does to space-time. This spiraled, folded, and pleated structure is described by a mathematical entity called a

---

21 Paracelsus, "The Aurora of the Philosophers," *Paracelsus and His Aurora & Treasure of the Philosophers, As also The Water-Stone of the Wise Men: Describing the matter of, and manner how to attain the universal Tincture. Faithfully Englished. And Published by J.H. Owen.* (London: Giles Galvert, 1659). Text may be found at www.levity.com/alchemy/paracel3.html, p. 27.

22 It also goes without saying that this principle is very much in evidence in the Great Pyramid as well! See my *Giza Death Star*, pp. 175–179, 217–221.

torsion tensor, of which there are two versions, a relatively weak one, known as the Einstein-Cartan tensor, and a stronger one, known as the Ricci tensor. We shall concentrate our attention in a moment on the Einstein-Cartan version.

1. Various Views

Besides two different versions of the torsion tensor, there are also slightly different views of what torsion actually represents, and to some extent, the two — the versions of the torsion tensor and the views of what it represents — are related. For example, we have illustrated the folding, spiraling, and pleating of space-time by the simple example of wringing a soda pop can, and indeed, the most common usage of torsion is to indicate precisely these things. However, in *some* cases not only is this in view, but the *degree or amount* of the spiraling and folding taking place.

But for Kozyrev and the Russian scientists who continued his work, torsion means all this, and something else as well. For them, torsion came to represent a kind of "sub-space" or "hyper-space" — a higher structured order — that *gives rise to* the fabric of space-time of the ordinary world.

2. The Strength of Torsion in the Einstein-Cartan Torsion Tensor

It was the mathematician Elie Cartan that first injected the idea of torsion into Einstein's General Relativity Theory. As researcher David Wilcock points out, however, the difficulty was that, in Cartan's version of the mathematics, "torsion fields would be some 30 orders of magnitude weaker than gravitation, and gravity is already known to be 40 orders of magnitude weaker than electromagnetic energy."[23] As a result of this extremely small amount of force, torsion fields "were basically an irrelevant footnote that would not make any noticeable contributions" to observed phenomena in the real world.[24] Moreover, in the Einstein-Cartan version, torsion *itself* remains a static field, accounting only for the spin orientation of a single system.[25]

3. The Ricci Torsion Tensor

But there is another, stronger version of the idea in mathematics, the Ricci tensor, named after the Italian mathematician who discovered it. In this version, the curvature of space-time is stronger and not a mere "footnote" in

---

23 David Wilcock, "The Breakthroughs of Dr. N.A. Kozyrev," from *The Divine Cosmos*, www.divinecosmos.com, p. 7.
24 Ibid.
25 Ibid.

an otherwise mathematically elegant theory. In fact, in some uses the Ricci tensor denotes not simply a static field, but an actual *flow* and structure to time, almost as if it possessed fluid-like properties, a point that will become quite important in our analysis of Dr. Kozyrev's work.

4. Dynamic Torsion, Vortex Mechanics, and Time

With the idea of the Ricci tensor we are confronted by the fact that torsion is a *dynamic* and *changing* phenomenon, and not merely a static field, for if the basic idea of torsion is that it is related to a rotating system, then it will be apparent that the universe is composed of rotating systems within other rotating systems, producing a continuously changing system with a changing *flow* of time. And if in addition those spinning systems are in turn emitting energy, such as a star pouring out its electromagnetic radiation from the fusion reactions within it, then that dynamic changes yet again, and with the constantly changing "spiraling and pleated" field of time, time *itself* takes on dynamic properties.

Such complex, interlocked systems of rotation may be thought of as "knots" of space-time that are so intensely concentrated that they form the objects observed in the physical universe. As such, all systems are in fact "space-time machines," and since they "contain" space-time they are not ultimately "constrained" by it, but rather, interact constantly, and in some cases, instantaneously, with it.

And this means that physics had to modify its mathematical modeling of time and space significantly. Or, as Dr. Kozyrev would imply, "time is not a scalar…"

D. Dr. Nikolai Kozyrev on the Nature of Time: The Physical Mechanics of Cause and Effect

In our examination of the thought and experiments of Dr. Kozyrev that now follow, we shall rely heavily on his own statements and mathematical argumentation, and equally upon the commentary of those who have studied his work, particularly Russians, in order to exhibit the fact that the full implications of his work were more than adequately understood by the Soviets, who, for obvious reasons, sealed it tightly behind a wall of high classification and KGB and GRU security.

Dr. Kozyrev outlined the fundamentals of his whole physics and philosophical approach in a paper first published in 1967: "The Possibility of Experimental Study of the Properties of Time." The title itself is suggestive, and breathtaking enough. But the contents of the paper — especially for one reared in the milieu of post-relativistic physics — as Dr. Kozyrev like all academic physicists was, is even more stunning. He announces his philosophy, and program, in no uncertain terms:

> In reality, *the exact sciences negate the existence in time of any other qualities other than the simplest quality of "duration" or time intervals....* This quality of time is similar to the spatial interval. The theory of relativity by Einstein made this analogy more profound, considering time intervals and space as components of a 4-dimensional interval of a Minkowski universe. *Only the…geometry of the Minkowki universe differentiates the time interval from the space interval. Under such a conception, time is scalar and quite passive. It only supplements the spatial arena,* against which the events of the universe are played out. *Owing to the scalarity of time, in the equations of theoretical mechanics the future is not separated from the past; hence, the causes are not separated from the results. In the result, classical mechanics brings to the universe a strictly deterministic, but deprived causality. At the same time, causality comprises the most important quality of the real world.*[26]

This densely packed, tightly-argued paragraph requires considerable unpacking in order for its truly breathtaking nature to be truly appreciated. Even then, as we shall see, it only hints at the breathtaking program and implications that Kozyrev outlines in the rest of his paper.

First, take note of the two basic points Kozyrev has made:

1) Time is not merely a "scalar" or "one-dimensional entity" in the geometry of space-time; it is not, therefore, to be viewed in the sense that the geometry of General Relativity — the Minkowski space — or for that matter, most physical theory, views it, namely, as *merely* duration; and,
2) That because physics has tended throughout the centuries to view time in only this way — as mere duration — modern physics in particular has no really adequate way to distinguish cause from effect *with formal, mathematical, explicitness.*

The last point may seem somewhat obscure, but it is directly related to the first point, and to what Kozyrev calls the "scalarity" of time, so it is best to have some basic understanding of what a "scalar" is in mathematical terms. By doing so, one will have an appreciation for its typical use in mathematical physics.

---

26 Nikolai A. Kozyrev, "Possibility of the Experimental Study of the Properties of Time," www.abyme.net/ , p. 1, emphasis added.

THE PHILOSOPHERS' STONE | 159

1. Time is Not a Scalar

A "scalar" is simply a term for what, in many mathematical languages, is simply a "dimensionless" entity, like a "point" in the kind of geometry one learned in elementary school. If one recalls that elementary school geometry, for example, a *line* required one dimension, a *plane* required two dimensions, and a real physical object required three dimensions, to describe them mathematically.

What Einstein did in his General Theory of Relativity, as many know, is that he made time a *fourth* dimension in the mathematical description of an object, since any physical object not only existed in space, but endured in time.

But this dimensionless "duration-only" description of time was completely inadequate, according to Dr. Kozyrev, for by pointing out the "scalarity" of time, he is simply pointing to the obvious fact that as a "dimensionless" entity it is not comprised of further "parts," so to speak. One cannot therefore adequately distinguish a "cause" from its "effect" within mathematical physics with any degree of formal, mathematical precision, since this idea of "mere duration" is incapable of further formal analysis. Time, on this physical view, was a passive player, and not an *active contributor*, to physical processes and forces. It was merely a backdrop or stage on which those processes and forces were played out. By viewing time as a non-scalar, Kozyrev has announced his philosophy, and his program: time is *active* and possessed of its own inherent parts and qualities, and experimental physics must investigate these with all the scientific rigor as it investigated *other* active forces and properties in previous centuries. With this insight, in other words, Kozyrev announced a wholesale assault on two of the foundations of modern physics and some of its hidden, and very counterintuitive, assumptions: Relativity, and Quantum Mechanics.

But only General Relativity has been mentioned thus far in this connection. What does the "non-scalarity" of time mean in terms of Quantum Mechanics? To answer this, one must delve a little further and deeper into his paper.

> The concept of causality is the basis of *natural* science. The natural scientist is convinced that the question "why" is a legitimate one...However, the content of the *exact* sciences is much more impoverished. In the *precise* sciences, the legitimate question is only "how?": i.e., in what manner a given chain of occurrences takes place. *Therefore, the precise sciences are descriptive. This description is made in a 4-dimensional world, which signifies the possibility of predicting events.* This possibility of prediction is the key to the power of the precise sciences.[27]

---

27 Nikolai A. Kozyrev, "Possibility of the Experimental Study of the Properties of Time," www.abyme.net/ , p. 1, emphasis added.

This short excerpt again belies the significance of what Kozyrev is actually saying by making the distinction between the *natural* sciences on the one hand, and the *precise or exact* sciences on the other.

2. A Theological Aside

With this distinction, Kozyrev has, in a way, revived a notion that was common currency in Western science prior to Descartes, and even more importantly, reached deeply back into the theologico-cultural past of Eastern Orthodox Russia. By the "exact" or "precise" sciences, Kozyrev means precisely what one ordinarily thinks of when one thinks of sciences such as physics or chemistry, and to some extent, genetics and biology. But what does he mean by "natural sciences"? And how does this relate to his conception of the non-scalarity and non-passivity of time?

The key to the answer to these questions may in fact lie in the work of an important seventh-century Byzantine Greek saint, Maximus the Confessor, a Church Father as important to the Orthodox Catholic East as Augustine of Hippo is to the Roman Catholic and Protestant West.

Briefly put, Maximus taught a basic "triadic structure" to cosmic history. Everything:

1) *began*, that is, came to be from a state of non-being; everything transitioned from non-being to becoming. Thus, "becoming" was the very essence of created being. This stage he called, aptly enough, *Genesis*, or "beginning (Γενεσις);
2) *moves*, that is, as a being whose nature is becoming, creatures have an inherent dynamism or motion, which he calls by the stage *Kinesis*, from the Greek word κινησις, meaning motion;
3) *comes to a state of rest, or to an end*, which he calls alternatively by the Greek words for rest — which is Stasis(στασις), where motion (of a particular *kind*) ceases — or by the word for "end" or "goal," Telos (Τελος).

Looking closely at the second stage, Kinesis, or motion, one perceives the link to Kozyrev's conceptual world, for in the patristic tradition of Eastern Orthodoxy, which suffuses Russian culture at a very deep level, time itself was understood to be a creature. Thus, time is an inherent and *active* property of motion, of *physical processes,* and not merely a passive, non-dimensional or mono-dimensional *backdrop* for such processes.

Consequently, while Kozyrev does not — to my knowledge — ever explicitly refer to such patristic theological conceptions, he nonetheless invokes some of its *consequences* in his own work. Moreover, it would be difficult to imagine such a highly educated and literate Russian of great intellectual subtlety and refinement, such as Kozyrev, being *unfamiliar* with the general outlines of such thought and its implications, even in the officially atheistic Soviet Union. By even suggesting such a mediaeval notion as "natural sciences" asking the question not of "how?" but of "why?" he is in fact invoking an old notion of the *end* or *goal* of a physical process, a notion popularly known as teleology, from the very same word by which Maximus and others referred to the "end stage" of creation: Telos. This Kozyrev refers to by his penetrating insight that without an understanding of time as *an active force within, rather than as a backdrop to,* physical processes, then perforce one has lost the ability to formally distinguish between cause and effect.

One of Kozyrev's modern interpreters, himself a Russian, understands the full implications of what Kozyrev's program entailed:

> ...when the number of facts in favor of existing of the so-called "fifth force" grows, when we fail to register gravitational waves and solar neutrinos are difficult to pick up in the necessary quantity, when *the definitions of life, consciousness, free will are absent in physics* **and an essential definition of time was not formulated**, when even a consistent theory of the electron was not created, (any) statement on the completeness of theoretical physics looks less convincing...[28]

In other words, the implications of Kozyrev's philosophical approach to the program of experimental and theoretical physics was nothing less than "catholic" in its scope, for it touched not only upon the "full range" of physical processes and of time itself as an active force within them, but also upon the intelligence and consciousness of their human observers.

3. Time and Kozyrev's Formally Explicit Definitions of Cause and Effect

But once again, what are the implications of this philosophical approach for Quantum Mechanics? In order to understand the answer to this question, one must bear in mind one of its principal features: the Heisenberg Uncertainty Principle. Named after its discoverer and formulator, the famous German physicist Werner Heisenberg, this principle states essentially that if

---

28 Dr. Lavrenty S. Shikhobalov, "N.A. Kozyrev's Ideas Today," p. 299, italicized emphasis in the original, boldface emphasis added. While Shikhobalov's comments may not make the connection between consciousness, free will, and time immediately clear, at least in the physics context, the connection was, once again, deliberately and explicitly made by St. Maximus the Confessor in the First of his *Opsucula Theologica et Polemica*.

one measures the position of an electron in an atom, one cannot measure simultaneously its velocity, and if one measures its velocity, one cannot measure its position.

As a consequence of this principle, it will be apparent that the mere act of determining what is measured is, to that extent, also an act of causing certain things to be. And this, of course, overturned all previous notions of physical causality, because reality to some extent was determined by the mere act of measuring it. It is this counterintuitive notion of cause and effect that Kozyrev's work addresses directly and head-on, for in quantum mechanics, since the act of measurement injected such notions of uncertainty into the physical process, ordinary cause and effect relations came to be dependent upon massive numbers of observations and *statistical* averages and probabilities.

Here too, however, Kozyrev points out that the "scalarity of time" once again enters the conceptual framework and clouds the actual picture that quantum mechanics, with its statistical probabilities, actually paints:

> (Even) in the statistical mechanics of an isolated system, under the most probable condition, the directivity of time will not exist. It is quite natural that in statistical mechanics, based on the conventional mechanics of a point, the directivity of time does not appear as a quality of time itself but originates only as a property of the state of the system. If the directivity of time and other possible qualities are objective, they should enter the system of elementary mechanics of isolated processes. However, the statistical generalization of such mechanics can lead to a conclusion concerning the unattainability of equilibrium conditions. In reality, the directivity of time signifies a pattern continuously existing in time, which, acting upon the material system, can cause it to transfer to an equilibrium state. Under such a consideration, the events should occur not only in time, as in a certain arena, but also with the *aid* of time. *Time becomes an active participant in the universe, eliminating the possibility of thermal death.* Then, we can understand harmony of life and death, which we perceive as the essence of our world. Already, owing to these possibilities alone, one should carefully examine the question as to the manner in which the concept of the directivity of time or its pattern can be introduced into the mechanics of elementary processes.[29]

To put it differently, since Kozyrev views time as an active participant in systems, he views the systems of physical processes in a profoundly different way from standard physical mechanics, in that for him they are (1) *open*

---

29 Nikolai A. Kozyrev, "Possibility of the Experimental Study of the Properties of Time," www.abyme.net/ , p. 2, emphasis added.

systems and (2) systems more or less far from *equilibrium*. More is involved, in other words, in the notions of cause and effect than the merely statistical probabilities of standard quantum mechanics.

Since time is *not,* on Kozyrev's view, merely a dimensionless scalar, this means in turn that it can have, like a vector of force in ordinary mathematical physics, a *direction, a movement from one point to another,* or what he calls, "the directivity of time." It is this non-scalar, almost vector-like quality that Kozyrev means when he states that time is in its own right a kind of physical force, hitherto not adequately understood by contemporary physics as an *exact* science, but once understood when in its infancy centuries ago as a *natural* science. On this more ancient "teleological" view, "causality is linked in the closest way with the properties of time, specifically with the difference in the future and the past."[30]

If that were all there were to Kozyrev's philosophical and experimental program, however, one might with some justification dismiss him as a rather belated revival of Aristotle. It is here that Kozyrev distinguishes himself in no uncertain terms from the New Age dilettante or presumed reviver of "natural science" in a merely mediaeval sense, for he outlines his program with three basic hypotheses that will guide him in the formal mathematical modeling needed to define Cause and Effect, Past and Future, with formal explicitness. "We will," he states, "be guided by the following hypothesis":

I. Time possesses a quality, creating a difference in causes from effects, which can be evoked by directivity or pattern. This property determines the difference in the past from the future.

The requirement for this hypothesis is indicated by the difficulties associated with the development of the Leibnitz idea concerning the definition of the directivity of time *through* the causal relationships. The profound studies by H. Reichenbach... indicate that one can never advance this idea strictly, without tautology. Causality provides us with a concept of the existence of the directivity of time and concerning certain properties of this directivity; at the same time, *it does not constitute the essence of this phenomenon, but only its result.*

Let us now attempt, utilizing the simplest properties of causality, to provide a quantitative expression of hypothesis I. Proceeding from those circumstances in which: 1) cause is always outside of the body in which the result is realized and 2) the result set in after the cause, we can formulate the next two axioms:

II. Causes and results are always separated by space. Therefore, between

---

30 Ibid.

their appearance there exists an arbitrarily small, but not equaling zero, spatial difference d_x.
III. Causes and results are separated in time. Therefore, between their appearance there exists an arbitrarily small, but not equaling zero, time difference d_t of a fixed sign.[31]

Note two very intriguing things from this passage.

First, Dr. Kozyrev refers very explicitly to the groundbreaking and painstaking research of Baron Reichenbach, a researcher well-known for his scientific experiments involving "sensitives," psychics, and what he called "Odinic" energy, or that we would now call the physical medium or aether, and its relationships to human consciousness and perceptions. This is the strongest possible indicator that Kozyrev knew full well the wide-ranging implications of his approach not only for the conventional relativistic and quantum mechanical programs of physical analysis, but also as a possible basis to plumb the depths of the physics of the paranormal.

But note secondly the astonishingly simple mathematical formalism by which Dr. Kozyrev has defined the distinction of cause and effect: it is some function of the *difference* in space and time of a system.[32] Lest the importance of this observation be lost, Kozyrev is actually implying that *space and time themselves must be quantized*. Without it, the notions of cause and effect, and the directivity of time, will be lost, and physics will become counterintuitive, as in fact it did in the early twentieth century with Relativity and Quantum Mechanics.

But what does quantization mean?

We are all accustomed to thinking of space as an empty void, infinitely divisible into ever smaller and smaller units, or infinitely multipliable into ever larger ones, without limit. The same with time: we are used to thinking of it as having no smallest unit of measure, nor a largest one. They are a *continuum* in the truest sense. But quantum mechanics changed all this with its idea that there is a smallest possible measurable unit of physical force or action — the Planck Constant — but what it did *not* do was go on to suggest that there was a smallest possible unit of *time*. However, with the idea of the quantization of space and time *itself*, we are closer not only to understanding

---

31 Nikolai A. Kozyrev, "Possibility of the Experimental Study of the Properties of Time," www.abyme.net/, p. 2–3. The notation is reproduced exactly as found in Kozyrev's paper, but his interpreter, Dr. Shikhobalov, indicates that Kozryev actually formulated the distinction in the form of Kronnecker deltas: The distances "between the cause and the effect (Kozyrev indicated them as δx and δt) and the course of time $c_2 (=\delta x/\delta t)$, namely to specify if they have statistic or deterministic character and if they are scalars or pseudoscalars... which does not correspond with the natural sense of the notion of distance." (Dr. Lavrenty S. Shikhobalov, "N.A. Kozyrev's Ideas Today," p. 300.)

32 Perhaps best symbolized as a simple function $f(d_x, d_t)$.

Kozyrev's work but also the equally breathtaking work of a little-known German physicist that we will encounter in part four. With this insight, Kozyrev is suggesting that space and time are quantizable, that they have their own inherent measurable structures, and thus, time *cannot be* simply a scalar, without a direction, an inert and passive dimensionless player on the physical stage, without orientation or an active force on physical processes.

Before proceeding with our examination of Kozyrev's work, it is worth pausing to take note of developments within Russian physics *after* Kozyrev, developments that were based upon it and that came to some astonishing conclusions of its own:

> …In physics (the notion of causality) appears only in the form of the so-called principle of causality. According to this principle, the Future cannot have any influence on the past (taking into consideration the theory of relativity it also means the impossibility of motion of bodies with the speeds which exceed the speed of light in vacuum). Thus, physics and the other exact sciences ignore the major part of causality notion aspects. N.A. Kozyrev also failed to formulate an irrefragable physical definition of causality.
>
> Obviously the first strictly formalized definition of causality can be found in the article by M.K. Arushabov and S.M. Korotaev and in following articles by S.M. Kortaev (member of the Institute of Earth Physics RAS…). Simply speaking, this definition *is based on the comparison of conditional probability of events: If we take two events, then the effect is the event with the probability of realization higher than analogous probability for another event.… With this second event will be the cause.*
> **(Editor's note: in first view it seems strange that *probability of cause is less than probability of the effect*.)** Such definition appeared to be consistent with Kozyrev's axiomatics.[33]

In other words, one of the implications of Kozyrev's work that was pursued by Russian physicists was Kozyrev's distinction of cause and effect, which, upon *statistical* analysis, indicated that *effects* were always of much higher probability than *causes,* and this, of course, was the whole reason Kozyrev also called his physics program by the name "causal mechanics," for it indicated the possibility of the engineerability of causes *via* the understanding of time and its spin orientation and role as an actual active participant in physical processes.

---

33 Dr. Lavrenty S. Shikhobalov, "N.A. Kozyrev's Ideas Today," p. 296, italicized emphasis added, boldface emphasis in the editor's original emphasized comment.

4. Quasi-Quantized Time-Space Spin Orientations

This consideration of the directivity of time, that it would have a definite *value or sign, positive or negative,* leads Kozyrev to one of his profoundest insights when he considers the nature of mirror worlds.

> From the pseudo-scalar properties of the time pattern, there immediately follows the basic theorem of causal mechanics:
> A world with an opposite time pattern is equivalent to our world, reflected in a mirror.
> In a world reflected in a mirror, causality is retained. Therefore, in a world with an opposite time pattern the events should develop just as regularly as in our world. It is erroneous to think that, having run a movie film of our world in a reverse direction, we would obtain a pattern of the world of an opposite time direction. We can in no way formally change the sign in the time intervals.[34] This leads to a disruption of causality, i.e, to an absurdity, to a world which cannot exist.[35]

A complete change of sign in the expression δt to -δt thus does not mean a "backward running movie" version of time, but rather, time running with an opposite spin orientation.

Thus, a "mirror world" is

> A world with an opposite time pattern, the heart in the vertebrates would be located on the right, the shells of mollusks would be mainly turned leftward, and in protoplasm there would be observed an opposite qualitative inequality of the right and left molecules. *It is possible that the specially formulated biological experiments will be able to prove directly that life actually uses the time pattern as an additional source of energy.*[36]

In other words, because of the directivity of time, its *spin orientation,* one should be able to experimentally verify whether or not changes in physical systems occur *on the basis of the rotation of the system, and the overall geometrical*

---

34 That is, in the interval δt. Thus, Kozyrev is saying that -δt and δt do not have opposite *direction in the linear sense,* but opposite *spin orientation.* Again, this strongly suggests that Kozyrev was implying a quantized view of space-time in which space and time had special spin orientation properties, a notion we shall again encounter in part four. A completely "reverse time" in the sense of a backward running movie would be, as he says, "an absurdity."
35 Nikolai A. Kozyrev, "The Possibility of the Experimental Study of the Properties of Time," p. 5.
36 Nikolai A. Kozyrev, "The Possibility of the Experimental Study of the Properties of Time," p. 5, emphasis added.

"context" *in which the experiment is performed, i.e., on the basis of* **when** *it is performed.*

Moreover, we've already encountered something like this before, for recall that the alchemists' alchemist, none other than Paracelsus, stressed over and over again that *timing* was an essential component to a successful alchemical operation! Recall also that David Hudson, in attempting to unravel the mystery of his anomalous material, discovered that some of its unique properties were the direct consequence of its high-spin states. Moreover, he also discovered that the spiraling geometry of the DNA helix itself had some relation to the ability of biological systems to transduce energy from somewhere.

Kozyrev has not only said *where,* but *how: in the spin orientation of the spiraling geometry of time itself:*

> Time enters a system through the cause to the effect. The rotation alters the possibility of this inflow, and, as a result, the time pattern can create additional stresses in the system. These variations produce the time pattern. From this it follows that *time has energy.*[37]

With the statement "time has energy" and the implication that "the time pattern can create additional stresses in the system," and actually *"enter"* a system, Kozyrev has signaled that his philosophical program has the ability to be experimentally verified, and this, as they say, is where it gets *really* intriguing, and where the alchemical parallels multiply like rabbits.

5. The Alchemical Implication

But before we go down that hole and visit that warren, one final point should be made about the implications of Kozyrev's work, lest it be missed. By stating that time *itself* "enters a system," Kozyrev is implying that time itself *is* the "transmutative *materia prima* or physical medium of the alchemists. In pointing out its dynamic torsional properties, Kozyrev has reinforced his view that "time is not a scalar." But in doing so, he has only deepened the mystery considerably, for he has really pointed out that the alchemists, too, knew this secret. After all, it was none other than Paracelsus — writing centuries before Kozyrev, and summing up the long chain of alchemists who had stressed the same thing before him in all the long centuries prior to the Renaissance — who wrote: "Since, then, our matter is our root for the white and the red, necessarily our vessel must be so fashioned that the matter in

---

37   Ibid., p. 8, emphasis in the original.

it may be governed by the heavenly bodies. For invisible celestial influences and the impressions of the stars are in the very first degree necessary for the work...."[38]

Kozyrev. Paracelsus.

It seems a bizarre and perhaps even surreal association to make.

And yet, as we shall now see, there are even more peculiar parallel transports to be made between Kozyrev's modern physics work, and ancient and mediaeval alchemy…

---

38 Paracelsus, "The Aurora of the Philosophers," *Paracelsus and His Aurora & Treasure of the Philosophers, As also The Water-Stone of the Wise Men: Describing the matter of, and manner how to attain the universal Tincture. Faithfully Englished. And Published by J.H. Owen.* (London: Giles Galvert, 1659). Text may be found at www.levity.com/alchemy/paracel3.html, p. 27.

Of Gyroscopes, Sponges, and Hydrogen Bombs
Kozyrev's Experiments and Wilcock's Analogies

•••

"Furthermore, Kozyrev proved decades ago that these fields travel at 'superluminal' speeds, meaning that they far exceed the speed of light. If you can have an impulse that moves directly through the 'fabric of space-time', travels at superluminal velocities and is separate from gravity and electromagnetism, you have a significant breakthrough in physics — one that demands that a 'physical vacuum', 'zero point energy' or an 'aether' must really exist."
David Wilcock[1]

"Rotating systems exist within rotating systems, producing ever-changing fields of dynamic torsion. Consequently, time is not a scalar."

Thus might one summarize the more than three decades of careful experimentation, meticulous observation, a veritable mountain of data, and the penetrating conceptualization and analysis that Dr. Nikolai A. Kozyrev performed. In the final analysis, however, any attempt to summarize the prodigious decades' effort of this Russian genius is doomed to failure. Here all that can be done is to trace the outlines of his thought as exemplified in some of his experiments, to exhibit its truly awesome implications, and its even stranger parallelisms with alchemy.

---

1 David Wilcock, "The Aether Science of Dr. N.A. Kozyrev," *Nexus Magazine,* Volume 14, Number 3, May-June 2007 (45–47), p. 47.

In order to find our way through this vast labyrinth of data, and the even more subtle labyrinth of Dr. Kozyrev's mind, some sort of systematic analogy will be helpful. Fortunately, with the signally important work of researcher David Wilcock, we have such an analogy ready to hand. More importantly, it is an analogy that lends itself quite well to tracing out some of the more obscure implications of Dr. Kozyrev's work.

A. Of Sponges: David Wilcock's Analogy

In an article in *Nexus* magazine summarizing his research into Dr. Kozyrev, author and alternative history and sciences researcher David Wilcock presents an extended analogy that we shall on occasion have recourse to in order to aid our presentation of Kozyrev's work:

> In order to truly grasp Kozyrev's work and related findings, certain new analogies for physical matter are required. Kozyrev's work forces us to *visualise [sic] all physical objects of matter in the Universe as if they are sponges submerged in water.* In all these analogies, we should consider the sponges as having remained in water for a long enough period of time that they are completely saturated. Bearing this in mind, there are two things we can do with such sponges underwater: we can decrease the volume of water that they contain or increase it by very simple mechanical procedures.
>
> 1. Decrease: If a submerged, saturated sponge is *squeezed, cooled or rotated,* then some of the water inside of it will be released into its surroundings, decreasing its mass. Once the sponge is no longer disturbed, the pressure on the millions of tiny pores is relieved, causing it again to absorb water and expand back to its normal resting mass.
>
> 2. Increase: We can also pump more water pressure into the sponge in its rest state, such as by *heating (vibrating)* it, thus causing some of the pores to expand with more water than they can comfortably hold. In this case, once we relieve the added pressure, the sponge will naturally release its excess water and shrink back down to its normal resting mass.
>
> Though it would seem impossible to most people, *Kozyrev showed that by shaking, spinning, heating, cooling, vibrating or breaking physical objects, their weight can be increased or decreased by subtle but definite amounts.* And this is but one aspect of his amazing work.[2]

Wilcock's analogy is very carefully chosen, and it is worth pausing to consider its *own* unique set of alchemical parallels.

---

2   David Wilcock, "The Aether Science of Dr. N.A. Kozyrev," *Nexus Magazine,* Volume 14, Number 3, May-June 2007 (45–47), p. 47, emphasis added.

*1. Submerged Sponges, Sub-Spaces and Hyper-Spaces:
An Alchemical Parallel*

Wilcock's analogy of sponges representing physical objects submerged in water is a very apt one, for as representing the manifold systems of the physical universe — from spinning spiraling galaxies, rotating masses of stars and planets and pulsars to molecules and atoms and sub-atomic particles — submerged in water, Wilcock is implying that our ordinary space-time is a subset of a much larger, hyper-dimensional "space-time," or alternatively, he is proposing a view of space-time in which hyper-spaces and sub-spaces all contribute to the formation of our physical universe. In this, as we shall see, he is adequately representing the actual thought and implications not only of Kozyrev's work, but, as we shall also discover in part four, the even more breathtaking work and thought of a little-known German physicist.

But in so doing, Wilcock is also pointing out another aspect of Kozyrev's work that is not without its own unique alchemical parallels, for if time, on Kozyrev's view, is not simply a scalar, but has qualities and dynamic properties capable of its own "multi-dimensional" descriptions, then one is reminded of esotericist Manly P. Hall's statement about alchemical operations having to be performed in more than one world simultaneously, by a common operation and common "recipe":

> As one of the great alchemists fittingly observed, man's quest for gold is often his undoing, for he mistakes the alchemical processes, believing them to be purely material. He does not realize that the Philosopher's Gold, the Philosopher's Stone, and the Philosopher's Medicine exist in each of the four worlds and that the consummation of the experiment cannot be realized until it is successfully carried on in four worlds simultaneously according to one formula.[3]

Such processes, as Hall notes, are not simply nor "purely material," implying that there are immaterial, i.e., hyper-dimensional, properties at work, and that the Philosophers' Stone itself is an object partly inhabiting this world-dimension, and partly inhabiting other worlds and dimensions.

2. Spinning, Vibrating, Heating, Cooling: Another Alchemical Parallel

There is another signally important point in Wilcock's analogical summary of Kozyrev's work, and that is the "list of processes" that Kozyrev employed

---
3 Manly P. Hall, *The Secret Teachings of All Ages* (Penguin), Reader's Edition, p. 508.

in his experiments: spinning, vibrating, heating, and cooling (among many others). This too sounds oddly familiar, not only recalling the processes employed by David Hudson, but as was seen from part one, the alchemists' recipes themselves.

To put it succinctly, not only do Kozyrev's views on the dynamic and qualitative properties of time and its non-scalarity parallel to an astonishing degree the views of the alchemists, but the parallels may be pressed to an even more detailed level, for the alchemists too subjected various materials to a variety of such physical stresses in their alchemical operations, and like the alchemists and David Hudson, Kozyrev is also noticing strange mass losses and gains when materials are subjected to a variety of physical stresses.

B. Of Gyroscopes and Other Things: The Experimental World of Dr. Kozyrev

As seen in the previous chapter, Kozyrev posited that time was actually a measurable physical force with properties of its own, manifest in the spiraling patterns one finds in nature from shellfish to DNA. Seizing on the principal feature of torsion — the rotation and spin orientation of a system — Kozyrev formulated a series of initial basic experiments to test and confirm his theoretical conceptions.

"The theoretical concepts indicate," he writes, "that the tests on the study of causal relationships and the pattern of time need to be conducted with rotating bodies: namely, gyroscopes. The first tests were made in order to verify that the law of conservation of a pulse is always fulfilled, and independently of the condition of rotating bodies."[4] With this observation, Dr. Kozyrev set up various experiments involving gyroscopes.

1. The Simple Gyroscope Experiments and Their Breathtaking Results

These gyroscope experiments were simplicity itself, and it is best to cite Dr. Kozyrev's own descriptions of the experimental apparatus to see how incredibly simple these actually were:

> These tests were conducted on lever-type weights (scales). At a deceleration of the gyroscope, rotating by inertia, its moment of rotation should be imparted to the weights (scales), causing an inevitable torsion of the suspensions.[5]

---

[4] Dr. N.A. Kozyrev, "Possibility of the Experimental Study of the Properties of Time," www.abyme.net, p. 9
[5] Ibid., p. 10.

The experiments, in other words, were deliberately designed to test the presence of *torsion*. Kozyrev continues:

> In order to avert the suspension difficulties associated with this, the rotation of the gyroscope should be held constant. Therefore, we utilized gyroscopes from aviation automation, the velocity of which was controlled by a variable 3-phase current with a frequency of the order of 500 (cycles per second). The gyroscope's rotor turned with this same frequency....During the suspension the gyroscope was installed in a hermetically sealed box, which excluded completely the effect of air currents. The accuracy of this suspension was of the order of (0.1 to 0.2 milligrams). With a vertical arrangement of the axis and various rotation velocities, the readings of the weights (scales) remained unchanged.[6]

But then comes an astonishing admission:

> The test was conducted for various directions of suspension and rotation masses of the gyroscope, at different amplitudes, and at an oscillation frequency ranging from units to hundreds of (cycles per second). For a rotating gyroscope, just as for a stationary one, the readings of the weights (scales) remained unchanged. We can consider that the experiments described substantiate fairly well the theoretical conclusion concerning the conservation of a pulse in causal mechanics. In spite of their theoretical interest, the previous experiments did not yield any new effects capable of confirming the role of causality in mechanics.[7]

If that had been all there was to it, however, Kozyrev probably would have done what any other scientist would have done: packed it in, admitted he was wrong, and gone on to other things.

But Kozyrev was no average scientist. For the experiments did do something *highly* unusual. The gyroscopes were, it will be recalled, *suspended* from a level balance scale, and thus were but only thinly connected with these finely calibrated balance scales which measured them. It is here the anomaly appeared:

> However, in their fulfillment it was noted that in the transmission of the vibrations from the gyroscope to the support of the weights (scales) *variations in the readings of the weights (scales) can appear, depending on the velocity and direction of rotation of the gyroscopes.* When the vibrations of the weights

---

6 Ibid.
7 Dr. N.A. Kozyrev, "Possibility of the Experimental Study of the Properties of Time," www.abyme.net, p. 10.

(scales) themselves begin, *the box with the gyroscope discontinues being strictly a closed system.*[8]

This was the first of many breakthroughs and anomalies that Kozryev's experiments would uncover.

2. Time Imparts a Spin Orientation to the Total System

Note that one aspect of Kozyrev's basic theoretical conception was in fact confirmed, namely, that the local space of a system itself appeared to have a *spin orientation*; it was not simply a "void with a curvature," as the post-Einstein popular imagination would have it, but it was a space with a *dynamic property: an orientation to rotation*. Additionally, as Kozyrev also noted, the system began to acquire this characteristic *when the gyroscope and the scales ceased being closed systems in themselves, and became open to each other.*

Kozyrev makes these points very clearly in an astonishing series of statements:

> With a certain type of vibration, which was chosen completely by feel, there occurred a considerable decrease in the effect of the gyroscope upon the weights (scales) during its rotation in a counterclockwise direction, if we examined it from above. During the rotation in a clockwise direction, under the same conditions, the readings of the weights (scales) remained practically unchanged. Measurements conducted with gyroscopes of varying weight and rotor radius, at various angular velocities, indicated that a *reduction of the weight...is actually proportional to the weight and to the linear rate of rotation.* For example, at a rotation of the gyroscope (D=4.6cm, Q= 90 grams, u=25 meters per second), we obtained the weight difference of -8 (milligrams). With rotation in a clockwise direction, it always turned out that (the weight difference) = 0. However, with a horizontal arrangement of the axis (of rotation of the gyroscope), in azimuth, we found the average value = -4 milligrams. From this, we can conclude that any vibrating body under the conditions of this experiment should indicate a reduction in weight. Further studies demonstrated that *this effect is caused by the rotation of the earth,* which will be discussed in detail later. Presently, the only fact of importance to us is that during the vibration there is developed a new zero reading relative to which *with a rotation in a counterclockwise direction, we obtain a weight reduction, while during a rotation in a clockwise direction we obtain a weight reduction, while during rotation in a clockwise direction we obtain a completely uniform increase in weight (± 4 milligrams).*[9]

---

[8] Ibid., p. 10, emphasis added.
[9] Dr. N.A. Kozyrev, "Possibility of the Experimental Study of the Properties of Time,"

Kozyrev leaves nothing to chance and spells out the implications of these observations and experimental data quite clearly: "Hence, *time possesses not only energy but also a rotation moment which it can transmit to a system.*"[10] With this, of course, one perceives an immediate connection to the Nazi Bell experiment, with its own rotating system and persistent allegations of time distortions and experimentation being explored in the Third Reich. And of course, once again, with this stunning conclusion, Kozyrev has also implied that the ancient alchemical observations that operational results varied with time have a basis not in mediaeval flights of fancy, but in real experimental science.

3. The "Intensity of Time" and Seasonal Variations of Experimental Results

Since time possesses these dynamic qualities, one may conclude that it can have a certain "density," i.e., that it can compress or rarefy; once again, time is "not a scalar" on Dr. Kozyrev's view. Indeed, he himself puts the case in so many words:

> In the case of time, *there also exists a variable property which can be called the density of intensity of time.* In a case of low density it is difficult for time to influence the material systems, and there is required an intensive emphasis of the causal-resultant relationship in order that the force caused by the time pattern would appear. **It is possible that our psychological sensation of empty or substantive time has not only a subjective nature but also, similarly to the sensation of the flow of time, an objective physical basis.**[11]

Note that what Dr. Kozyrev is calling "the time pattern" means precisely the degree of spiraling, folding, and pleating that enters a system due to the amount of torsion it is experiencing, and that this may indeed be the actual physical basis for why people experience the flow of time now more, and now less, "quickly" under certain conditions.

All this leads to the inevitable conclusion that his experiments involving gyroscopes, pendulums and various compounds would themselves vary according to the local celestial geometries and the amount of torsion his experiments transduced. This indeed was what Kozyrev found:

> Evidently many circumstances exist affecting the density of time in the space surrounding us. *In late autumn and in the first half of winter all of the tests can be easily managed. However, in summer these experiments become difficult*

---

www.abyme.net, pp. 10–11, emphasis in the original.
10   Ibid, p. 11, emphasis in the original.
11   Ibid, p. 18, italicized emphasis in the original, boldface emphasis added.

to such an extent that many of them could not be completed. Obviously, the density of time changes within broad limits, owing to the processes occurring in nature, and our tests utilize a unique instrument to record these changes.[12]

However, confirming this "alchemical insight" was not enough for Kozyrev. He immediately continued by formulating the theoretical possibility of alchemy itself:

> If this be so, *it proves possible to have one material influence another through time.* Such a relationship could be foreseen, since the causal-resultant phenomena occurred not only in time but also with the aid of time. Therefore, in each process of nature time can be extended or formed. **These conclusions could be confirmed by a direct experiment.**[13]

But in what way is this statement alchemical? After all, Kozyrev does not mention the term "alchemy" itself, nor does it appear that it was in his mind.

The answer, once again, is rather simple. Recall that torsion is a higher-dimensional entity; that its "space" *encompasses* that normal space-time of ordinary physical reality. To recall Wickock's analogy once again, it is the hyper-dimensional water in which the sponge of this world is submerged. Thus, by means of the "pattern of time" impressed in an object, "it proves possible to have one material influence another through time," that is, by *means of* it. Kozyrev has once again come to the same basic insight and agreement with the alchemists: taking note of the proper *time* is an essential component in the confection of the Philosophers' Stone, for time itself is an essential quality of the *materia prima,* of the Philosophers' Stone itself.

As if that were not enough, it should be noted how often in the previously cited quotations that Dr. Kozyrev mentioned the "causal-resultant" relationship, or phenomenon. By stating that having "one material influence another through time" could be "confirmed by a direct experiment," Kozyrev has signaled the fact that he believes the cause-effect relationship could be *deterministically engineered,* and thus, Kozyrev ceases speaking of the "causal-resultant" relationship, and starts to defer to it as "the causal-resultant dipole."

---

12 Dr. N.A. Kozyrev, "Possibility of the Experimental Study of the Properties of Time," www.abyme.net, pp. 18–19, emphasis in the original.
13 Ibid., p. 19, italicized emphasis in the original, boldface emphasis added.

4. The "Cause-Effect Dipole," Torsion, "Pre-Action" and Superluminal Information Transfer

But what is a "dipole" and what is the significance of Kozyrev's deliberate and carefully chosen change of language in its regard?

A dipole is best illustrated by what everyone knows about magnets and electricity, for both are dipolar phenomena: magnets of course have two poles, a "north" and a "south" pole, which give rise to the magnetic field that "moves" between them. Electricity, similarly, has a dipolar nature, since electricity moves between a positive and negative polarity or charge, or, if one prefers more generalized engineering terms, between a power source and its load end. In turn, these dipolar phenomena are a result of fundamental broken symmetries, of a *non-equilibrium condition, a stress, within the system itself.* They are two mutually important aspects of a common underlying reality without which those physical forces and phenomena simply would not exist.

Consequently, by stating the cause-effect relationship in terms of a *dipole*, Kozyrev was stating that this, too, is a broken symmetry, and that both components were essential conditions to physical action, and moreover, that they could be *engineered* and in some cases, their roles *could be reversed by changing the physical and geometrical parameters and relationships within an experimental system.*

He states all this by outlining a series of experiments performed with pendulums and electric motors:

> Since we are studying the phenomenon of such a generality as time, it is evident that it is sufficient to take the simplest mechanical process in order to attempt to change the density of time. For example, using any motor, we can raise and lower a weight or change the tension of a tight elastic band. We obtain a system with two poles, a source of energy and its outflow; i.e., the causal-resultant dipole. With the aid of a rigid transmission, the pole of this dipole can be separated for a fairly extensive distance. We will bring one of these poles close to a long pendulum during the vibrations of its point of suspension. It is necessary to tune the vibrations in such a way that the full effect of southward deflection would not develop, but only the tendency for the appearance of this effect. It turned out that this tendency increases appreciably and converts even to the complete effect if we being near to the body of the pendulum or to the suspension point that pole of the dipole where the absorption of energy is taking place. However, with the approach of the other pole (of the motor), the appearance of the effect of the southern deflection in the pendulum invariably became difficult. In the case of a close juxtaposition of the poles of the dipole, their influence

on the pendulum practically disappeared. It is evident that in this case a considerable compensation of their effects occurs. It turned out that *the effect of the causal pole* does not depend on the direction along which it is installed relative to the pendulum. Its effect *depends only on the distance (spacing)*. Repeated and careful measurements demonstrated that *this effect diminishes,* **not inversely proportional to the square of the distance, as in the case of force fields,** but *inversely proportional to the first power of the distance*. In raising and lowering of a 10-kg weight suspended through a unit distance, its influence was sensed at a distance of 2–3 meters from the pendulum. *Even the thick wall of the laboratory did not shield this effect....*

The results obtained indicate that nearer the system with the causal-resultant relationship the density of time actually changes. *Near the motor there occurs a thinning (rarefaction of time), while near the energy receiver its compaction takes place. The impression is gained that time is extended by a cause and, contrariwise, it becomes more advanced in that place where the effect is located.*[14]

In other words, the behavior of the pendulum would vary according to which part of the dipole, the cause (the motor) or the effect (the load end), were brought into proximity with the point of suspension of the pendulum, *or* to the outermost point of its swing.

That time itself would gate energy into the system of the pendulum depending on the geometrical arrangement of the power-load end dipole in connection with it, led Kozyrev to formulate a bold conclusion, one which tied his experimental observations directly into the non-local nature of the underlying *materia prima*, time, itself:

> The tests proved the existence of the effects through time of one material system upon another. This effect does not transmit a pulse (momentum), *meaning it does not propagate but appears simultaneously in any material system.* In this manner, in principle it proves possible to have a momentary relationship and a momentary transmission of information. Time accomplishes a relationship between all phenomena of nature and participates actively in them.[15]

By the "momentary" appearance of an effect from one isolated system in another isolated system, Kozyrev means that the affect is transmitted, via the non-local medium — the "water in which they are submerged" — at faster-than-light velocities.

---

14 Dr. N.A. Kozyrev, "Possibility of the Experimental Study of the Properties of Time," www.abyme.net, p. 19, italicized emphasis original, boldface emphasis added.
15 Ibid., p. 23, emphasis added.

5. Kozyrev's "Torsion Telescope"

Time, *torsion,* thus constituted a kind of "pre-action" determinant on systems. It was itself an energy traveling at much greater than light velocities.

According to one Russian researcher into his ideas, in order to test this new idea, namely, that one could detect *future* action, Kozyrev constructed a "torsion detector telescope:"

> From the mid-50s to the late 70s, professor N.A. Kozyrev (with V.V. Nasonov) conducted astronomical observations using a receiving system of a new type. When the telescope was directed at a certain star, the detector (designed by N.A. Kozyrev and V.V. Nasonov) positioned within the telescope registered the incoming signal even if the main mirror of the telescope was shielded by metal screens. This fact indicated that electromagnetic waves (light) had some component that could not be shielded by metal screens.[16] *When the telescope was directed not at the visible but at the* **true** *position of a star, the detector then registered an incoming signal that was much stronger. The registration of the true positions of different stars could be interpreted only as a registration of star radiation that had velocities billions of times greater than the speed of light.* N.A. Kozyrev also found that the detector registered an incoming signal when the telescope was directed at a position symmetrical to the visible position of a star relative to its true position. *This fact was interpreted as a detection of the* **future** *positions of stars.*[17]

This astronomical "pre-action" was also reflected in their laboratory experiments as well. As Kozyrev's assistant of many years, V.V. Nasonov noted, "Already in the torsion-balance experiments it was repeatedly noticed that, at optimum experimental conditions, a small deflection of the system readings, directed oppositely to the expected effect, *appeared prior to the action.*"[18]

---

16 Recall Kozyrev's explicit mention of the research of Baron Von Reichenbach. It is significant that Reichenbach also conducted experiments where normal light was totally shielded, and yet, his "sensitives" and other equipment he designed picked up similar radiations. Kozyrev's explicit mention of Reichenbach indicates that at least *some* aspect of his own thought was deliberated connected with the research of a man most would consider to be in the realm of the esoteric or paranormal.

17 Yuri V. Nachalov, "Theoretical Basis of Experimental Phenomena," www.amasci.com/freenrg/tors/tors3.html, p. 2, italicized emphasis added, boldface emphasis in the original. Nachalov cites the paper by N.A. Kozyrev, and V.V. Nasonov, "On Some Properties of Time Discovered by Astronomical Observations" (In Russian), *Problemy Issloedovaniya Vselennoi* 1980, #9, p. 76.

18 V.V. Nasonov, "Physical Time and the Life of Nature: A Talk at the Seminar on the Problems of Time in Natural Science," (Moscow: University of Moscow, 1985), cited in A.P. Levich, "A Substantial Interpretation of N.A. Kozyrev's Conception of Time," www.chronos.msu.re/EREPORTS/levich_substan_inter/levich_substan_inter.htm, p. 12.

a. Kozyrev's Progeny and The Reverse Engineering of Time: The Alleged Russian "Scalar Spheres"

If the cause-effect dipole and time itself could to some extent be engineered and even reversed, as was implied in Kozyrev's pendulum experiments, then the possibility of an even more breathtaking engineering occurs. But this possibility is not without its critics as we shall see.

But first, the allegation itself:

In a paper entitled "Time and Its Physical Relationships," Andrew Michrowski notes one of the possibilities Kozyrev opened in the Pandora's Box of his causal mechanics:

> In preventive time engineering, one could delay the approach of a known cause and to artificially close the consequence loop, and thereby annulling it from ever achieving an effect. In other words, one can make the effect "happen" before its normal time, disrupting space structure with its related "speed of the course of time." One can also make it "happen" after its expected deadline. This technology could have interesting implication in strategic situations (preventing an extra-planetary body from attaining a collision course).
>
> According to the Russian experience, when the spatial structure is disrupted by time-engineered causal mechanics, the affected region undergoes relative greater entropy (or less order). The volume of space if forced out to somewhere else, generating torsion fields...[19]

And this, according to Michrowski, led to actual Russian experiments with time machines in the 1990s, curiously, in the same time frame as the emergence of the whole "Red Mercury" nuclear terrorism scare of the same period:

> For over 15 years, a Russian association of scientists has conducted experiments with acceleration and deceleration of time with 4 prototypes of time machines. Light-heartedly, the units are called "muskrat traps" since the experiments conducted in remote forests were disguised as a high-tech electromagnetic technology for trapping muskrats. The time machine units are sphere ranging in diameters of 30cm, 1 meter, and 2.1 meters. The shells are encased with coils designed to produce convergent waves. Team leader Vadim A. Chernobrov describes converging electromagnetic waves as moving from a periphery to a central point. They are observed when a hoop is thrown into the water and inside the hoop the waves converge. If a

---

19 Andrew Michrowski, "Time and Its Physical Relationships," The Planetary Association for Clean Energy, Inc. (No date given), p. 4.

potential is applied to do work and to initiate the energy differential process, the other reverse direction scalar (the reverse-time energy flow) must react. Thus, compensation of time — in the form of the deceleration or acceleration of the rate of time — can take place.

The first trials involved mice, in which most (25 out of 31) died. Eventually there were successful 2-hours runs of time travel. An experiment with a dog that was clearly frightened also showed no ill effects. This led to experimentation with humans, the first being Ivan Konov who, on August 26, 2001, decelerated into the past by 3% of planetary time during a half-hour trial. Dozens of others have experienced the phenomenon and report such sensations as: quicker pulse, giddiness, itching skin, body twisting, numbness at extremities and a case of an out-of-body experience. Harmful effects on living systems do not appear to be linked to the change of the rate of time, but rather to the variations of the time rate value among regions of a living organism.[20]

There is something strange about this list, for we've seen it before:

Readers of my previous books, *Reich of the Black Sun* and *The SS Brotherhood of the Bell*, will recall that some of the effects of the Nazi Bell, when in operation, produced similar results: a pricking sensation in the skin, or an "itching" sensation, giddiness and muscular effects. Other effects were noted in observers outside of the apparati, including headaches.[21]

The question, as we shall see below, is not over whether such strange apparati were ever constructed or tested, but rather, whether or not they and the tests represent real science, or, as some within Russia are now charging, pseudo-science.

b. Torsion, System Memory, and Psychometry

The charge indeed has its roots in some aspects of Kozyrev's own thought, so before investigating the charges, it is essential to understand the philosophical and conceptual soil in his thought from whence the science, or pseudo-science, sprang.

One of Kozyrev's most crucial observations in the formulation of his approach was that the prized Second Law of Thermodynamics in physics does not, in actuality, square with any observed phenomena of nature:

Kozyrev pointed out the sharp contradiction between the second law of thermodynamics which brings nearer the thermal death of the Universe,

---

20 Ibid., p. 5.
21 Andrew Michrowski, "Time and Its Physical Relationships," p. 6.

and the absence of any signs of equilibrium in the diversity of the Universe. He stressed that "the attempts to explain the absence of thermal death have been quite apart from the real Universe observed by astronomers. The point is that the celestial bodies and their systems are so well isolated from each other that their thermal death must have occurred much sooner than any external system could interfere. Therefore degraded states of systems ought to dominate, whereas they are almost never met. And the task is not only to explain the non-equilibrium state of the whole Universe, but also to gain an understanding why separate systems and celestial bodies themselves continue to live despite the short relaxation times."

Various hypotheses are possible attempting to "save" the second law of thermodynamics. For instance, the one asserting that the Universe is isolated indeed but the present moment of cosmological time is not very far from the "initial" fluctuation (singularity, cataclysm), so that the signs of degradation cannot be too clear, i.e., the "death" is postponed to a remote future. N.A. Kozyrev suggested an alternative version: the Universe and its subsystems are not isolated, i.e., the necessary condition for the second law of thermodynamics is not valid: "there are permanently acting causes in nature, preventing entropy increase." A necessary factor, violating the isolated state of systems, is just the Kozyrev flow (of time).[22]

Entropy is not observed, nor is any heat death observed in the Universe, on a large scale, simply because of the fact that time and torsion actually *interface* what otherwise are apparently isolated systems. It is the very fact that such large-scale entropy is *not* observed in the Universe that indicates for Kozyrev that such linkage *must* be the case.

And it is this, precisely, that gives rise to the "paranormal possibility" with all the allegations of "pseudo-science" that this implies, for as has been seen, torsion not only can but does affect the transfer of information from one system to another at greater than light velocities, and by its means, one system can impress itself on another. Materials can "influence" each other through time, to recall Kozyrev's own words. Torsion induces a stable "system memory" than can impress itself from one system to another.

For the budding esotericist, this was a conceptual basis within formal physics considerations not only for the possibility that actual memories could be impressed into physical objects — a concept known as psychometry — but also that there may indeed be a real physics basis for that most hated of all

---

22 A.P. Levich, "A Substantial Interpretation of N.A. Kozyrev's Conception of Time," p. 1, citing N.A. Kozyrev, "Causal Mechanics and the Possibility of Experimental Studies of the Properties of Time," *History and Methodology of Natural Science,* 2nd issue, Physics, Moscow, pp. 91–113; and N.A. Kozyrev, "Causal of Nonsymmetric Mechanics in a Linear Approximation," Pulkovo, 1958, respectively.

the pseudo-sciences: astrology, with all its lore of the influences of celestial geometries on personality.

This was, in fact, one of the results of Kozyrev's actual experiments, for as he himself stated, "A body placed for a certain time near a process and then brought to a torsion balance, produced the same effect on them as the process itself. Memorizing the action of processes is a feature of different substances, except aluminum."[23] Note the huge implications of Kozyrev's remarks, for they imply that so vast and thorough was his experimentation into various materials, that some were found to be torsion process memory preservers, while aluminum was discovered to shield against it! He had, in other words, through his experiments assembled a large catalogue of the torsion-amplifying or damping properties of various elements and compounds.[24] And these properties in turn are a result of their lattice structure, a fact that will come to have even more significance in the next part of this book.

c. Peculiar Parallels: Quantized Results

One of the most significant results that Kozyrev obtained in his various experiments involving balances was the fact that the weight gains and losses were *quantized*, i.e., they occurred in discrete steps or increments, and were not continuous and smooth:

> In the vibration experiments on a balance the weight reduction $^\Delta Q$ occurs stepwise, beginning with a certain vibration power. As vibration frequency is further increased, the weight reduction $^\Delta Q$ at first remains the same and then again grows stepwise by the same value. Thus it has been observed that, apart from the basic step $^\Delta Q$, good harmonic oscillations make it possible to see a series of quantized values: $^\Delta Q/2$, $^\Delta Q$, $2^\Delta Q$, $3^\Delta Q$..., corresponding to continuous vibration frequency increase.... So far a realistic explanation of this phenomenon has not been found....Similar quantum effects have been observed with pendulums. Afterwards it turned out that effect quantization takes place in almost all the experiments.[25]

In a rare lapse for Kozyrev, he seems to have missed what his own results, coupled with his previous insights concerning the nature of time and its torsion effects, implies, namely, if these quantized steps are manifestations of

---

23 N.A. Kozyrev, "Astronomical Observations Using the Physical Properties of Time," *Vzpykh-ivayaushchiye Zvezdy*, Yerevan, pp. 209–227 (In Russian), cited in A.P. Levich, "A Substantial Interpretation of N.A. Kozyrev's Conception of Time," p. 12.

24

25 N.A. Kozyrev, "On The Possibility of the Experimental Investigation of the Properties of Time," cited in A.P. Levich, "A Substantial Interpretation of N.A. Kozyrev's Conception of Time," p. 13.

the nature of time, then this would seem to imply *that space-time itself has a quantized structure, and is not a smooth infinitely divisible continuum.* As will be seen in the next part of this book, this crucial insight forms the basis of a breathtaking theory by one of Germany's most brilliant and least-known physicists.

Before continuing, it is necessary to pause and consider the two basic features and implications of Kozyrev's experiments:

1) Time and space are quantized and not continuous; and,
2) Experimental operations vary in their results dependent upon the season they are performed, i.e., depending on the ever-changing local geometry of space and time.

These two results recall the work of yet another physicist, working across the ocean, in America, and finding some very similar results.

(1) Quantized Results and the Gravitor Experiments of Thomas Townsend Brown

That physicist is Thomas Townsend Brown, well-known in alternative literature to have been allegedly involved in the Philadelphia Experiment,[26] and even more well known for his experiments in "electro-gravity" and a device which he called a "gravitator." These were dielectric blocks suspended from a roof, like a pendulum, and charged to high electrical potential. When this happened, the devices would lose mass, and actually lift, tilting upward. But more importantly, Brown observed that the devices would regain mass, and return to their starting position, in a series of discrete quantized steps, rather than float smoothly back down into their starting position. Moreover, Brown, like Kozyrev, also observed that the effect and the efficiency of his device varied with the seasons of the year and the phases of the Moon. In other words, Brown, like Kozyrev, had experimentally implied the quantization of space and time, since the results varied with the local celestial and ever-changing geometry.[27]

---

26 See my *Secrets of the Unified Field: The Philadelphia Experiment, the Nazi Bell, and the Discarded Theory* (Kempton, Illinois: Adventures Unlimited Press, 2008), pp. 132–141.

27 For a much fuller and more detailed discussion of Thomas Townsend Brown's gravitator experiments, see my *The Giza Death Star Deployed: The Physics and Engineering of the Great Pyramid* (Kempton, Illinois: Adventures Unlimited Press, 2003), pp. 205–212, especially pp. 209–212.

I am indebted to Mr. Richard C. Hoagland for pointing out to me the parallels of Kozyrev's and Brown's respective experimentations.

(2) Quantized Results and the Alchemists

Even more peculiarly, Kozyrev's work, with its discovery of the time-varying and quantized results of torsion experiments, not only echoes the observations of the alchemists that alchemical operations had to be performed at certain times, but also recalls their own suggestions that the results of their operations were quantized.

Consider only the implications of the following quotation from the mediaeval alchemist Roger Bacon, whom we have already encountered:

> The second multiplication is an Augmentum quantitatis of the stone with its former power, in such a way that it neither loses any of its power, nor gains any, but in such a manner that *its weight increases and keeps on increasing ever more, so that a single ounce grows and increases to many ounces.* To achieve this increase or Multiplication one has to proceed in the following manner: Take in the Name of God, your stone, and grind it to a subtle powder, and add as much Mercurii Solis as was taught before. Put these into a round vial, seal with sigillo Hermetis, and put it into the former oven exactly as taught, except that the time has to be shorter and less now. Whereas before you previously used ten (alii thirty) days, you may now not use more than four (alii ten) days.[28]

Note the typically ambiguous alchemical language, for the implication of quantization is both clear, yet vague, in the reference to "multiplication." Moreover, note as well the role which time plays in the operation.

An even more suggestive hint that alchemists also observed some sort of quantization in the results of their operations is made even clearer in the following quotation from Paracelsus:

> No precise weight can be assigned in this work of projection, though the tincture itself may be extracted from a certain subject, *in a defined proportion,* and with fitting appliances. For instance, that Medicine *tinges sometimes thirty, forty, occasionally even sixty, eighty, or a hundred parts* of the imperfect metal.[29]

Why Paracelsus is careful to elaborate the projection in terms of multiples of ten — thirty, forty, sixty, eighty, or a hundred — and does not mention fifty,

---

28   Roger Bacon, *Tract on the Tincture and Oil of Antimony,* www.levity.com/alchemy/rbacon2.html, p. 12., emphasis added.
29   Philippus Theophrastus Areolus Bombastus von Hohenheim, a.k.a. Paracelsus, "Concerning the Projection to be Made by the Mystery and Arcanum of Antimony," *The Aurora of the Philosophers,* www.levity.com/alchemy/paracel3.html, p. 20, emphasis added.

seventy, or ninety, is very suggestive that he has possibly observed quantized effects in his alchemical operations. And of course, no one stressed the need not only for attention to, but the embodiment of, temporal geometries, of time itself, in the alchemical apprati and operations.

*(3) Torsion, System Memory, Crystal Growth and Defects:
A Philosophers' Stone in the Desert of Egypt*

There is yet another strange parallel, and to see how strange a parallel it is, one must refer to two statements made by some of Kozyrev's progeny. In a rather innocent-sounding paper entitled "Experimental Detection of the Torsion Field," Yuri V. Nachalov and E.A. Parkhomov make the following observation:

> If any substance (or even the physical vacuum in general) is subjected to the influence of an external torsion field, then this influence causes a transverse spin polarization of that substance. Since this transverse spin polarization can be retained as a metastable state, *then a torsion field of a given spatial configuration can be "recorded" upon any physical object.*[30]

We have encountered this before, as the phenomenon of "psychometry."

But these authors have a specific substance in mind: "Torsion fields are able to change the rate of any physical process, for instance, they significantly alter the oscillation frequency of quartz crystals."[31] If this be true, then it is likewise possibly equally true that torsion fields *also* affect the very growth of such crystals themselves, and thus, *the standard defects of lattice structure that one finds in crystals are the result of their growth patterns and defects recording the torsion fields to which they are subjected when grown.* Indeed, the torsion tensor itself is capable of describing such lattice defects!

And this brings us to the veritable "Philosophers' Stone in the Egyptian Desert," the Great Pyramid of Giza, for in my book *The Giza Death Star Destroyed*, I suggested precisely this torsion and hyper-dimensional basis for the placement of the interior chambers of the structure:

> Very obviously, the Great Pyramid is itself a crystal, composed of innumerable amounts of smaller crystals of quartz embedded in its limestone and granite. That is, if one wishes to couple a harmonic oscillator to the medium

---

30   Yu. V. Nachalov, E.A., Parkhomov, "Experimental Detection of the Torsion Field," www.amasci.com/freenrg/tors/doc15.html, p. 4.

31   Ibid., p. 3.

itself, and if this medium be a kind of "crystal"...then the most logical choice for an efficient oscillator of such a medium would be to ... give it an overall crystalline structure, right down to the "defects" all crystals possess.[32]

I then elaborated on the idea in the footnote:

> I am suggesting, in other words, that the placement of the inner chambers inside the Great Pyramid may not be accidental, but **deliberately designed... defects**, since the energy needed to produce such defects is precisely calculable....Of course, to suggest that the interior chambers of the Pyramid might be precisely placed, and have some function of coupled oscillation of the medium deriving from their placement, is an *extremely* speculative suggestion...[33]

In other words, the placement of the interior chambers may ultimately prove to be based on a precise knowledge of torsion fields in the local celestial system, on a precise knowledge of the nature and physical properties of time itself.[34]

d. The Charge of Pseudo-Science: Curious Echoes of the "Red Mercury" Story

Intriguing and suggestive as all these parallels are, however, it is in the parallel of the more recent attacks on Kozyrev and his progeny as "pseudo-science" coming from within Russia itself that one encounters the most peculiar parallel of them all: the resemblance of the attacks to similar attempts to denounce "Red Mercury" as a hoax that also emanated from within Russia, during the same time period. The first attacks began, in fact, in 1959 with a series of articles in Pravda, and hence Kozyrev's work had to wait several years before publication.[35] Notably, this attack occurs in more or less the same time period that Soviet Russia ceased open discussion and publication of its work in producing "pure fusion bombs," a connection we will explore more fully in a moment. With the fall of the Soviet Union, a series of virulent attacks began to appear in the Russian press denouncing those pursuing the lines of research implied by Kozyrev's work as being "pseudo-scientists." While it is probably

---

32 Joseph P. Farrell, *The Giza Death Star Destroyed: The Ancient War for Future Science* (Kempton, Illinois: Adventures Unlimited Press, 2005), p. 188.
33 Ibid., n. 23, italicized emphasis in the original, boldface emphasis added.
34 If such precise knowledge could be regained, and if my hypothesis ever be verified, then this placement might also function as a way of precisely dating the Great Pyramid. It is my belief that, if and when this happens, the structure will be shown to be far older certainly than standard Egyptological views maintain, and even older than the oldest dates proposed in much alternative science and history literature.
35 Dr. Lavrenty S. Shikhobalov, "N.A. Kozyrev's Ideas Today," p. 298.

true that much of this was indeed the case, with charlatans capitalizing on the confusion of the times to gain funding from the Russian government for their fraudulent projects, the possibility that it was also an attempt to cast a pall of suspicion over the whole legitimate development of "causal mechanics" in Kozyrev's hands must also be entertained, for his work, at that point, was no longer a secret.

However, the question is: Why would the Russian government possibly have been trying to cast a pall of suspicion over Kozyrev's work?

To answer that, we must turn once again to the Red Mercury legend of the early 1990s, a legend which the Russian press was equally at pains to denounce as a hoax and as "pseudo-science."

C. Back to "Red" Mercury: Of Hydrogen Bombs, Far-from-Equilibrium Systems and Torsion

As has been noted, Kozyrev's basic insight against the Second Law of Thermodynamics was simply that the Universe exhibited far too much non-equilibrium in order for it to be true in the sense most physicists understood it to be. Indeed, at the very core of his theoretical edifice, the idea of systems far-from-equilibrium is paramount.[36] And no systems could be farther from equilibrium than stars, and their man-made counterparts, hydrogen bombs.

In fact, it will be recalled that the whole principal motivation, by his own reckoning, for his decades' long investigation of torsion, and of the torsion amplifying and shielding properties of various elements and compounds, was precisely the fact that there were simply not enough neutrinos being emitted by stars for the standard model of stars as gigantic perpetual fusion reactors — essentially perpetual hydrogen bombs — to be true.[37]

As a recent Russian commentator on Kozyrev's work has noted,

> Long-term experiments by R. Davis on registration of solar neutrinos lead to a *conclusion that the temperature of the central part of the Sun is lower than the temperature, which is necessary to provide its radiance due to the thermonuclear reactions only. This result fully matches to the conclusion made by N.A. Kozyrev.* He came to this conclusion on the basis of the analysis of observational astronomical data. According to this conclusion, the *processes of thermonuclear synthesis cannot serve as a main source of stellar energy.*
>
> At present time the presence of numerous and various solar-earth and moon-earth connections, which cannot be explained from positions of traditional physics, is firmly determined. The given circumstance makes us to regard Kozyrev's

---

36   A.P. Levich, "A Substantial Interpretation of N.A. Kozyrev's Conception of Time," p. 2.
37   Q.v. A.P. Levich, "A Substantial Interpretation of N.A. Kozyrev's Conception of Time," p. 1.

hypothesis on the interrelation of all world phenomena by means of physical properties of Time with attention.[38]

But what has all this to do with the Red Mercury legend? Why even *associate* Kozyrev's work with it?

The association at one level lies in what may be the more than coincidental attack on Kozyrev in *Pravda* in 1959, effectively preventing him from open publication of his experimental results and theoretical conceptualizations, and the similarly-timed disappearance of discussion of clean fusion bombs in the open Soviet literature. Let us speculate a bit.

We have already encountered the fact that the earliest atmospheric hydrogen bomb tests far exceeded their pre-test calculated yields. In other words, *just as in stars, some other energy source was being tapped into, and transduced by, the thermonuclear detonation itself.* And if we extend this line of speculation, the Russians most likely encountered the same phenomenon in their hydrogen bomb testing. Moreover, in Kozyrev, they had an astrophysicist who thought he knew *why stars*, also implicating thermonuclear processes, appeared *not* to be radiating neutrinos energy for the thermonuclear model to be true.

One may reasonably and logically conclude, therefore, that the 1959 *Pravda* attack on Kozyrev was really a cover story to denounce his work, to *de-legitimize* it to anyone in the West who may have been paying attention to it, while Kozyrev, and his work, disappeared — as they did — into the highest reaches of classification within the Soviet Union, for his work provided the necessary key to understand why H-bombs were returning such anomalous yields, yields that, moreover, most likely varied with the time of their detonation. Kozyrev knew why: it was because the bomb itself became, for that brief brilliant nanosecond of the initial explosion, a dimensional gateway, a sluice-gate, opening the spillway to a hyper-dimensional cascade of torsion into the reaction itself.

And with his subsequent investigation of the torsion shielding *and amplifying properties* of various materials, we have another clue, perhaps, to what the Soviets were really after as well, for his investigations also opened up the possibility of precisely the creation of a "ballotechnic explosive," of a conventional explosive powerful enough to initiate a thermonuclear fusion bomb without the necessity of an A-bomb for a "fuse," for torsion, as noted in the many quotations cited throughout this chapter, could *impress its processes into a material,* just like the Philosophers' Stone, making such an explosive possible, at least in theory.

Indeed, if one now goes back and reconsiders the Red Mercury legend, and in particular the peculiar "recipes" alleged for the confection of the substance

---

38 Dr. Lavrenty S. Shikhobalov, "N.A. Kozyrev's Ideas Today," p. 297, emphasis in the original.

— the "alchemical" presence of antimony in some of its alleged chemical compositions, and more importantly, the use of mercury and radioactive isotopes, and its synthesis by immersion in a reactor core under high heat and pressure — all of this suggests not only the alchemical parallel, but that the Russians, taking to heart Kozyrev's work, had hit upon the very "Paracelsan" idea of impressing the motions of the heavenly bodies, of stars, of time, of torsion, into its confection.

Small wonder, then, that with the collapse of the Soviet Union which sponsored that work, and the emergence of the Red Mercury scare of the 1990s, that the Russian government, through its considerable media organs, made every effort to denounce the one as a hoax, and the other as leading to pseudo-science.

But as has been seen, Kozyrev's work is hardly the latter, and that casts the whole Red Mercury episode into a very different light indeed.

Perhaps, just perhaps, it was, after all, a code name for the Soviet Philosophers' Stone, the red tincture of the Great Elixir, the Red "Mercury" able to draw down the fires of the stars and of time itself, for time, as Kozyrev pointed out, was what ultimately fires them to begin with....

Conclusions to Part Three

∴

"For invisible celestial influences and the impressions of the stars are in the very first degree necessary for the work."
Philippus Theophrastus Areolus Bombastus von Hohenheim,
a.k.a. Paracelsus[1]

"What they found, was that the nucleus of these atoms deforms, goes to a high-spin state.... high-spin nuclei pass energy from one atom to the next with no net loss of energy..."
David Hudson[2]

"Time possesses not only energy but also a rotation moment which it can transmit to a system."
Dr. Nikolai A. Kozyrev[3]

What conclusions may be drawn from our examination throughout parts one through three that can guide us into the belly of the brutish beast of Nazism and the Third Reich's own alchemical pursuits? To answer this question, it will be helpful to reprise verbatim the conclusions of the two previous sections, adding those we have discovered in this one.

---

1 Paracelsus, "The Aurora of the Philosophers," *Paracelsus and His Aurora & Treasure of the Philosophers, As also The Water-Stone of the Wise Men: Describing the matter of, and manner how to attain the universal Tincture. Faithfully Englished. And Published by J.H. Owen.* (London: Giles Galvert, 1659). Text may be found at www.levity.com/alchemy/paracel3.html, p. 27.
2 www.asc-alchemy.com/hudson.html, p. 31.
3 N.A. Kozyrev, "Possibility of the Experimental Study of the Properties of Time," *Joint Publications Research Service #45238,* www.abyme.net, p. 11.

From Part One, we concluded the following:

1) The Stone comes in at least two or, more especially, three "parts," a material "body" and a more ethereal "soul," implying that whatever properties were being observed and referred to by the alchemists, that some of them appeared to exist in a state *beyond that of the three dimensions of normal matter itself;* in short, and in more modern terms, there was a *hyper-dimensional* aspect to the Philosophers' Stone, points which shall be taken up again in parts two and three;
    a) This aspect is made more apparent by the later references in the Middle Ages to the Stone having at least one constituent which "occupies no space," a peculiar quality easily explainable by reference to higher dimensions;
    b) In one very questionable instance, that of the Englishman Digby, it appears to be able to effect action at a distance;
2) The Stone appears to exist in at least three states:
    a) As a stone or mineral;
    b) As a reddish-purple Elixir, Tincture, or Liquid; and,
    c) As a powder or granular particulate;
3) The successful confection of "Alchemical Gold" is signaled by a distinctive progression through the electromagnetic optical spectrum, with the salient points being a progression from white to reddish-purple, the latter of which denotes the acquisition of the Great Elixir;
4) Throughout all versions of the confection of the Stone, an elaborate process is implied, involving the use of heat at every step, and in one case — that of Moses and the Golden Calf — involving very intense heat able actually to *consume or burn* gold itself, an observation that in turn requires a rather sophisticated furnace or forge technology;
5) It is also claimed that the Stone, Great Tincture, or Grand Elixir is able to effect cures, healing, and to prolong human life far beyond its natural life span;
6) It is also claimed that the Stone is indestructible, partaking directly of the indestructibility of the underlying primary transmutative medium;
7) In the suggestive references to be found in the Chinese alchemist Go-Hung, the "Pill of Immortality" must be fashioned on a great and famous mountain, an allusion, perhaps, to the distant connection between alchemy and pyramids; and finally,
8) Throughout alchemical literature not only are certain minerals associated with certain celestial bodies but also the very processes of alchemical operations are with those bodies as well, implying that

these processes are wrought most efficiently when those bodies are in certain alignments. Moreover, the association of astrology with certain gems — with *crystals* — also implies an association of lattice structures of crystals with celestial geometries, a significant insight as we shall see in future chapters;
9) Finally, as noted in the examination of Paracelsus, some alchemical experiments and operations could only successfully be performed at specific times and seasons, a point which will become a primary focus of part two.

The following conclusions and implications were developed in part two (noting that the numbering of points has been changed in this context):

10) The basic principles and assertions of certain alchemical texts are verified, namely:
    a) *with respect to the color sequence indicating successful confection of the Philosophers' Stone:* Hudson, via known and standard chemical and physical techniques, was able to replicate the overall color sequence of the derivation of the Philosophers' Stone. As was seen, two of these colors, a cranberry-red color, almost the color of grape juice, signified a stage in the process which ended with a fine "white powder of gold";
    b) *with respect to the composition of the Philosophers' Stone:* As was seen from chapter one, many alchemical texts stressed the *powder form* of the Philosophers' Stone, a form amply demonstrated by Hudson's material;
    c) *with respect to the Philosophers' Stone "occupying no space yet being confined in matter":* As was seen, one of the most significant anomalies exhibited by Hudson's white powder was its unusual mass loss anomaly, an anomaly that physicist Hal Puthoff explained by maintaining that some of the material was actually existing in a wholly different space and time, a "sub-" or "hyper-space." Thus, alchemical texts that indicate the existence of the Philosophers' Stone in different "worlds" would appear to be capable of interpretation along hyper-dimensional physical models.
11) In turn, these anomalous properties appear to be based upon:
    a) Extraordinary or non-ordinary geometries and shapes of atomic nuclei in elements within the platinum-group to mercury range in the periodic table of the elements; these are in turn the result of,
    b) extremely high-spin states of the nuclei; furthermore,

> i) in *some* cases these states are excited at a spin frequency of 19.5, and deexcited at a spin frequency of 21.5, in apparent correlation of the "tetrahedral physics" model of popular Mars anomalies researcher Richard C. Hoagland; that is to say, energy appears to gate, in some elements, into the system at 19.5, and to exit or deexcite at 21.5, indicating the possibility that Hoagland's model might be scale-invariant;
>
> c) the total angular momentum of the system, which is composed of two sub-systems:
>> i) the individual particles in the nuclear shells; and,
>> ii) the total angular momentum in the "proto-nuclear cluster" of the superdeformed nucleus;
>
> 12) These superdeformed atoms apparently bond in "quasi-molecular fashion" via a resonance phenomenon related to the spin frequencies and angular momentum of their various nucleus shells;
> 13) These superdeformed atoms likewise can undergo spontaneous asymmetrical fission *without* neutron bombardment, and often can yield extremely high bursts of gamma radiation when deexciting from their high-spin state. The weaponization potential and implications of this phenomenon will be explored in the next two parts of the book. Similarly, superdeformed nuclear isotopes in a high-spin state do not decay at standard rates of radioactive decay.

From this part of the work, and summarizing the preceding points, the following conclusions may be drawn:

> 14) Kozyrev's work implies that space and time themselves are quantized phenomena, and this in turn verifies the indications in alchemical texts that alchemists observed similar quantized phenomena in their operations;
> 15) Similarly, Kozyrev's work also demonstrated that the experimental operations performed at different times yielded surprisingly different results, verifying similar observations by alchemists, and placing their assertions that celestial geometries — i.e., the proper times — were an essential component to the successful performance of their work;
> 16) Kozyrev also demonstrated that the spiraling geometry of torsion, with all its hyper-dimensional properties, was essential to a proper understanding of time and its influence on physical processes, to the extent that torsion imparted a spin orientation to physical systems, verifying, from a different perspective, the observations David Hudson made concerning the high-spin states of his anomalous platinum group materials;

17) Likewise, Kozyrev's work also demonstrated that torsion fields could impress themselves into a physical system, such that after the removal of the original fields from that system, that it in turn could impress those dynamics on other isolated physical systems, thus confirming the alchemical view that the medium is non-local and that information transfer can occur at superluminal velocities between systems, and that the Philosophers' Stone could impress its dynamics on other material at a distance;

18) With his observations concerning the anomalous energy output of stars and the inability of the thermonuclear fusion model of stellar energies to account for all their energy production, Kozyrev also obliquely confirmed the alchemical view that celestial geometries, and in particular the position of the sun, were crucial to their work;

19) With his own observations of weight gains and losses depending on the rotational orientation of his experimental systems, Kozyrev similarly confirmed the anomalous weight gains observed by Hudson in his materials — which were due to their high spin state — and the observations of the alchemists themselves which suggested that they too observed anomalous patterns of weight gains and losses in their experiments;

20) With the implication, suggested by Kozyrev himself, that hyper-dimensional torsion effects could also account for the variations of perception in conscious observers of the faster or slower pace of the flow of time, Kozyrev likewise provided a physical basis for alchemical insistence that the consciousness of the alchemist himself be adequately prepared, implying that they too had observed similar phenomena;

21) By dint of his extensive testing of various elements and compounds for their torsion amplifying and shielding properties, and by subjecting such substances to a variety of processes of stress — heating, cooling, vibrating, spinning, and so on — Kozyrev paralleled the similar operations of the alchemists, and of David Hudson, who similarly subjected their materials to a variety of intense and stress-inducing processes. In all of this, Kozyrev was laying the foundations for a new science of "causal mechanics" and the engineering of exotic materials able to take on certain impressed dynamics, and impress them — or unleash them — on other materials. In this, too, he parallels and to a certain extent confirms the insights and observations of alchemy over the centuries.

22) Finally, we saw that the theoretical possibility of a "torsion substance," that is, of a compound impressed with the very properties of torsion

itself and therefore able to some extent to *induce or transduce it* into our normal three-dimensional space, was possible, giving some possible foundation to the Red Mercury legend as having some actual basis in truth. It may indeed have been a substance calling down the extreme fires and energies of the stars, itself, as seen repeatedly in the previous pages, an alchemical conception.

We may, therefore, with some confidence, and some trepidation, now turn to the most alchemical material of them all: The Nazi Serum of the Alchemical Reich.

Part Four
THE NAZI "SERUM" AND THE ALCHEMICAL REICH

⋯

"The German alchemist Franz Tausend began to produce gold from mercury in the 1920s. He began to work in association with General Ludendorff in 1925, and eventually produced artificial gold for the Nazis."

Robert A. Nelson,
Adept Alchemy,
www.levity.com/alchemy/nelson2_7.html, p. 7

Final Farm Hall Farce
Wirtz, Diebner, and the Mysterious Photochemical Process

❖❖❖

"Now, the beta synchrotron sends the electrons through this magnet which bends the course of them down to the reaction vessel.... Now, this is a tunable excimer laser. It's tuned to the exact resonance of the plutonium-239 that's in the reaction vessel at that end...."
The character Dr. John Mathewson, portrayed by actor John Lithgow,
from the movie *The Manhattan Project*[1]

The movies may seem an odd place to begin any discussion of the Nazis' secret research projects into exotic matter in particular, or for that matter of the alchemical nature and dalliances of the Third Reich in general. But in this case, it allows a direct entry into the resolution of some outstanding mysteries, and raises new questions. The movie in question is the 1986 release *The Manhattan Project*, starring well-known actor John Lithgow. The central character of the movie, however, is a precocious teenager with a taste and talent for science, Paul Stephens, played by actor Christopher Collet. Stephens steals some highly enriched plutonium from a local laboratory to build his own home-designed atom bomb. But what Stephens does not know, however, is that the laboratory's director, Dr. John Mathewson, played by Lithgow, has perfected an entirely new method of isotope enrichment involving tunable lasers. This new process makes the stolen plutonium only a few hundredths of a percent less than completely pure. With such pure plutonium, the high

---

[1] The opening lines from the movie *The Manhattan Project* (1986), starring John Lithgow, Christopher Collet, and Cynthia Nixon.

school student's homemade A-bomb has a far larger yield and destructiveness than he imagined, and the military spares no effort to track Stephens, and his bomb, down.

What is of interest here is the technology for isotope separation and enrichment only very briefly described in the first initial minutes of the movie. In point of fact, the description is so brief that one might miss it entirely if one did not realize its significance, nor its horrifying real world existence. This brief description, given by Lithgow's character Dr. John Mathewson, cited in the epigraph that began this chapter, even more oddly finds a peculiar — and indeed horrifying — echo in the real world almost exactly *forty-one years earlier.*

And as if to magnify the already horrifying implications of that fact, it was also made by a small group of Nazi physicists...

A. The Farce at Farm Hall: Wirtz and Diebner on the Photochemical Process of Isotope Enrichment

The group that made the comments were, of course, some of the German A-bomb scientists interred at Farm Hall, England, after the war. And indeed, some of their comments are the source of a persisting mystery, which I first mentioned in my book *Reich of the Black Sun: Nazi Secret Weapons and the Cold War Allied Legend.* But the mystery did not stop there. As many have discovered, entering research into Nazi secret weapons presents mysteries capable of several different types of analysis.

And the statements of the German scientists at Farm Hall are certainly one of them. To see why, it is once again necessary to see what I wrote before about them, in my book *The SS Brotherhood of the Bell,* for that background is needed to understand the new analysis that occurs in this chapter, courtesy of Mr. Richard C. Hoagland, and discussions we had over several days in the final months of 2007.

With that in mind, here is what I wrote in *The SS Brotherhood of the Bell:*

"When approaching the more extreme claims advanced in some exotic literature for Nazi secret weapons developments, one has the impression of some dark mediaeval alchemists' laboratory, with the alchemists dressed in the black uniforms of the SS, cracking whips on the backs of emaciated concentration camp slaves to perfect their dreadful machines of power and annihilation. It has all the elements — were the scope of human suffering involved not so real and so enormous — of a bad Hollywood "B" movie, with a gaunt Boris Karloff orchestrating an oddball cast that includes Bela Lugosi, Peter Lorre, Vincent Price and Sydney Greenstreet, all playing larger-than-life villains conspiring to conquer the world.

"But the situation is not helped by the strange quotations from *reliable* sources which, if one *really* pays attention to them, should give one pause. Consider two examples from the Farm Hall Transcripts, the transcripts of the conversations of the German atom-bomb scientists interred at Farm Hall, secretly recorded by the British and declassified by them only in 1992. The two examples, which I cited without extensive commentary in my previous book on Nazi secret weapons, *Reich of the Black Sun*, concern artificial rubies, and an unknown "photochemical process" of isotope enrichment. I will cite both sets of quotations and my remarks concerning them from that book directly.

1. The "Artificial Rubies" Passage from The Farm Hall Transcripts (Reich of the Black Sun, *pp. 142–132*)

"On pages 142 and 143 of *Reich of the Black Sun* I observed that one of the interred Farm Hall scientists made a rather astonishing statement, a statement made astonishing not only for its "matter-of-fact" passing nature and brevity, but also because it called forth almost no comment from the editor of the transcripts:

Then, on July 21, 1945, the handsome and cynical Horst Korsching, discussing the prospects for making a living with Diebner and Bagge, offers a curious observation:

"BAGGE: For the sake of the money, I should like to work on the uranium engine; on the other hand, I should like to work on cosmic rays. I feel like Diebner about this.

"KORSCHING: Would you both like to construct an uranium engine?

"DIEBNER: This is *the* chance to earn a living.

"KORSCHING: Every layman can see that these ideas are exceedingly important. Hence there won't be any money in it. You only make money on ideas which have escaped the general public. If you invent something like *artificial rubies* for the watch making industry, you will make more money than with the uranium engine."[2]

"I then commented as follows:

---

[2] Bernstein, *Hitler's Uranium Club*, p. 99, emphasis added.

THE PHILOSOPHERS' STONE | 207

Artificial rubies? Of course, such things were used in watchmaking before the invention of quartz movement. But in 1945, the idea was fantastic. Of course, by the time of the declassification of these transcripts, the world's first laser, which did in fact use an artificial ruby as the main component of the lasing optical cavity, was history, having been invented in 1961. But in July 1945 the idea was more than a little ahead of its time. Is this another possible, though slight, indication that something else was going on inside Nazi Germany? Later in the conversation, Korsching expresses his desire to return Hechingen to collect his telescope, lenses and prisms, an indication that he was perhaps involved in optical as well as nuclear research.[3]

"The mention there of the connection between lasers and artificial rubies was not accidental, for on page 104 of *Reich of the Black Sun*, I cited the following quotation from former British intelligence officer-cum-journalist Tom Agoston, who first broke the story of the *Kammlerstab*, the SS's secret weapons think tank, to the West in the 1980s:

> Its purpose was to pave the way for building nuclear-powered aircraft, working on the application of nuclear energy for propelling missiles and aircraft; *laser beams*, then still referred to as "death rays": a variety of homing rockets, and to seek other potential areas for high-technology breakthrough. In modern high-tech jargon, the operation would probably be referred to as an "SS research think tank." Some work on second-generation secret weapons, including the application of nuclear propulsion for aircraft and missiles, was already well advanced.[4]

"What is interesting is the juxtaposition of these two quotations, for while Agoston offers no *evidence* for his assertion that the SS was working on the development of lasers during the war, he nonetheless clearly states that this was an avenue of research being pursued.

"Thus, the *subsequent* appearance of the Farm Hall Transcripts in the early 1990s oddly corroborates Agoston's assertions with Korsching's very curious reference to artificial rubies, a then costly and time-consuming process that surely would have merited more than merely making rubies for watches. Since the transcripts had not yet been declassified by the British government when Agoston wrote his book, we can only assume that Agoston did *not* know the contents of the still secret Transcripts and was basing his assertions on his confidential talks with Dr. Wilhelm Voss, who first disclosed the story of

---

3   Joseph P. Farrell, *Reich of the Black Sun*, pp. 142–143.
4   Agoston, *Blunder! How the U.S. Gave Away Nazi Supersecrets to Russia* (New York: Dodd, Mead and Company, 1985), p. 65, emphasis added.

Kammler's black projects secret weapons think tank.

"This juxtaposition argues very strongly, though only circumstantially, that the SS was indeed involved in researching *lasers*. While we do not yet know how far they pressed this research, nor to what degree of success, it is in any case not of immediate concern, since the *theoretical possibility* of lasers already existed within quantum mechanics at that time.

"But what is of *real* interest is the *type* of physics that Agoston's and Korsching's remarks — made decades apart and independently of each other — indicate about the nature of the physics *concepts* the Germans were investigating, in this case, aspects of *quantum mechanics and coherence*. As was seen in *Reich of the Black Sun*, Nazi ideology itself, with its rejection of "Jewish" relativistic physics, would have naturally turned to the home-grown, "purely Aryan" and equally successful, quantum mechanical theory as a conceptual basis from which to pursue its advanced projects. We shall return to this all-important though brief clue, *coherence*, in a moment. For now, let us turn our attention to a second odd quotation, another weird glimpse into the possible areas of physics the SS was investigating.

2. The Farm Hall Transcript's Indications of a German Photochemical Process of Isotope Separation and Enrichment

"As recounted in *Reich of the Black Sun*, once the interred German scientists had learned of the Allied A-bombing of Hiroshima, they then begin to debate how the Allies could have "done it so soon," a discussion that naturally quickly turns to the question of separating and enriching enough uranium-235 isotope. In one short exchange between Karl Wirtz and Otto Hahn, the discoverer of nuclear fission, the deduction is quickly made by Hahn that the Allies could only have achieved the production of a uranium A-bomb with such processes, a sentiment quickly echoed by Wirtz with his comment "They have it *too*,"[5] a short admission pregnant with implications that methods of isotope separation were clearly known and available to the Germans during the war.

"But as I noted in *Reich of the Black Sun*, 'a short, but astonishing, exchange between Hahn, Weizsäcker, Harteck, Wirtz, and Diebner' then follows:

HAHN: I think it's absolutely impossible to produce one ton of uranium 235 by separating isotopes.

WEIZSÄCKER: What do you do with these centrifuges?

---

[5] Bernstein, *Hitler's Uranium Club*, p. 144, cited in *Reich of the Black Sun*, p. 144.

HARTECK: You can never get pure "235" with the centrifuge. But I don't believe that it can be done with the centrifuge.

WIRTZ: No certainly not.

HAHN: Yes, but they could do it with mass spectrographs. Ewald has some patent.

DIEBNER: *There is also a photochemical process.*[6]

As *Reich of the Black Sun* goes on to indicate, this little exchange is a sign that perhaps the scientists are playing out a farce on their British captors, and that the farce may even be continuing by the British themselves in their declassification of the transcripts.

"What do I mean by this? What I mean is that the British declassification is significant for *when* it occurs, and I only explored a few of the possible reasons in the previous book:

> Note that the transcripts are declassified by the British *after* the German reunification in 1989, an oblique admission, perhaps, that there was no more purpose in maintaining whatever secrets they still held, since there would now be other sources available to tell the story that had been long suppressed: that the Nazis had been either perilously close to, or had *actually acquired* the atom bomb before the Allies.[7]

To see what the "farce" being played out by the interred German scientists and the British government's declassification may be, one needs to examine a significant question that I only hinted at in various places in *Reich of the Black Sun:* What is this "photochemical process"?

"A second reference to this mysterious unknown photochemical process occurs a little later in the Transcripts, during a brief but very suggestive interchange between Hartek and Wirtz:

HARTEK: They have managed it with mass spectrographs on a large scale or else they have been successful with a photochemical process.

WIRTZ: Well I would say photochemistry or diffusion, ordinary diffusion. They irradiate it with a particular wavelength (all talking together). [8]

---

    6 Bernstein, *Hitler's Uranium Club,* p, 118, emphasis added, cited in *Reich of the Black Sun,* pp. 144–145.

    7 Joseph P. Farrell, *Reich of the Black Sun,* p. 145.

    8 Bernstein, *Hitler's Uranium Club,* p. 148, cited in *Reich of the Black Sun,* p. 148.

I then commented as follows:

> At this juncture, Bernstein[9] again observes that "it is not clear" what this photochemical process is.[10] In any case, whatever the process was, Wirtz's mention of it and of irradiation with a particular "wavelength" appears to have provoked a burst of conversation from the other scientists. Were they intentionally trying to drown him out and mask his statements so as not to be recorded? We will never know.[11]

But perhaps we really *do* know what this mysterious though unknown photochemical process of isotope enrichment is.

"A clue is again afforded by the *timing* of the declassification of the Farm Hall Transcripts: after German reunification. But their declassification also occurs *after* another significant event, the discovery and publication of *the first cold fusion experiments by Pons and Fleischmann*. Most do not associate cold fusion with isotope separation and enrichment, but the fact of the matter is, *by very easily engineered processes that might best be called precisely "electrolytic" or "photochemical,"* nuclear transmutation of elements has been observed in laboratories all over the world since then, and using materials relatively easily obtained, and able to be engineered in any competent high school or in a garage laboratory bench. The literature covering these cold fusion transmutations is easily available on any internet search.[12]

"If this or something similar is indeed what the Germans meant by a 'photochemical process,' then the contents of the Farm Hall Transcripts are nothing less than revolutionary, for it means that we have a *second* indicator, in a very different way, that the Germans were researching the phenomenon of quantum coherence of the medium in a variety of systematically pursued ways.[13] On this view the timing of the release of the transcripts is significant in a dual way, as occurring (1) *after* German reunification and (2) *after* the first public release of cold fusion experimental data.

---

9 Jeremy Bernstein, the editor and annotator of the transcripts.
10 Referring to Bernstein's comment on p. 120, n. 38, of *Hitler's Uranium Club*.
11 Joseph P. Farrell, *Reich of the Black Sun*, p. 148.
12 Q.v. the crucial paper on permeation methods for transmutation by Yasuhiro Iwamura, Mitsuru Sakano, Takehiko Itoh, "Elemental Analysis of Pd Complexes: Effects of D_2 Gas Permeation," *Japan Journal of Applied Physics,* Vol. 41 (2002) pp. 4642–4650, Part I, No 7A, July 2002. As for the SS and fusion research, German researcher Karl-Heinz Zunneck states that fusion energy was one area of the SS's investigations, though he does not indicate how far it had pressed this research (see Karl Heinz Zunneck, *Geheimtechnologien, Wunderwaffe und die irdischen Facetten des UFO-Phänomens,* Schleusingen, Germany: Amun Verlag, 2002), p. 151.)
13 I realize that at this juncture the term "quantum coherence" seems to have little to do with cold fusion, but will expand the term to a more accurate description later. Suffice it to say that the only viable explanations for the anomalous excess energy present in cold fusion experiments is that some aspect of the vacuum energy has been cohered and entered the experiments via mechanisms not yet fully understood.

"The implications of the transcripts' two allusions to such a photochemical process by interred German scientists in 1945 is *staggering*, for if they had actually discovered and conducted such experiments — and there was certainly ideological impetus from the Nazi government in its pursuit of 'energy independence' to do so — and if they had advanced from an initial 'Pons and Fleischman' apparatus to a more sophisticated version that accomplished the transmutation of elements at room temperature, as in the recent Japanese experiments, then we have a second, though still tenuous, indicator that the SS was pursuing some very extraordinarily advanced physics conceptions indeed. And there would have been *little* if anything required in such experiments that did not exist in some form available to the Germans during the war.

"Small wonder then, that when Wirtz begins to talk of irradiation at particular wavelengths that the other scientists appear to begin talking all at once, as if to drown out his comments from the British recorders, an effort that, if one takes Bernstein's editorial mystification as any indicator, was entirely successful.[14]

"What do we have, then, at this juncture?

"First, we have allusions, from the same source, to two *different* phenomena — lasers and a 'photochemical process' of nuclear transmutation — involving quantum coherence at some level. Second, in the case of the reference to artificial rubies and their implied use in lasers, we have further corroboration from an independent source of research, Agoston, whose corroboration of the "laser" side of the story comes *prior* to the release of the Transcripts themselves and Korsching's passing remark about artificial rubies. Finally, the declassification and release of the Farm Hall Transcripts themselves may be significant with respect to timing in a dual sense, as occurring after German reunification, and after the first public release of cold fusion laboratory data by Pons and Fleischmann."

And there, more or less, is where my analysis in *The SS Brotherhood of the Bell* ended, but a new possibility emerged.

After *SS Brotherhood* was published, I was invited to be interviewed by Mr. Tim Ventura for his well-known website American Antigravity.[15] The interview was, of course, about the Nazi Bell. Mr. Ventura in turn was a friend and associate of well-known space science and Mars anomalies researcher Richard C. Hoagland.

Unbeknownst to me, Mr. Ventura had mentioned my books to Mr. Hoagland, and out of that mention developed an initial contact and series of

---

14 Bernstein would certainly know of cold fusion, so his mystification may reflect rather his own adherence to that school of "public consumption physics" that rejects a priori any validity to cold fusion claims, since they represent an "open systems" paradigm and approach to COP > 1 systems.

15 www.americanantigravity.com

emails beginning in the autumn of 2007. Over the course of these emails, Mr. Hoagland asked for copies of all my books, which I arranged to send him.

After reading *SS Brotherhood*, Mr. Hoagland began a series of emails with me, concerning a different possibility of analysis of the above texts concerning the mysterious "photochemical process" than I mentioned in the book. To be honest, I had considered, and rejected, the idea for inclusion in my analysis of this mystery of the Farm Hall transcripts in *SS Brotherhood*, for a variety of reasons, not the least of which were the technological difficulties to be encountered at the time, and hence I opted for a less technologically sophisticated interpretive speculation — cold fusion — in the book.

Mr. Hoagland's analysis and argumentation, however, cause me to reconsider the idea. By presenting his analysis here, and offering what I believe to be some *additional* circumstantial information that possibly corroborates this new, far more radical, interpretation, I hope to exhibit how this speculation actually resolves *other* mysteries from the war *not* previously mentioned in any of my books on Nazi secret weapons research. It is also to be noted that Mr. Hoagland's interpretation sticks very closely to the transcript text itself, and additionally, provides yet another rationale in addition to the ones offered in *SS Brotherhood* for the reason that the transcripts were kept classified for so long.

But first, Mr. Hoagland's analysis.

B. Richard C. Hoagland's Analysis and Hypothesis

On Tuesday, November 20, 2007, around nine-thirty in the evening, I received the following email from Mr. Hoagland, and it began a series of discussions between us.

1. Hoagland's Analysis and Hypothesis: The Emails

Joseph,
My second point concerns certain "enigmatic" references within the "Farm Hall Transcripts."

(Remember, we had already been emailing back and forth discussing other topics of no relevance here. He continues):

Specifically, the (apparent) mystery of what was meant when one of the captive German scientists (Diebner) referred to as a "photochemical (isotope separation) process...." And later, when another one (Wirtz) says—

"Well, I would say photochemistry or diffusion, ordinary diffusion. They irradiate it with a particular wavelength—"

And then, everyone in the room (according to the transcriber...) starts "talking at once" — as if they *know* they're being recorded by their British "hosts" and want to drown out any further tell-tale references...

OK, this process is *not* "mysterious." It's well-known — now.

It's called "laser isotope separation."

...

Now, lasers (supposedly) did *not* exist in the early 1940s.

However, if the previous reference in these same Farm Hall transcripts you quote, to "artificial rubies," was accurate, it doesn't take a rocket scientist (!) to put these references together... and plausibly propose that these oblique statements to "photochemical isotope separation" *could* actually refer to a possible Nazi *laser project* for precisely such isotope separation! At least — theoretically.

Broadband, high-intensity lights — even with "narrowband" filters — simply *could not* do the job. It takes the ultra-narrowband (hundredths of an Angstrom wide!), *tunable* lasers (a second or third generation laser development...) to even *begin* to do this properly. But —

If they had the idea (based on the German development of quantum mechanics), the rest was "only" money ... and time.

Thus, those two (supposedly unconnected) references would seem to argue that the Germans *had* thought of "lasers" a *long* time before Townes made it practical... and may have even made some working systems!

...once you're on the trail — especially, if you have essentially *unlimited* (Reichsmarks), a lot of later laser developments (like, the invention of "tunable dye lasers") could have been *radically* compressed in time; nothing like a war...

Anyway, just thought you'd like to know...[16]

And there, Mr. Hoagland's remarkably succinct and breathtaking email ended, leaving my mind running a thousand different directions.

As I stated previously, I had considered the idea, and rejected it, largely for the technological degree of sophistication required to make it work, a sophistication Mr. Hoagland himself mentions in his email. But this was not what really caught my attention. What caught my attention were two things I had *not* considered, and that Mr. Hoagland very evidently *had*.

In order to see what things really caught my attention in his email, it would be helpful to summarize his argument:

---

16   Richard C. Hoagland, personal communication with the author, Nov. 20, 2007, 9:29 p.m.

1) The German scientists clearly imply a photochemical process of enrichment by references to irradiating material with "a particular wavelength";
2) Such a process is well-known today, though in World War II it is *decades* in advance of the thinking of the day (after all, the Manhattan Project had not thought of, nor pursued, its development);
3) The scientists all begin talking as if to drown out further commentary, as if they are protecting a great secret, and confirming the suspicions of many analysts that the Nazi scientists knew they were being recorded;
4) Korsching's reference to the manufacture of "artificial rubies" — the lasing cavity for the first *known* optical laser a decade and a half later — *is thus corroboratory of some sort of laser project in Nazi Germany, "at least — theoretically" as Mr. Hoagland puts it.*
5) Having thought of the concept of lasers, and laser isotope enrichment, the Nazis would then have realized that in order to separate various isotopes of various elements, they would need not a solid state lasing optical cavity, like a ruby crystal, but a gaseous one, one that was *tunable* to the wavelengths of particular isotopes over an extremely narrow bandwidth no wider than a few hundredths of Angstrom units;
6) They "may have even made some working systems!" and finally,
7) They may have done so *if there were adequate funding — a pile of Reichsmarks and enough labor and technical staff — to do it.*

Note very carefully what Mr. Hoagland is *not* saying. He is *not* saying that they actually *did* it, but only acknowledging the *possibility* that they may have done so, and pointing to the specific remark of Korsching mentioning artificial rubies as an indicator that thinking along the lines of lasers, and perhaps the first steps on the technology tree to the tunable laser — the first-generation solid state optical lasing cavity, a ruby laser — may have been actually taken.

But it was precisely Korsching's remark, and Mr. Hoagland's reference to what might have been accomplished with enough money, that stuck in my mind. I clung (rather desperately, I might add) to the technological difficulty, and responded to Mr. Hoagland's email, all the while knowing certain odd facts and mysteries from the war that were *not* mentioned in *SS Brotherhood of the Bell*, and therefore were presumably unknown to Mr. Hoagland, knowing full well that they not only tended to corroborate his analysis, but even more besides:

Richard,
　　Well, laser isotope separation...*by Nazis!?!?!?* I thought I was pressing

> the point too much with the little cold fusion excursion…! Even mentioning lasers in connection with it seemed to me to be crawling way out on a limb…. Yes I knew of this process (of laser isotope separation)…, but thought I'd be crucified badly enough for what I *did* say…
>
> But…as you say…**nothing like a *war* and unlimited Reichsmarks to light the way**…they *had* Hartek's centrifuges, they had Von Ardenne's mass spectrometer version of Lawrence's beta calutrons…so why not lasers?

As can be seen, what had sparked my mind to consider the possibility more seriously was Mr. Hoagland's reference to "enough money," implying a *serious crash program to develop the technology needed for massive enrichment of **very pure** fissile material in an atom bomb project.* More about this in a moment.

As can also be seen, my chief difficulty — the one that had prevented me from mentioning the laser isotope separation possibility in *SS Brotherhood* — was precisely the difficulty of the technological sophistication in lasers required to pull it off:

> *Interesting observation* you have about tunable lasers and the remarks of "irradiating it with a particular wavelength…" Now that *does* really open a Pandora's Box if indeed this *is* what Wirtz is referring to! It is, if you're correct, just downright *scary.*
>
> If you're right (and from the purely scientific point of view I see no reason to think you're *wrong* here), then there is even more research to do on what the Nazis were really up to. **The technological aspect here is the only down side…could they have really pulled that off??**[17]

In other words, it was not for lack of funding, for the SS, as I had established in *Reich of the Black Sun* and in *The SS Brotherhood of the Bell,* would have been responsible for such a project, and it was awash with money, and able to draw on an almost limitless supply of slave labor and skilled technicians and engineers.

Hoagland had, in other words, by his emphasis on money and the wartime situation, forced me to reconsider the possibility. And as my email response to him indicates, I was clinging doggedly to the technological difficulty of such a project.

Having fired off this email response, I thought surely the technological difficulty would persuade him of the difficulty of this suggestion ever having been *an actual possible project* inside the Third Reich.

---

17   Personal communication of the author to Richard C. Hoagland, Nov. 20, 2007, italicized emphasis in the original, boldface emphasis added.

Not so:

Joseph,
(The) *major* point in favor of Nazi laser development, or at least, *theoretical* explorations of the same:
It was *not* contrary to existing quantum physics. In fact, many science writers and physicists, over the past decades, have wondered again and again why it took *so long* to actually "invent" the laser — given the long history of quantum mechanics before Townes finally built one...
"Cold Fusion," on the other hand, was a "lucky accident."
And, because its explanation is *not* standard quantum physics...its full development has been (so far) delayed by *almost a full generation* (at least — publicly).

So far, so good, the science was there. We both knew that. And, interestingly enough, Hoagland is *implying* yet another possible alternative and "suppressed physics" explanation for recent history, namely, the reason that the laser was not publicly invented much sooner was that *the Nazis had already done so as a component of isotope separation for their atom bomb project.* Now, while he has not stated this in so many words, it is the clear implication of his remarks.

And indeed, as I had pointed out in my comments in *The SS Brotherhood of the Bell* which were cited previously, one of the things the British journalist Agoston had in fact *explicitly stated* was that they had actually *built* lasers within the super-secret black projects think tank of SS *Obergruppenführer* Hans Kammler! So, the *science* was there, as I had clearly stated in *SS Brotherhood* and as Mr. Hoagland had also outlined in his next email to me, and the *allegation* was there in Mr. Agoston's statements that they had actually built lasers.

The case was beginning to build, yet the technological sophistication still weighed in my mind. Ruby lasers certainly, at the minimum, seemed to be implied by the totality of the comments in the Farm Hall Transcripts and the wider context of information known about the Kammler Group (*Kammlerstab*). And let us recall what the mission brief of the *Kammlerstab* was: it was to brainstorm its way to second-, third-, and even fourth-generation weapons and technical systems by mapping out the necessary steps of the technology trees to acquire them, and then *actually do it*.[18]

Then Hoagland reminded me of one more interesting remark in the Farm Hall Transcripts, a remark that pointed in a certain direction, a direction that Mr. Hoagland at that point in time did not know of, because he had not yet read my book *Reich of the Black Sun:*

---

18 See my *Reich of the Black Sun* (Kempton, Illinois: Adventures Unlimited Press, 2004), pp. 100–107.

So, no, lasers would have been a *lot* easier for the Nazis to imagine… and develop.

And, those references to "artificial rubies" and "photochemical isotope separation" in those transcripts are *major* clues; you can *only* do photochemical isotope separation with *lasers* — because of the ultra-narrow, ultra-stable frequency control that's needed. No other "photochemical light source" will do the job. *Period.*

So, why mention it…unless you know of the *one way* it *can* be done — because **you've seen it work?** [19]

This last comment was really what made me consider the idea seriously, and to present it here. For though he had not yet read *Reich of the Black Sun,* and hence did not yet know about what this comment suggested to me, as we shall see, his thinking *independently* was headed the same direction as mine now was, but without the aid of knowing what I knew…

2. What Hartek Saw at I.G. Farben and What Mr. Hoagland Did Not Then Know: Back to the Auschwitz "Buna Factory"

What Hoagland's remark about "seeing it work" reminded me of was a short exchange in the Farm Hall Transcripts that I had cited in *Reich of the Black Sun*. Having not yet received that book nor read it, Mr. Hoagland could not have known why I found that remark so significant. Here is the exchange, placed in the context in which it occurs in *Reich of the Black Sun.* The interred German scientists have just heard of the American A-bombing of Hiroshima, and are debating how the Allies could have enriched so much uranium so quickly. I am indenting the block quotation of my own words from *Reich of the Black Sun,* and placing the there-cited quotations from the Farm Hall Transcripts in quotation marks:

> For Hartek and the other Farm Hall scientists, the problem was not means or methods, it was simply a labor shortage, a shortage the SS was not experiencing.
>
> Later, Hartek is even more specific:
>
> "Considering the figures involved I think it must have been mass spectrographs (that the Allies used to enrich uranium). *If they had had some other good method they wouldn't have needed to spend so much. One wouldn't have needed so many men."* [20]

---

19   Richard C. Hoagland, second personal communication with the author, Nov. 20, 2007, italicized emphasis in the original, boldface emphasis added (i.e., the last comments Mr. Hoagland emphasized, and I am doubly emphasizing here).

20   Joseph P. Farrell, *Reich of the Black Sun: Nazi Secret Weapons and the Cold War Allied Legend*

The standard reading of this passage according to the "postwar Allied Legend" is, of course, that Hartek, like all the German scientists, was bemoaning the fact that they simply did not have the money nor labor resources to enrich enough uranium on the scale of the Manhattan Project.

But in *Reich of the Black Sun* I argued that this point within the Allied Legend is simply absurd in the extreme, since the SS with its enormous slave labor pool *did* have such labor and monetary resources available. Indeed, following the lead of Carter P. Hydrick and other researchers into the matter, I suggested that this whole aspect of the Allied Legend of Nazi nuclear incompetence was a deliberate postwar misdirection to conceal a very large and *very successful* Nazi enrichment process.

However, if the misdirection were also being done *to keep attention away from the Nazi enrichment technology itself* — if indeed tunable lasers were one of the technologies involved — then there is yet another reason for the misdirection, for it would mean a laser technology that is at least second- or third-generation technology had been developed to enrich *very pure* fissile isotopes, which would have allowed for *smaller* critical masses, and an A-bomb of much lighter weight and smaller dimensions, able to be used as warheads on their rockets. Mr. Hoagland's suggestion, in other words, *fit the overall context and practical requirements of the Nazi A-bomb program quite well.*

It is in this context, then, that I believe one should read the following exchange between the Farm Hall scientists. Once again, I am citing the context from *Reich of the Black Sun*; the scientists' actual conversation is noted by the appearance of their names prior to their individual comments:

> Korsching responds (to Hartek's previously cited comments), and a small debate ensues, in which a sensitive topic is barely touched upon by Hartek, and Bernstein's editorial comment becomes either an exercise in ignorance, or deliberate omission:
>
> KORSCHING: "It was never done with spectrographs."

(Remember, this is the same Korsching who made the remark about artificial rubies!)

> HEISENBERG: "I must say I think your theory is right and that it is spectrographs."
>
> WIRTZ: *"I am prepared to bet that it isn't."*

(Kempton, Illinois: 2004), p. 149, citing Jeremy Bernstein, *Hitler's Uranium Club: The Secret Recordings at Farm Hall,* Second Edition (New York: Copernicus, 2001), p. 122, emphasis added in the present book.

This is a highly significant comment, and it is emphasized here. It is highly significant because, as I argued in *Reich of the Black Sun,* Heisenberg and the high-profile scientists were probably kept *out* of Nazi Germany's real A-bomb project, and used as fronts for the Allies to worry about, while the lesser-known but equally capable scientists, such as Korsching, Deibner, Wirtz and so on, were involved with the real one being run by the SS. Thus, Heisenberg *would not have known,* with his own abysmally small laboratory efforts, of any technologies of isotope enrichment being developed in uttermost secrecy by the SS. Hence, Wirtz's comment to Heisenberg suggests that *he* knows of a technology for isotope enrichment vastly less labor-intensive, and much less costly, than mass spectrographs, a technology that *Heisenberg* does *not* know about, or at least, would not have considered to be practical.

And let us not forget, *this is the same Wirtz who made the remarks about "irradiating" an isotope with the proper wavelength to begin with!*

Thus, Mr. Hoagland's suggestion of laser isotope enrichment begins to offer a way of interpreting the statements of the Farm Hall scientists *that makes perfect sense of the transcripts themselves, which the Allied Legend of Nazi nuclear incompetence ultimately utterly fails to do.*

But Heisenberg then continues with his own statements that imply very clearly that some sort of labor-intensive isotope enrichment project was underway in Germany. However, note carefully the remarks of Korsching and Hartek that Heisenberg's comments provoke:

> HEISENBERG: "What would one want 60,000 men for?"
>
> KORSCHING: *"You try and vaporize one ton of uranium."*
>
> HARTEK: *"You only need ten men for that. I was amazed at what I saw at I.G."*[21]

Before analyzing Korsching's and Hartek's comments, it is necessary to cite the commentary I made in *Reich of the Black Sun* immediately after this passage:

> Bernstein's[22] only comment here is to note the obvious, that "I.G." means "I.G. Farben," nothing else is said. Either Bernstein is unaware of the Farben "Buna plant" and its mysterious properties of consuming more electricity than Berlin and producing no Buna, or he has intentionally omitted any

---

21 Joseph P. Farrell, *Reich of the Black Sun,* p. 150, citing Bernstein, *Hitler's Uranium Club,* p. 122. Italicized emphasis added in the present book, boldface and italicized emphasis added in *Reich of the Black Sun* and doubly emphasized here.

22 Bernstein, i.e., the editor and commentator on the Farm Hall Transcripts.

further clarification of Hartek's remark. The Allied Legend, in so far as Bernstein is concerned, is intact.[23]

But it isn't intact at all. In fact, in the light of Mr. Hoagland's extraordinary proposal about laser isotope enrichment, it would appear to be altogether shredded.

The reasons why are extraordinarily simple, and extraordinarily breathtaking at the same time.

Note the comment of Horst Korsching, the physicist, once again, who made the comment about artificial rubies: "You try and vaporize one ton of uranium." Of course, one enrichment technology for which this process of reduction to a gas would be necessary would be for thermal gaseous diffusion, a relatively primitive technology. It was certainly a technology the Germans possessed, but the resulting purity of U-235 acquired from it would have been rather low, and required its deployment *en masse* in a manner similar to the Manhattan Project. *But the other technology which would have utilized it would be precisely uranium separation by lasers.*

However, the thermal process, since it would have required massive amounts of equipment and labor to run it, would have been labor-intensive, *whereas laser enrichment would not.* It is *this* fact that makes nuclear chemist Paul Hartek's comment which follows so breathtaking: "You only need *ten* men for that," ten men, not thousands! In other words, Hartek's brief comment *by the nature of the case* implies a very *advanced and sophisticated enrichment technology* in existence and being *used* by the Third Reich to enrich uranium! Even the use of Von Ardenne's mass spectrographs — a technology very similar to Lawrence's beta calutrons in the American Manhattan Project — would have required more manpower than that, especially if employed *en masse.*

The clincher is Hartek's final statement: "I was *amazed* at what I saw at I.G.," that is, at I.G. Farben.[24]

Why is this so significant?

It is significant because, first, it points to the so-called synthetic rubber or "Buna" plant I.G. Farben built at Auschwitz, relying upon the massive amounts of slave labor both to build and man it. However, as I pointed out in *Reich of the Black Sun,* none of its signatures are those of a Buna plant at all. Its close proximity to water, its need for a massive labor pool and technical staff, and its enormous electrical power consumption — it used more electricity than the entire city of Berlin, the eighth largest in the world at that time — and the fact that in the entire four years of its operation *it produced not one ounce of synthetic rubber* all point to the fact, as I argued in *Reich of the Black*

---

23   Farrell, op. cit., p. 150.
24   Emphasis added.

Sun, that it was not a Buna factory at all, but a massive Oak Ridge-sized isotope enrichment facility.[25]

But there are two additional factors that must now be mentioned and considered, for they corroborate the possibility that laser isotope enrichment may in fact have been one of the technologies employed at that plant. The first of these is that the Buna plant itself was plagued by constant technical difficulties and breakdowns. This would be inconceivable for a technologically sophisticated and experienced company like Farben if the technology were only that of synthesizing rubber, or even, for that matter, the technologies of uranium enrichment by diffusion or by mass spectrometers, both rather simple technologies. Farben's track record of accomplishments in sophisticated industrial plants would seem to rule this possibility out.

So how to explain the constant difficulties and breakdowns Farben experienced in its "Buna" plant operations at Auschwitz? If one of the technologies employed was indeed a *primitive tunable laser technology utilizing gases rather than solid state crystals as its lasing optical cavity, a rather sophisticated technology,* then the breakdowns make more sense.

But the *other* reason Hartek's mention of I.G. Farben is so significant is that Farben was a *chemical cartel;* its expertise was in the manufacture of *chemicals* and in particular, *gases.* If the Auschwitz "Buna plant" had only employed centrifuge or mass spectrographs as its primary enrichment technologies, then the selection of an electrical engineering firm such as Siemens to build the plant at Auschwitz would have been far more logical.

It is the fact that it was *Farben* which built the plant that points to the probable use of *other* technologies implying sophisticated expertise in the handling of *gases* that point in turn to the use of gaseous diffusion as one probable technology in use at Auschwitz *en masse.* This would explain the high electrical consumption and enormous labor requirements.

It is here, however, that Hartek's remark must be recalled once again, for he clearly states that (1) he was *amazed at what he saw, implying something in actual use* by Farben, and (2) that what he saw required only *ten* people to operate! The gaseous and thermal diffusion techniques known and available to the Germans (and everyone else), would hardly have been a technology to call forth "amazement" on Hartek's part. It is this fact, plus the facts of its extremely *low* labor requirements, and I.G. Farben's peculiar track record of accomplishments in high-technology chemical processes and its expertise in gases, plus the fact of its consistent *difficulties* in operating the Auschwitz plant, that point to the possible use of some sort of extraordinarily *sophisticated and, for the day, advanced technology of isotope separation and enrichment in use there.*

---

25   Joseph P. Farrell, *Reich of the Black Sun,* pp. 25–43.

It points to, and corroborates, the possible use of tunable lasers.

Reductions of a metal to gas. Use of light of specific frequency, or color, to accomplish a transmutation and purification..

It sounds a lot like alchemy...

But more of that later.

For now, let us return to Mr. Hoagland for a moment, and recall that when he was writing his emails to me, he had not yet read the previously outlined facts in *Reich of the Black Sun*. This makes the comments contained in his *next* email to me even more stunning, for they independently *confirm* the previously outlined argumentation and analysis:

> Joseph:
>
> I guess I need to define "tunable."
>
> ...
>
> I mean, like mixing "a ... *lot* of chemical compounds, made out of organic dyes,...

And, lest it not be remembered, *dyes* were precisely how Farben and indeed, the whole world-renowned German chemical industry got its start!

> ...made out of organic dyes, placing them between parallel, reflecting telescope half-silvered mirrors (a *standard* optical technology in the 1940s...), and exciting these compounds with simple light to *laser* emission...

In other words, *the technology existed*. All that was needed was the science, which was there, and the combination of this technology with the appropriate chemical know-how, which was also there in the form of Farben's expertise and the *Kammlerstab's* well-known utilization of various technologies in new combinations! And before we resume with Mr. Hoagland's email, it should be noted that Horst Korsching — the scientist who made the cryptic comment about artificial rubies — was also recorded at Farm Hall as stating his desire to return to Germany to collect his crystals and *telescope*. [26]

> Once the *right* compound was found (by brute trial and error, if no other way...)

Which would be one way to account for the difficulties Farben was experiencing in operating its "Buna plant."

---

26   See my *Reich of the Black Sun*, p. 143.

> ...then *that* specific compound could be *scaled up to an industrial level laser separation system...and voila, a working laser separation system.*

"Scaled-up" certainly the Auschwitz Farben plant was!

> So, the "tunable" part was in the (trial and error?) discovery and preparation of the *right* "lasing chemistry compound." Once *that* was accomplished, separating the actual isotopes using this technology would be pretty straightforward...for a society "making synthetic rubber," making "synthetic oil," etc. etc.

And Nazi Germany was certainly synthesizing both to an astonishing degree. In the final analysis then, there is nothing unusual nor really so extraordinary about Mr. Hoagland's proposal from the technological standpoint. Indeed, as his remarks clearly imply, though he did not know all the details at the time, he has certainly seen that the same case could be argued and advanced on a more detailed basis here, as has been done.

> Money and political will. The only things ultimately required — if the *theoretical* basis for the laser was known by some German physicists early in the War.
> Oh, and an absolute dictatorship — to organize the men and materials to make it work!
> Nothing else was needed; the *science* was built into the German quantum mechanics...from the beginning.[27]

Once again, while Mr. Hoagland is *not* suggesting that the Nazis actually *did* this, he is outlining the necessary parameters *for what to look for* if the possibility is to be entertained. And as I have presently argued the case thus far, *a reasonable circumstantial case can be developed that this, indeed, may actually have taken place, and that such technologies may actually have been in use by I.G. Farben. There are few other plausible ways of interpreting* **all** *of the strange remarks in the Farm Hall transcripts.*

And if this argument actually be the case, it would afford yet another rationalization for why the Farm Hall Transcripts were kept classified for so very long after the war; they implied a technology so far in advance of anything in use or even envisioned by the Allies in the Manhattan Project, that if anyone were the nuclear bunglers, it was the Allies, not the Germans.

---

27 Richard C. Hoagland, personal communication to the author, Wed. November 21, 2007, 2:23 a.m.

But is there any *other* evidence that would tend to corroborate the use of such a sophisticated technology inside the Third Reich? Indeed there is. But before we look at those possible corroborations, we need to have a closer look at the technology itself. The closer look will afford a glance at why I originally balked at this interpretation of the Farm Hall Transcripts and the Auschwitz Buna plant.

3. *The Modern Version of Tunable Lasers*

When Mr. Hoagland first emailed me about his insights into the implications of the Farm Hall Transcripts, he included a link to a website where the laser isotope technology is outlined, shown, and explained for a general audience. It is referred to here as it contains excellent pictures and good generalized descriptions of the actual technology and technique used in laser isotope separation.

The website is that of the Lawrence Livermore National Laboratory, and the article is entitled "Laser Technology Follows in Lawrence's Footsteps."[28] The basic idea of Laser Isotope Separation, or LIS as it is sometimes called, is simplicity itself:

> The technique is based on the fact that different isotopes of the same element, while chemically identical, absorb different colors of laser light. Therefore, a laser can be precisely tuned to *ionize only atoms of the desired isotope*, which are then drawn to electrically charged collector plates.[29]

Note two things here that were already paralleled and implied in the Farm Hall Transcripts.

The first is the reference to *ionization*, a process that would in fact *vaporize* the element, i.e., turn it into a gas. And this, of course, *might* be the process that Korsching had in mind when he referred to vaporizing a ton of uranium. The second is the reference to electrically charged collector plates. If such a technology were employed by Farben at its "Buna plant" at Auschwitz on a large scale and *en masse,* then this would possibly also account for the plant's high electrical consumption just as well as would massive numbers of centrifuges, diffusion machines, or mass spectrographs.

One gets an idea of the scale of such a laser from this picture which accompanies the article:

---

[28] www.llnl.gov/str/Hargrove.html
[29] Stephen Hargrove, "Laser Technology Follows in Lawrence's Footsteps," www.llnl.gov/str/Hargrove.html, p. 1, emphasis added.

The three large squares one sees in the picture are the actual separation tanks containing the element to be separated, while the large tube leading into and out of them are the actual tubes conveying the tuned laser light into the tanks.[30] In the light of the German nuclear chemist Paul Hartek's remarks cited above that referred to some technology of enrichment in use at I.G. Farben that only needed ten men, it is interesting that one sees just two men barely visible in the laser isotope separation facility at Lawrence Livermore Laboratories!

But what is the most intriguing thing mentioned in the article is something that tends to corroborate the argumentation I made in the previous section. The comment occurs beneath the following picture:

Plant-Scale Tunable Dye Laser at Lawrence Livermore National Laboratories[31]

The caption beneath this picture states the following information:

> LIS plant-scale dye laser chains absorb *green light from solid state lasers and reemit it at a color that can be tuned to the isotope of interest.* For uranium enrichment, the green light was converted to red-orange light of the three different wavelengths that are absorbed only by uranium-235.[32]

---

30   Stephen Hargrove, "Laser Technology Follows in Lawrence's Footsteps," www.llnl.gov/str/Hargrove.html, p. 2.
31   Ibid., p. 4.
32   Stephen Hargrove, "Laser Technology Follows in Lawrence's Footsteps," www.llnl.gov/str/Hargrove.html, p. 4, emphasis added.

Notice that the *tunable gas dye laser* is being stimulated by ordinary *solid state lasers* to emit its lasing light. This is significant because the solid state laser is precisely what is implied in Korsching's remarks about artificial rubies.

In other words, fully *five decades* before the technology at Lawrence Livermore, the Farm Hall Transcripts of the interred Nazi A-bomb scientists are making remarks that imply the following technologies

1) solid state lasers (Korsching's artificial rubies comment);
2) photochemical isotope separation and enrichment (Wirtz's remark about irradiating a substance with specific wavelengths), which imply a tuneable chemical gas dye laser;

And now, from Lawrence Livermore, we see why the Nazis apparently were entertaining *both* ideas, for in the Lawrence Livermore version of the technology

3) the use of the first, the solid state laser, to stimulate the lasing activity of the second, the chemical tunable gas dye laser.

In other words, the comments of the Nazi scientists at Farm Hall — which clearly in Hartek's case imply an *actual technology in use* — would seem to indicate that they had rationalized the whole process out very thoroughly, and that in turn implies once again, an actually existing though rudimentary tunable gas laser technology.

There is another highly intriguing comment made in the Lawrence Livermore article:

> Development efforts to achieve high enrichment efficiency centered on *improving laser beam uniformity and uranium vapor conditions*. Eighty percent of the plant's enrichment efficiency goal was achieved in several tests, *including the 290-hour demonstration tests in March 1999 that had laser systems operating at record power levels.*[33]

One may infer from these observations that Lawrence Livermore was encountering difficulty in achieving the right vapor conditions for uranium, in controlling the beam frequency, and also consuming enormous amounts of electrical power. Such difficulties would readily explain the otherwise inexplicable difficulties I.G. Farben was experiencing operating its so-called "Buna plant" at Auschwitz, for if indeed it *was* a Buna plant, it should have posed no difficulty to Farben whatsoever. But controlling a precisely tuned laser beam from a chemical gas dye laser being in turn stimulated to lasing action by solid state lasers, all with crude 1940s technology, *would* explain it.

But the real advantage of such a technology lies not only in its cost effectiveness, but also in the *much greater purity of isotope obtained for significantly reduced passes through the system,* a point which will become extremely important in the next section of this chapter:

> ...the *LIS process uses only 5 percent of the electricity consumed by existing gaseous diffusion plants, and LIS facilities would cost substantially less to build than those for other enrichment techniques such as centrifuge technology.*
>
> The *enrichment of uranium, from natural levels of 0.7 percent uranium-235 (235U) to between 3 and 5 percent 235U, is achieved in a few passes with LIS, a great improvement over the hundreds to thousands of passes required by other processes.* This translates into a much smaller plant and production costs substantially below those of either gaseous diffusion or gas centrifuge technology....
>
> Indeed, the system is remarkably compact. A vacuum chamber holding one separator unit produces output equivalent to that of several thousand of the best commercial centrifuges. *A commercial LIS plant would use 84 enrichment units, compared to more than 150,000 centrifuge machines.*
>
> ...Instead of using uranium hexafluoride, the starting material required by other processes, LIS uses uranium metal, which is less hazardous.

---

33   Stephen Hargrove, "Laser Technology Follows in Lawrence's Footsteps," www.llnl.gov/str/Hargrove.html, p. 5, emphasis added.

Compared to centrifuge or gaseous diffusion, the laser process requires about 30 percent less natural uranium ore to produce a comparable amount of enriched product, which also minimizes the amount of uranium tailings by about 30 percent.[34]

There are several points to be made here, each of which would have been of supreme importance to the Nazis, *if* they were aware of the theoretical possibilities of such a technology — and as has been seen there is every indication that they were — or *if* they were in possession of some rudimentary version of it — and as has been seen a strong circumstantial case can be made that they were.

Note first something that seems to contradict our argument, and that is that such a plant would consume far *less* electricity than a plant relying on massive banks of diffusion gates or centrifuges. But I have never argued that only *one* type of technology was in use at Auschwitz. However, suppose for the sake of argument that only one type was in use, and that it was some rudimentary form of LIS technology. As the above quotation makes clear, even a commercial plant would require several such laser units, and this would — especially given a much cruder form of it — still conceivably account for the Auschwitz plant's anomalously high electrical consumption, *especially if that plant were also producing the gases for the lasers.*

Secondly, not only are *much fewer passes* necessary through the LIS plant — in fact, the required number of passes is reduced several orders of magnitude — but the result of such passes is *much greater amount and purity of isotope material.* For the Nazis, in a hurry *to miniaturize their A-bombs to mount on their rocket warheads, this technology is, in other words, an **essential and necessary component to that goal.** From the standpoint of the exigencies of the war, then, the Nazi development of this technology seems not only logical, but is in a certain sense **mandatory.*** At this stage of the argument, what would rather need to be explained is why they did *not* develop it *if indeed they had known of its theoretical possibility,* which, once again, as seems evident from the Farm Hall Transcript remarks of Korsching, Wirtz, and Hartek, they did.

Note also, in conjunction with the previous point, that this also means *less uranium ore is needed* to achieve the same results. Again, this would have been another selling point.

The Lawrence Livermore article then produces the following diagram to illustrate the basic process of laser isotope enrichment:

---

34 Stephen Hargrove, "Laser Technology Follows in Lawrence's Footsteps," www.llnl.gov/str/Hargrove.html, pp. 7–8, emphasis added.

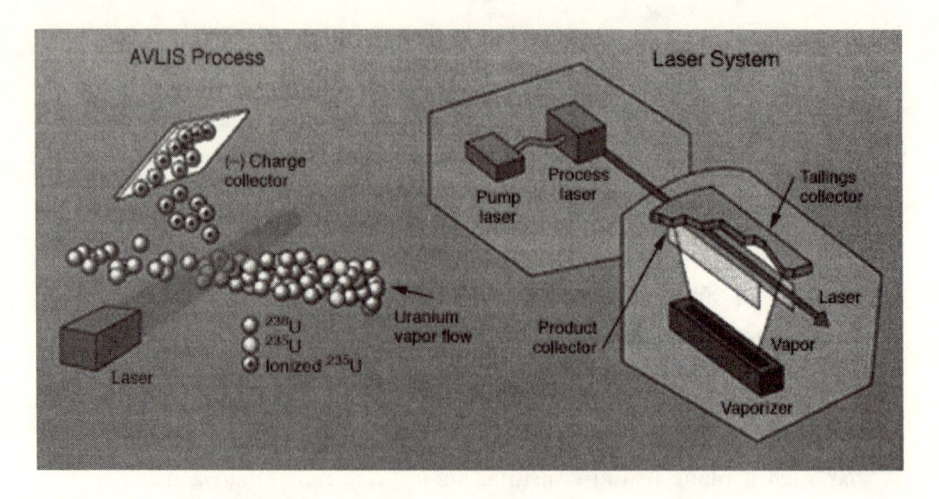

Basic Process of Laser Isotope Enrichment[35]

The following commentary precedes this diagram:

> In LIS enrichment, uranium metal is first vaporized in a separator unit contained in a vacuum chamber. The vapor stream is then illuminated with laser light tuned precisely to a color at which 235U absorbs energy.
>
> The generation of laser light starts with diode-pumped, solid-state lasers providing short, high-intensity pulses at high repetition rates. *This green light from the solid-state lasers travels via fiber-optic cable to energize high-power dye lasers....*
>
> Given the several kilowatts of high average power of the dye laser beam, it's a significant achievement that the wavelengths are stable to better than 1 part in 10 million and that the beam's ability to travel long distances is nearly perfectly preserved.
>
> Because the ionized 235U atoms are now "tagged" with a positive charge, they are easily collected on negatively charged surfaces inside the separator unit. *The product material is condensed as liquid on these surfaces and then flows to a caster where it solidifies as metal nuggets.* The unwanted isotopes, which are unaffected by the laser beam, pass through the product collector, condense on the tailings collector, and are removed.[36]

Note the reference to one final technology involved with LIS, and to the collection of the refined and separated 235U isotope as a *liquid* when first

---

35 Stephen Hargrove, "Laser Technology Follows in Lawrence's Footsteps," www.llnl.gov/str/Hargrove.html, p. 12.

36 Stephen Hargrove, "Laser Technology Follows in Lawrence's Footsteps," www.llnl.gov/str/Hargrove.html, p. 12, emphasis added.

separated. Additionally, surely the first reference to fiber optic technology would be beyond anything the Nazis dreamed of!

Unfortunately, however, as we shall see in the next section, the idea of 235U — or of any isotope for that matter, existing in a very pure liquid condensate state, and the idea of fiber optics, all in the historical context of Nazi Germany, may, in fact, be just possible.

C. Other Mysteries that Tend to Corroborate Hoagland's Proposal

Were it only for Hoagland's suggestions concerning the reading of the Farm Hall Transcripts as implying some sort of rudimentary laser isotope separation, and were it only for the more detailed "contextual case" based upon his idea presented here, one might be inclined to grant the hypothesis a *slight* possibility. After all, any such case as might be made for Nazi knowledge, much less *use*, of laser isotope separation technology must remain circumstantial if for no other reason than if that were the case, the postwar Allied Powers would be loathe to admit that they had entirely overlooked the possibility, and spent billions, and wasted *years* in the development of other technologies, and the building of huge and costly enrichment plants based upon them, while a much less costly, much more efficient, but much more sophisticated and therefore hard-to-develop technology was inherent in quantum mechanics all along, right along with nuclear fission itself.

But there are, in fact, *other* mysteries from Nazi Germany and World War II that, examined within the context and by means of the lens of laser isotope enrichment, begin to make a great deal of sense. And these things in turn tend to corroborate more fully the circumstantial case being advanced for its actual existence in a rudimentary form within Nazi Germany.

1. Vast Quantities of Enriched Uranium in Nazi Germany Implied by the Evidence

The first of these mysteries is the apparently huge amount of uranium ore that the Nazis not only *possessed,* but the large amount of 235U which they actually *enriched.* In my book *Reich of the Black Sun* I noted that as late as December of 1944, an official Manhattan Project document from the chief metallurgist as Los Alamos indicated that American stocks of fissile weapons-grade uranium-235 were still far short of the needed critical mass for a workable atom bomb.[37] Nor would America, according to the memorandum,

---

37 Farrell, *Reich of the Black Sun,* pp. 57–58.

have enough of the material for a uranium-fueled bomb at its then current production rates until *November* of 1945!

But curiously, within a matter of mere weeks after the German surrender, the stocks and output of Oakridge of the precious material *doubled*. As I pointed out there, this tore yet another gaping hole in the postwar Allied Legend about American nuclear physics mastery and German nuclear engineering incompetence:

> If the stocks of weapons grade uranium ca. late 1944-early 1945 were about half of what (the Americans) needed to be after two years of research and production…how then did the Manhattan Project acquire the large amount of uranium-235 needed in the few months from March to the dropping of the Little Boy bomb on Hiroshima in August, only five months away? How did it accomplish this feat, if in fact after some three years of production it had only produced less than half of the needed supply of critical mass weapons grade uranium? Where did its missing uranium-235 come from?…
>
> Of course the answer is that if the Manhattan Project was incapable of producing enough enriched uranium in that short amount of time — months rather than years — then its stocks had to have been supplemented from external sources, and there is only one viable place with the necessary technology to enrich uranium on that scale… That source was Nazi Germany.[38]

In support of this contention I produced evidence that the German U-boat, U234, which had surrendered to American authorities, contained eighty *gold-lined cylinders* of what the evidence suggested was most likely highly enriched uranium-235.

There were two significant facts that supported this contention. First, the two Japanese officers accompanying the precious cargo from Germany to Japan were seen painting paper labels on each of the cylinders, which simply read "U235." And secondly, gold was used to shield only *highly enriched* U235, since, as a non-corrosive metal, it would not have corrupted or reduced the purity of the stock.[39] In would thus appear that the U234 was transporting the entire critical mass of uranium-235 to Japan for a bomb, for the cylinders contained about 560 kg of the uranium oxide, more than enough for a critical mass for several bombs![40]

This means that Nazi Germany was enriching uranium on a massive scale, and to great purity. The question is: *how?*

---

38 Farrell, *Reich of the Black Sun*, p. 58.
39 Ibid., pp. 60–61.
40 Ibid., p. 61.

But there is an even deeper mystery surrounding Nazi Germany's uranium stocks. As I also noted in *Reich of the Black Sun*:

Germany too appears to have suffered the "missing uranium syndrome" in the final days prior to and immediately after the end of the war. But the problem in Germany's case is that the missing uranium is not a few tens of kilos, but several hundred *tons*. At this juncture, it is worth citing Carter Hydrick's excellent research at length, in order to exhibit the full ramifications of this problem:

"From June of 1940 to the end of the war, Germany seized 3,500 tons of uranium compounds from Belgium — almost three times the amount (Manhattan Project chief General Leslie) Groves had purchased.... and stored it in salt mines in Strassfurt, Germany. Groves brags that on 17 April, 1945, as the war was winding down, Alsos recovered some 1,100 tons of uranium ore from Strassfurt and an additional 31 tons in Toulouse, France.... And he claims that the amount recovered was all that Germany had ever held, asserting, therefore, that Germany never had enough raw material to process the uranium either for a plutonium reactor pile or through magnetic separation techniques.

"*Obviously, if Strassfurt once held 3,500 tons and only 1,130 were recovered, some 2,370 tons of uranium ore was unaccounted for — still twice the amount the Manhattan Project possessed and is assumed to have used throughout its entire wartime effort....* The material has not been accounted for to this day....

"As early as the summer of 1941, according to historian Margaret Gowing, Germany had already **refined 600 tons of uranium to its oxide form, the form required for ionizing the material into a gas,** in which form the uranium isotopes could then be magnetically or thermally separated or the oxide could be reduced to a metal for a reactor pile. In fact, Professor Dr. Riehl, who was responsible for all uranium throughout Germany during the course of the war, says the figure was actually much higher....

"To create either a uranium or plutonium bomb, at some point uranium must be reduced to metal. In the case of plutonium, U238 is metalicized. Because of uranium's difficult characteristics, however, this metallurgical process is a tricky one. The United States struggled with the problem early and still was not successful reducing uranium to its metallic form in large production quantities until late in 1942. *The German technicians, however... by the end of 1940 had already processed 280.6 kilograms into metal, over a quarter of a ton.*"[41]

---

41 Farrell, *Reich of the Black Sun*, pp. 58–59, citing Carter P. Hydrick, *Critical Mass: The Real Story of the Atomic Bomb and the Birth of the Nuclear Age,* Internet published manuscript, www.3dshort/nazi-

I then noted that

> These observations require some additional commentary.
>
> First, it is to be noted that Nazi Germany, by the best available evidence, was missing approximately two thousand tons of unrefined uranium ore by the war's end. Where did this ore go?
>
> Second, it is clear that Nazi Germany was enriching uranium on a *massive scale*, having refined 600 tons to oxide form for potential metalicization as early as 1940. This would require a large and dedicated effort, with thousands of technicians, and a commensurately large facility or facilities to accomplish the enrichment. The figures, in other words, tend to corroborate the hypothesis…that the I.G. Farben "Buna" factory at Auschwitz was not a Buna factory at all, but a huge uranium enrichment facility. However, the *date* would imply another such facility, located elsewhere, since the Auschwitz facility did not really begin production until sometime in 1942.[42]

There are two significant things to note here: first, Hydrick's reference to the vaporization of uranium to an *ionized gas*, and secondly, the fact that large-scale uranium separation is occurring inside of Nazi Germany *prior to* the Auschwitz facility coming on line in early 1942.

The problem is, there is *no other such large facility in all the Third Reich or its conquered territories* prior to this period. The *absence* of such a large facility thus implies the existence of a sophisticated technology, capable of producing such stocks of enriched uranium *rather quickly* and *efficiently* without the need for deployment in large numbers.

We already know what the most likely technology that would be capable of doing that is…

2. The Alleged Test of a Small Critical Mass High-Yield Nuclear Device at Ohrdruf in March, 1945

There is, however, yet another piece of evidence that points to the existence of a highly sophisticated technology of isotope separation inside of Nazi Germany during the war, and this is the alleged detonation of a small critical mass — an *extremely* small critical mass — atom bomb of highly efficient levels of yield at the troop parade ground of Orhdruf, in the Harz Mountains of Thuringia in south central Germany on March 4, 1945. Significantly, Adolf Hitler himself was present in the area during the period of the test.

---

bomb2/CRITICALMASS.txt, 1998, p. 23, emphasis added in original citation, double emphasis added here. Hydrick's internet manuscript was subsequently published in paperback, and is well worth reading.

42 Farrell, op. cit., pp. 59–60.

What concerns us here is that the test, according to some German sources, involved the detonation of a device with a critical mass of a mere *100 grams*. This fact led me to speculate in *Reich of the Black Sun* that at least one of the processes that this test, if it really occurred, had to have employed was "boosted fission," a technique whereby some neutron emitting source is doped into the critical mass to yield an extra burst of fast thermal neutrons into the reaction cascade that would *not* result from the splitting of the critical mass itself. This process served several purposes, for it allowed (1) a lower critical mass, (2) a higher yield for a smaller investment of material, or, to put it more bluntly and clearly, more bang for the Reichsmark. Under its ordinary uses it also would allow for a slightly lower level of purity of the critical mass of fissile material to be used.[43]

But the extraordinarily small mass for the Orhdruf test — a mere 100 grams! — was far below any weight, even with ultra-pure 235U, that could cause an atomic explosion. As I pointed out in my next book on Nazi secret weapons, *The SS Brotherhood of the Bell,* this fact meant that the bomb had to have been fueled by extraordinarily pure *plutonium*, and moreover, plutonium that was boosted via the boosting process. However, this introduced a new problematical factor into the equation, for plutonium can only be synthesized in a nuclear *reactor,* and according to the Allied Legend and the known evidence, all known German attempts to construct an atomic pile were highly unsuccessful, due either to failures in the moderators, or design factors, or both.

This led me to speculate that if the Orhdruf test did take place — and the evidence on the ground to this day in Germany suggests clearly that it did — then there had to have been a *hidden and unknown* functional reactor technology inside of Nazi Germany *or its allies* at a fairly early date, a technology that the postwar Allies have kept hidden to this day, or that they simply did not ever discover. And this technology was producing *plutonium.* I then speculated on two possible sources.

The first was that the Germans actually built a *methane*-moderated reactor, for such had actually been proposed *in 1941(!)* by German physicists Fritz Houtermans and Manfred Baron Von Ardenne for the express purpose of manufacture of "element 94," plutonium. And at that point in history, plutonium had only *one* practical use: as fuel for an atom bomb.

But there is another source for a possible functioning reactor technology using graphite as a moderator, and I only briefly mentioned it in *The SS Brotherhood of the Bell,* and that source is Fascist Italy. It was Enrico Fermi's home university, after all, the University of Milan, which had taken out the Italian patent for just such a reactor in 1938, and it was indeed on the

---

43 Farrell, *Reich of the Black Sun,* pp. 80–88.

basis of *that* design that Fermi built his first atomic pile in the squash courts of the University of Chicago for the Manhattan Project! While there is to my knowledge no evidence to support the notion that Fascist Italy had a functioning atomic pile, by the same token, if it did, once again the victorious Allies would have been loathe to disclose the fact.

A functioning Italian atomic pile also presents certain possible resolutions to other operational mysteries of the war, which must be mentioned here before resuming our circumstantial case of some rudimentary laser isotope enrichment taking place inside of Nazi Germany. First, as many military commentators have noticed, the entire Allied invasion of Italy and the subsequent long, slow, hard-fought campaign against Field Marshal Kesselring's forces seem militarily inexplicable. A direct Allied invasion of the Balkans — as Churchill himself advocated — would have paid the Western Allies far larger operational dividends. So, might the ultimate rationalization for what is otherwise a militarily inexplicable invasion be, in fact, a very successful Italian atomic pile program? Possibly.[44]

There is one other thing that a successful Italian atomic pile program might also help to explain, and that is simply the case of the several hundreds of tons of missing uranium from Nazi Germany. One place it *might* have gone would have been precisely Italy, the nation whose scientists had designed the first workable large-scale atomic reactor.

All this speculation, however, leaves one problem about the Ohrdruf test to be explained. That *something* unusual was exploded there, there is no doubt. The question is whether or not it was, as some German researchers now maintain, a boosted fission device, for even if one admits that possibility, it not only requires *plutonium* and hence a functioning *reactor,* but once one has synthesized plutonium, one must then *separate enough Pu239, the isotope needed for a bomb, to extraordinarily high purity* in order for a boosted fission bomb of only 100 grams critical mass of fuel to be possible. One must have, in other words, a technology capable of separation of isotope vastly more efficient than diffusion gates and mass spectrographs.

Something like laser isotope enrichment.

---

44 The problem with this explanation is, however, that any such Italian program would much more likely have been situated in the industrial north of the Po valley around urban and university centers such as Milan, Genoa, Trento, rather than in the more rural and inaccessible southern part of Italy. Hence, if this was the ultimate rationalization of the invasion, the Allies would have been more likely to invade further north than they did.

In support of the idea of *some* sort of Italian atomic research, however, is the peculiar and curious fact that Fascist Italy had begun to modify several of its four-engine long-range bombers for possible eventual use on the USA in one-way atom bomb suicide attacks. Any conventional arms attack on the USA would have only been a propaganda victory for Mussolini's government, and thus, militarily indefensible *unless* the *Commando Supremo* had something else up its sleeve that made the venture worthwhile.

3. The Thorium and Radium Mysteries

An even greater mystery from World War II is the fact that when the Allies' "Alsos" intelligence teams entered the Reich, antennae pulsing with suspicion, snouts to the ground sniffing out any clue or lead into the nature of Germany's atom bomb project, one of the most unusual things they encountered was the extreme German interest in *thorium*, and the extreme lengths the Nazis went to scour all of occupied Europe for every last granule of the element. It was — and is — a mystery, because no one knows what the Germans were *doing* with it. As Nazi secret weapons researcher Igor Witkowski puts it: "The U.S. Alsos mission was unable to explain the high role of thorium in the German research,"[45] a statement calling into question once again the postwar Allied Legend concerning their own nuclear superiority, for if they were so superior, why can they not explain this mystery?

Witkowski suggests a wider context, however, in which to view it, one which will lead us, finally, back to the Nazi Bell project and its mysterious and "alchemical" Serum 525. In a personal communication to this author, Mr. Witkowski graciously shared with me the latest results of his continuing research on Nazi secret weapons, research that will be incorporated as supplements to the pending German publication of his magisterial book in English on Nazi secret weapons, *The Truth About the Wunderwaffe*. In an extended supplemental commentary to be added to the end of his chapter on Nazi nuclear weapons research, he appends the following new information for the German edition. It is presented here, with thanks to Mr. Witkowski for allowing me to share it with an English-speaking audience:

> Compartmentalization was one of the reasons why the subject of German nuclear activities was not comprehended, but there was another one too — because some of the most important facilties were located in the East — in the future Soviet sphere of interest, which proved to be a very significant "obstacle" for Western researchers....Recent archival searches have revealed however the existence of other, similar places. Even more important ones, for in the case of the two presented below, the allied intelligence reports described them not as "working on nuclear technology" etc., but *expressis verbis* as the ones in which the nuclear weapons were manufactured or developed.
>
> In one of the reports there is a description of a secret camouflaged underground facility near the town of Sagan, in Niederschlesien:

---

45 Supplement to be appended at p, 219 of the English edition of his *The Truth About the Wunderwaffe* for its pending German publication, from a personal communication of Igor Witkowski to the author.

"German soldiers reported on Aug. 1 that 15km from Sagan, on a bearing of 120° there is an underground factory in the woods with an aerodrome. Factory presumably manufacturing an atom bomb, rumored to be a new weapon. An aerodrome near Sagan is reported (to be) in process of conversion to experimental production."

Another report seems to be even more interesting, this time pertaining to a previously unknown research team under the control of Werner Heisenberg (!), formed on the basis of scientists from Dresden:

"*University of Dresden: Said to be the heart of German development of the secret weapons. At the University of Dresden a group of chemists and engineers, and professors are working under the direction of Heisenberg…on by-products of radium at the Schicht mines in Aussig near radium mines at Jachymov in the Protektorate.*"

Jachymov (St. Joachimsthal) was, on the other hand, one of the main exploitation areas in the former Reichsprotektorat der Böhmen und Mahren[46] in which the Red Army teams worked later on. In this case, an entire battalion was dispatched by Smersh,[47] supported by unspecified technical groups.

Only recently has it turned out that their aim was to obtain the German nuclear bombs. Apart from the "revealed" role of Heisenberg and his Dresden team, the mention of "by-products of radium" may be equally important. It's hard not to set it up with another bit of information: the cargo transported by a Japanese submarine (I-29) which left the base in Lorient on the 16th of April 1944, heading to Penang in today's Indonesia. According to accessible sources it transported, among others, an "amalgam of mercury and radium."

….It's hard to resist a suspicion that these two cases are manifestations of some high-priority and still largely unknown nuclear project — German as well as Japanese.[48]

This important quotation contains a number of highly significant clues and bits of information.

First, as Witkowski indicates, more recently discovered archival material indicates that Heisenberg was leading a team of *chemists,* engineers and professors

---

46   i.e., the Reich Protectorate of Bohemia and Moravia, i.e., the north "bulge" part of Czechoslovakia, at the center of which was, of course, Prague. Hitler and Himmler basically turned all of Bohemian Czechoslovakia into a private SS preserve, more or less turning the whole country into an "Area 51," with SS General Hans Kammler's "black projects think tank" headquartered in Pilsen.

47   The N.K.V.D's and Soviet Military Intelligence counterintelligence and assassination squads.

48   Igor Witkowski, "Supplements to the English Edition of *The Truth About the Wunderwaffe* to be added to the Pending German edition," personal communication with the author. Comments to be added to page 219 at the end of the chapter on nuclear weapons.

at the University of Dresden. Such a team, with Heisenberg in charge, one of the founders of quantum mechanics itself, would be precisely the type of team needed to develop *and deploy* a laser-based isotope enrichment technology.

Secondly, the fact that it is the University of Dresden that is involved is highly significant. For as the only major university besides that of Breslau in Lower Silesia that is close to the rats' run of huge SS underground factories and bunkers that stretched from the Harz region of south-central Germany down through the curve of Silesia to Breslau, the presence of such a team there under Heisenberg's control — and the fact that its existence if not its function were at least known to the Allies — finally gives a military rationalization to one of the most barbarous atrocities ever committed: the firebombing of Dresden and the horrifying loss of civilian life, under brutal circumstances, toward the end of the war. The Allied bombing, otherwise morally reprehensible and indefensible, has the inexplicable air of desperation about it, one that Witkowski's information implies might have had a very secret and covert military reason behind it.

Thirdly, note that with the mention of radium and its "by-products" we have an indicator of Nazi interest in yet *another* radioactive material. Putting all the previous information together, we are now confronted by a massive dilemma: the Nazis are showing extreme interest not only in uranium-235 and plutonium-239, a normal interest for any nation pursuing an atom bomb, but in thorium and radium "by-products", i.e., isotopes, as well. And that wide range of interest implies once again the existence of a technology *flexible* enough to separate isotopes of several *different* elements, *if* that is indeed what the Nazis were up to with these "by-products." That technology is, once again, laser isotope separation. Even if they were only interested in these materials for the purpose of a speedy production of a "dirty" radiological bomb, the means to enrich *enough* deadly material still had to exist in order for a radiological bomb to be a practical weapon. This too implies the necessity for a flexible technology of isotope separation.

4. *The Laser Mystery*

As we have seen, the Farm Hall Transcripts contain suggestive remarks indicating the possibility of the development of a laser technology inside the Third Reich either before or during the war. The remarks furthermore found not only corroboration in the research of British journalist Tom Agoston, but he explicitly stated that this was one area that the SS was actually experimenting in. But is there any other corroborating data?

There not only is corroboration, but it too reveals the depths to which Nazi science had actually penetrated, going far beyond the accomplishments of the Western Allies or the Soviets even long after the end of the war.

On the 16th of November, 1944, a secret war department document addressed to Major F.J. Smith from Lt. Colonel Merillate Moses, stated the following:

> g) Experiments with "Death Rays" were conducted by AEG Seimensstadt Berlin at Tempelhof *in 1939*. Guinea pigs were killed at a distance of 200 meters.
> h) An individual employed "on electrical matters," not named, or otherwise described, told the prisoner that *the Germans had for years been experimenting* with these death rays.
> i) It is believed that it is a *sort of magnetic beam*, capable of stopping the motors of planes at a great distance.
> j) Experiments of which he knew, (1) destroying the functions of aircraft motors with induced magnetic fields, (2) exploding aircraft in the air by direct ultra-violet ray.
> k) This weapon which emits waves or rays *based on piezoelectricity is a development of the death ray.*[49] It is known that tests were made *in 1938* and that at a distance of eight hundred meters an automobile motor was successfully stopped.[50]

Observe that according to this American War Department document, actual *experiments* began on automobiles and such ca. 1938 and 1939, indicating that even *earlier* proof of concept experiments were probably performed.

While the secret weapon project being described here is obviously *not* a laser, there is a suggestive reference to "piezoelectrics" which *might* once again indicate the use of a crystal for the solid state optical cavity of a laser. Shades of Horst Korsching's remarks about artificial rubies in the Farm Hall transcripts!

Finally, notice the firm involved: AEG, or the *Allgemeine Elekstrisitäts Gemeinschaft*, the very *same firm* that built the power plant of the Bell, under the direction of future NASA administrator Dr. Kurt Debus![51]

But it gets worse.

According to Witkowski, Nazi research was researching *X-ray and gamma ray lasers:*

> …among other things work was carried out on some kind of "X-ray laser"
> — a source of coherent X-ray or gamma radiation….

---

49 Readers who have read *all* of my books will permit me to say "Hmmm!"
50 The text, plus an actual photocopy of the document, is reproduced and cited here from Igor Witkowski, *The Truth About the Wunderwaffe*, p. 88, emphasis added.
51 See my *SS Brotherhood of the Bell*, (Kempton, Illinois: Adventures Unlimited Press, 2006), pp. 154–157, and Witkowski, *The Truth About the Wunderwaffe*, pp. 256–257.

Kurzer Bericht über die Arbeiten, welche vom Unterzeichneten und seinen Mitarbeitern im Institut für röntgenologische Roh- und Werkstofforschung im Rahmen des Forschungsauftrages DE 6224/0109/43 seit 20. April 1943 durchgeführt wurden.

1. Ausarbeitung der Grundlagen des Projektes für die Besprechung mit dem Herrn Generalfeldmarschall am 20.4.43.
2. Verschiedene Denkschriften und Berichte an GL/St mit Einzelheiten meiner Planung und Vorschläge für die Durchführung.
3. Mitarbeit bei den technischen und organisatorischen Vorarbeiten über die Auswahl des Platzes und der Einrichtung der Versuchsstelle Groß Ostheim.
4. Exposé über die Bedeutung des Wideröeschen Strahlentransformators für die vorliegenden Pläne und maßgebende Beteiligung an Verhandlungen mit Herrn Dr. Wideröe.
5. Ausarbeitung der Pläne betr. Auf- und Ausbau der großen Halle in Gr. Ostheim, gemeinsam mit dem Büro Prof. Tamms (Arch. Sander) und den zuständigen Herren vom Luftgau XII.
6. Aufstellung der Pläne für Installation, Telefonanlage usw. für die zunächst vorgesehenen Laboratoriumsbaracken und den Ausbau der Unterkunftsräume.
7. Versuch zur Einrichtung wissenschaftlicher Arbeitsräume nebst Feinmechanikerwerkstatt, zunächst mit dem Inventar des Instituts für röntgenologische Roh- und Werkstofforschung zum Zwecke der Vornahme dringlicher Vorversuche. Diese Einrichtungen wurden auf Veranlassung der Forschungsführung inzwischen wieder aus Ostheim entfernt.
8. Schulung der aus der Truppe herausgezogenen wissenschaftlichen und ingenieurtechnischen Mitarbeiter.
9. Zusammenstellung der für die Arbeiten wichtigen Literatur und Ausarbeitung eines zusammenfassenden kritischen Berichtes über die bisherigen Ergebnisse auch auf dem Gebiete der Physik durchdringen der Röntgen- und Gammastrahlen, sowie Elektronen.
10. Aufstellung von Inventarlisten und Geräten für die erste Einrichtung von Laboratorien und Werkstätten.
11. Beschaffung der 1,2 Mill.Voltanlage vom Hamburger Staat.
12. Bestellung einer 2,2 MV-Röntgenanlage bei der Fa. C.H.F.Müller in Hamburg.
13. Entwicklungsauftrag an die Firma C.H.F.Müller-Hamburg über einen 15 MV-Strahlentransformator Bauart Wideröe..
14. Beschaffung von Laboratoriumsbedarf und Geräten aus Beutestellen der Luftwaffe.
15. Wissenschaftliche Arbeit theoretischer Natur über die Ausbreitung und Schwächung harter Röntgenstrahlen in Luft variabler Zusammensetzung, Dichte und Temperatur für verschieden hohe Primärenergien
16. Projektierung einer Großanodenröntgenröhre nach eigenen Vorschlägen.
17. Projektierung einer Anlage zur Erzeugung hoher Spannung und großer Stromstärke nach dem Kaskadenprinzip auf Grund eines eigenen DRP.

Leipzig, den 4. Mai 1944.   Prof.Dr.E.Schiebold.

German Luftwaffe Document Referring to X-Ray and Gamma Ray Laser Research: Courtesy of Igor Witkowski and The Truth About the Wunderwaffe

Thanks to searches carried out in German archives, it has been possible to establish that in 1944 a special Luftwaffe research establishment received the task to develop such a weapon, situated in the town of Gross Ostheim. Materials relating to this work are currently located in a civilian establishment — the Karlsruhe research centre (*Foruschungszentrum Karlsruhe*) and were disclosed several years ago... The third and most mature version of the weapon assumed the irradiation of a target 5 kilometers away at a rate of 7 rads a second for 30 seconds which, as affirmed in the report, was completely sufficient to totally paralyze the aircraft's crew.[52]

But what is more interesting is something that Witkowski does not mention, and it occurs in the actual document itself.

I have displayed on the previous page the actual document itself that Witkowski refers to, because its contents are so astounding that one might be inclined to believe I made it up, unless the actual document were shown. Before getting to the actual contents of the document itself, there are two important points to remember when reading it.

The first is that in the 1940s, the term *laser*, standing for "light amplification by stimulated emission of radiation," had not yet even been coined. So the Germans, if they are indeed researching lasers, will have to describe the *concept and principle operative in a laser, namely, a sudden cascade of photon radiation emission*. The second point is equally crucial. Recall that Richard C. Hoagland, when he proposed his "laser isotope separation" interpretation of the Farm Hall Transcript comments, also indicated that some sort of documentary proof was needed that the Germans *had worked out the theory*.

As this document indicates, they not only had worked out the *theory*, but had gone beyond mere optical solid state lasers, i.e., a first generation of the technology, but were already thinking in terms of lasing in the high-frequency range of deadly X-rays and even deadlier gamma rays.

With these points in mind, on to the translation of the document itself:

> Brief Report concerning the Works, which was performed by the undersigned and his co-workers at the Institute for X-Ray-Related Natural and Raw Material Research, since April 20, 1943, under the scope of the Research Directive DE 6224/0109/43.
>
> 1. Completion[53] of the foundation of the project for the conference with the Field Marshal on 20/4/54 (April 20, 1943).

---

52 Witkowski, *The Truth About the Wunderwaffe*, p. 92. The actual German document is pictured on p. 99.
53 Or "development;" the German word is *Ausarbeit*.

2. Distinct memoranda and reports to GL/St particulars of my plan and proposal for its realization.
3. Cooperation with the technical and organizational preparation concerning the selection of places and the arrangement of the research area Gross Ostheim.
4. Exposition concerning significance[54] the Wideröe *beam-transformers*[55] for the existing plans and the decisive participation of discussion with Herr Dr. Wideröe.
5. Completion of the plan concerning the construction and conversion of the great hall in Gross Ostheim in common with Bureau Professor Tamms (ach. Sander) and the appropriate personnel of Air District XII.
6. Formation of the plane for installation, telephone exchange, etc., for the crucial provision of laboratory barracks and conversion of the billets.
7. Development of the scientific equipment and workrooms in addition to precision instrument-making components, above all with the inventory of the Institute for X-Ray-Related Natural and Raw Material Research with the object of the *previously mentioned urgent pilot tests...*
8. Training of the scientific and engineering staff from the unit.
9. Collection of the important literature for the work and development of a comprehensive critical report concerning the results thus far for the area of the physics of the penetration of X-rays and gamma rays, as electrons.
10. Disposition of fixtures and equipments for the first layout of laboratories and workshops.
11. Procurement of the 1.2 million volt installation from the State of Hamburg.
12. Placement of the order for a 2.2 million volt X-ray equipment at the C.H.F. Müller factory in Hamburg.
13. *Commission of a 15 million volt ray-transformer of the Wideröe model at the Firm of C.H.F. Müller, Hamburg.*
14. Procurement of laboratory requirements and equipments surplus stocks[56] of the Luftwaffe.
15. *Scientific work of a theoretical nature concerning the propagation and weakening of hardened*[57] *X-rays in air under variable conditions, density and temperature for various high primary energies.*
16. Proposal of a large high tension X-ray tube of unique fluxes.

---

54 Or "meaning" or even in this context, "implication." The German word here is *Bedeutung*.
55 Wideröeshen Strahlentransformators, apparently a technical designation named, perhaps, after its inventor, whose surname appears at the end of the statement. The name is not typical nor perhaps even German, but is more suggestive of Norwegian origin.
56 Literally, the "booty stocks," i.e., supplies captured as war booty.
57 Or "hard" or "harder"; the German appears to be a typographical error, having "harter" rather than "härter."

17. Proposal for an installation for the manufacture *of high voltage and large current intensity according to the Cascade Principle as a basis of a single DRP.*

Leipzig, May 4, 1994 /s/ Prof. Dr. E. Schiebold

What is evident from the document, regardless of how one translates certain nuances — which is difficult without the rest of the context! – is that certain passages are starkly unambiguous in their implications, not the least of which is the final point which mentions the "Cascade Principle," the very principle utilized in lasers. Notably, the way the point is phrased, taken in the context of the rest of the preceding document, indicates that some proof-of-concept design and testing had already taken place, and that the project was moving into full-scale equipment design and testing.

Additionally, point number 16 indicates where the German project is heading, for atmospheric propagation and distortion effects would certainly be of concern to the Luftwaffe in the deployment of *any* type of laser.

What is also highly significant is the *name* of the inventor ascribed to the "ray-transformer": Wideröe. As mentioned before, the name is Norwegian, not German, and is in fact the name of one of the pioneers in particle accelerator design and development: Rolf Wideröe. Wideröe had in fact built a first linear resonance particle accelerator at the Rhenische-Westfalen Technische Universität in Aachen, Germany, in 1928. Wideröe's work in particle accelerators, which he called the "ray-transformer," was crucial in the technology tree leading to the cyclotron.[58]

The document thus seems to suggest that the Nazis were attempting some combination of the linear resonance particle accelerator developed by Wideröe as possibly the pumping mechanism for an X-ray laser based upon the cascade principle operative in lasers. Thus, *the Germans had not only apparently done the theory, but were developing large scale equipment.* Moreover, the use of a resonance accelerator suggests the highly speculative possibility that the Germans were also thinking in terms of a *tunable* laser.

There is one final important point about the document that suggests that the project was more than theoretical, but concerned with the practical development and implementation of equipment on a large scale based on ideas and prototypes that had already *passed* the proof-of-concept phase of development, and that is that the document is a dull, boring *"talking points" briefing paper* outlining the points to be covered in a briefing with an unnamed Luftwaffe field marshal! Clearly, if one is shopping for installations of 1.2, 2.2,

---

58 See Rolf Wideröe, *The Infancy of Particle Accelerators,* ed. Pedro Waloschek, www.waloschek.de/pedro/pedro-texte/wid-e-2002.pdf, pp. 25–40.

and ultimately 15 million volts, one is beyond the small laboratory efforts and into a more serious phase of the project.

D. A Tentative Conclusion

We have already noted Witkowksi's reference to that curiously "alchemical" cargo being carried by a Japanese submarine, a cargo consisting of *an amalgam of radium and mercury!* This is the strongest circumstantial clue as to what might have comprised the chemical recipe of the Nazi Bell's mysterious "IRR Xerum 525." Might the heavy interest of the Nazis in radium and thorium — an interest that seems at the minimum as great as their interest in the fuels for an atom bomb, uranium and plutonium — be because these and possibly other elements were necessary ingredients in the chemical recipe of the Bell's "Xerum 525"?

This question is a crucial one, because Xerum 525, as described in SS general Jacob Sporrenberg's war crimes trial affidavit in Poland, resembles no compound more than it does "Red Mercury" and the alchemists' "tincture of antimony," both in its alchemical "reddish-purple" color, and in its state as a very dense, liquid *goo* that, like the various materials tested in Kozyrev's experiments, was subjected to a variety of stress by being spin and electrically pulsed.

But that is the story of another chapter. For now, we may conclude as a highly speculative though plausible possibility, that Mr. Hoagland's hypothesis of some version of laser isotope separation has a great deal of circumstantial evidence to support it. And the strange and almost "alchemical" combination of technologies suggested by the Luftwaffe document prompts an important question:

What, exactly, *was* the role and influence of alchemy inside of Nazi Germany, and more importantly, inside of the Nazi party itself. And what role, if any, might it have played in the conceptualization of the Bell?

Get a strong cup of coffee, breathe deeply, sit down, and brace yourself...!

The Greater German Alchemical Reich
The Goldmaker, the Gold, Gerlach, and "Himmler's Rasputin"

∴

> "Primal law: 'Above as below, below as above!' Therefore in the middle
> there is a neutral force — (i.e., generational-) field!"
> Jarl Widar, a.k.a. Weisthor, a.k.a. SS Brigadier General Karl Maria Wiligut[1]

During the 1920s and the early 1930s prior to the Nazi assumption of power, Germany was a veritable cauldron of alchemical ferment and other occult and esoteric activity. This phenomenon has spawned a plethora of books on the subject of Nazism and the occult, some more, and some less, scholarly, but all more of less coming to the same conclusion: there was, to whatever debatable extent, at least some degree of influence of esoteric and occult doctrines in the formation of the Nazi Party and later of the Nazi Reich.

Curiously, few of those studies have paid much more than passing attention to the role that *alchemy* as an esoteric discipline played in those formations. The omission is peculiar because, as we shall see, one rather infamous alchemical episode involving high-level and influential Nazis even caused something of a major scandal in the ailing Weimar Republic, a scandal that even drew international attention and a few articles in *Time* magazine and the *New York Times*. The affair is made even more curious because it occurred

---

1 Jarl Widar (Karl Maria Wiligut), "The Creative Spiral of the 'World-Egg'!", *Hagal* 11 (1934), Heft 9, pp. 407, cited in Stephen E. Flowers and Michael Moynihan, *The Secret King: The Myth and Reality of Nazi Occultism* (Los Angeles: Feral House, 2007), p. 106.

in more or less the same time period that one of Germany's, and indeed the world's, most respected and capable physicists was writing a newspaper article for the general public in which the possibility of alchemy is mentioned quite explicitly. These facts not only make the lack of serious interest in the subject of alchemy inside of Weimar and Nazi Germany curious, but make more curious the even *greater* absence in the literature of any serious attempt to draw connections between alchemy and physics.

But the question demands our attention.

Why *would* senior and influential Nazis be drawn into a self-evidently fraudulent scheme to turn base metals into gold? And why *would* a prominent, internationally-known and respected German Nobel Laureate physicist be writing an article in a widely-circulated German newspaper about alchemy being a serious reality? And even more importantly, what are the connections, if any, between the two?

To answer these questions, it is best to begin the story in the middle, and work our way backward to the scandal, and forward to Nazi Germany's most highly classified secret weapons project: the Bell, and its precisely alchemical "Serum 525."

A. The Strange Case and Alchemical Beliefs of "Himmler's Rasputin": SS Brigadier General Karl Maria Wiligut

Many people are vaguely aware that the Nazis had at least some sort of occult and esoteric influence at work not only in the formulation of their ideology, but in the actual structuring and policy formation of the Nazi state, the Third Reich. Fewer people have actually studied the matter, but those who have tend to assign this influence to a now typical cast of characters and secret societies moving in the background, helping to midwife the Nazi Party into existence: the *Thulegesellschaft* and its own alleged roots in the esoteric, occult, and rabidly pan-Germanic groups that flourished in pre-World War I imperial Germany, and more importantly, imperial Austria.[2] On the basis of a considerable circumstantial case, many of these in turn have speculated that Hitler himself was initiated into one or more or these groups. Yet, the documentary proof of this has never been forthcoming.

One of the primary sources for this belief was, of course, Baron Rudolf Von Sebottendorff's publication prior to World War II of his now somewhat infamous book, *Bevor Hitler Kam* (*Before Hitler Came*), a book that so infuriated the Nazi government once it was in power that they put it on their "index," prohibited publication, and attempted to round up as many copies as they could.

---

2 See for example Dusty Sklar, *The Nazis and the Occult*.

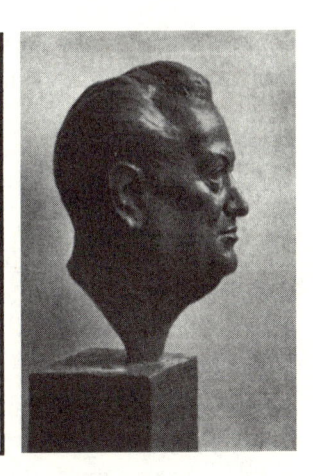

Cover, Inside Cover, and Frontispiece from
Baron Rudolf Von Sebottendorff's Bevor Hitler Kam

Part of the reason may lie in the fact that, according to Von Sebottendorff, the influence of the *Thulegesellschaft* on the Nazi Party's formation and *ritual* was pervasive, even to the use of its symbols, including the swastika, as the pictures on the previous pages demonstrate, as well as its forms of greeting, including, of course, the Nazi "Sieg Heil!" salute and extended arm.[3]

A more important reason for the ban on the work, however, must surely lie in the lists of its members that it contains, not only of prominent figures in the future Nazi State, including Nazi Party *Reichsleiter* and occultist dabbler Rudolf Hess, but prominent, very wealthy and powerful, backers, such as Heila Countess Von Westrapp, then Secretary of the society, and the powerful and wealthy Von Thurn-und-Taxis family, in the person of Gustav Franz Maria Prince Von Thurn-und-Taxis, as shown on the "honor roll" list of the society's "martyrs and confessors" on the next page.

Sebottendorff also provides the clue, however, to something else, and that is the possible exact relationship between the SS, its esoteric "culture," and the highly advanced paradigm of physics it was investigating with the Bell. That link is "Himmler's Rasputin," SS Brigadier General Karl Maria Wiligut, a.k.a. Weisthor. With recent scholarly publications, we are now also in a position to see the possible connection of alchemy to the SS' esoteric culture and its advanced physics projects via the various conceptions entertained by Wiligut, and passed on to Himmler.

In order to appreciate the significance of these speculations, however, one must place them against the background of the basic conclusions

---

3 Rudolf Von Sebottendorff, *Bevor Hitler Kam* (Munich: Deustula-Verlag Grassinger & Co, 1933, 1 Auflage), p. 190.

concerning the relationship of the Nazis to the pre-Third Reich secret societies, a connection that has been the more or less basic prevailing view within scholarship since the end of the war. This view maintains, on the basis of the numerous circumstantial connections between personnel, beliefs, and rituals of *Thulegesllschaft* and other such societies in the pre-Nazi era and the Nazi Party itself, and its rituals, beliefs and policies, that the influence was direct and immediate. Such studies often point in turn to the influence on the *Thulegesellschaft* of the pre-World War I pan-Germanic esoteric orders and teachings originating from Vienna, and the Order of New Templars, or as it is also known, the Order of the New Temple (ONT) of Lanz Von Liebenfels, his rabidly racist views and racially-based reconstructions of history in his *Ostara* magazine. While this is not the place for a detailed rehearsal of all these scholarly views,[4] one impression they invariably convey is that this relationship ran directly to Hitler, and on that basis, such researchers often speculate that he was actually an initiate into one or more of these three esoteric societies and influences.

But as will now be shown, the influence is at one and the same time both more indirect and more direct, for it is mediated not by any one of these societies to Adolf Hitler, but by one man — Karl Maria Wiligut, "Himmler's Rasputin" — and not to *Hitler* and the Nazi Party as a whole, but to *Himmler* and the SS as an institution within the Nazi Party and State. As such, it is less accurate to speak of an occult influence on the entire Nazi State, as it is to speak of an esoteric influence at the uppermost levels of the command structure of the SS. One is, so to speak, dealing with a Black Reich within the Reich, and at the uppermost reaches of the SS, with a very secret esoteric, and specifically alchemical, belief system.

1. Himmler's Rasputin
a. Brief Biography

Wiligut was born on the 10th of December 1866 to an Austrian army officer, and, following in his father's footsteps, enrolled in the Imperial Cadet School at Vienna-Breitensee in 1880, and by 1913, one year before the outbreak of World War I, had been promoted to the rank of major, reaching the rank of colonel by the war's end after serving with some distinction in the Imperial army on the Eastern Front. On the breakup of the Austro-Hungarian Empire in 1919, he took up residence in Salzburg.[5] It was during the period prior to World War I that Wiligut also began the publishing of esoteric works

---
 4 For the views, see my *Reich of the Black Sun*.
 5 Stephen E. Flowers and Michael Moynihan, *The Secret King: The Myth and Reality of Nazi Occultism* (Los Angeles: Feral House, 2007), p. 44.

that would make him a well-known esoteric scholar and garner him a certain amount of respect and renown in German-speaking occult circles.

Around 1908 Wiligut came into contact with Lanz Von Liebenfels' Order of the New Temple, most likely being introduced to it "by his cousin, Willy Thaler, who was a member of the Liebenfels circle"[6] and thus entered the milieu of "esoteric nationalism" permeating the occult secret societies of postwar Austria and Germany.

A brief word is necessary about Von Liebenfels and his historical and racial ideas, for they are to some extent paralleled in those of Wiligut, though the degree of influence on the latter is a matter for debate. According to Sebottendorff and his exposé of Thule Society influence on the Nazis, Von Liebenfels, a former Roman Catholic monk, had — through his knowledge of early Christian patristic literature — reconstructed what he believed to have been the actual early text of the New Testament[7] as a component of a larger program of historical revisionism based in part on the Theosophical beliefs of Helena Petrovna Blavatsky, in which various "root races" gradually declined from a golden age past to mankind's current decadent condition. In Von Liebenfels' hands, this decline was due to the racial corruption of the "pure Aryan" lineage of a white Teutonic race by intermingling of the cultures and blood of inferior races, such as — predictably enough — the Jews. Von Liebenfels published his views in a notorious book entitled *Theozoologie* which had, as its racially slurring subtitle, *On Sodom's Monkeys*. While most scholars of the subject point out the parallel between Von Liebenfels' racial beliefs and those of the Nazis, few take note of the larger view of history in which they are cast, namely, that there was, eons ago and predating the known classical civilizations of Sumer and Egypt by millennia, a previously existing Very High Civilization. It is this view, as will be seen in a moment, that is paralleled in the esoteric system of Wiligut, with its own peculiar twists. Thus it is *Wiligut*, not Von Liebenfels, who actually exercises a direct influence on Himmler, the SS, and its leadership.

In any case, after his retirement from the Austrian army and move to Salzburg in 1919, Wiligut "seems to have immersed himself in esoteric studies."[8] It was during this period that Wiligut came to his own version of esoteric "higher biblical criticism" and "reconstruction" of Christian history, maintaining that "the Bible had originated in Germany, and through mistranslation and intentional misrepresentation it had been revised into its present form."[9] In this he parallels Von Liebenfels.

---

6 Ibid., p. 45.
7 Rudolf Von Sebottendorff, *Bevor Hitler Kam,* p. 32.
8 Flowers and Moynihan, *The Secret King,* p. 46.
9 Ibid.

Wiligut quickly became involved in the post-World War I right-wing politics of Austria and Germany, editing a journal called *Der eiserne Besen (The Iron Broom)*, a journal that like Baron Von Sebottendorff's later Thule Society-sponsored *Völkischer Beobachter (People's Observer)* — the newspaper that eventually became the Nazi Party's "official newspaper"! — had as its goal to expose "the conspiracies of the Jews, Freemasons and Roman Catholics (especially the Jesuits)."[10] During this period his marriage began to fail, leading to his wife ultimately, and successfully, getting him declared mentally incompetent and confined to a mental institution in 1924.[11] Even during this period of confinement, however, Wiligut managed to maintain contact with members and acquaintances within the Order of the New Temple.

Eventually released from the asylum in 1927, Wiligut remained in Salzburg and received visitors both from Austria and Germany anxious to seek his esoteric expertise, especially from the esoteric German Edda Society. It was during one of these visits during 1932 that he received Fräulein Frieda Dorenberg, a member of the National Socialist Party, and herself "deeply involved in esoteric matters" as well as being a member of the Edda Society.[12] Smuggled into Germany later that year, Wiligut assumed the pen name Jarl Widar and published important esoteric articles — about which more later — in the esoteric journal *Hagal*. These contacts and articles gained him a great deal of notoriety and respect in German occult circles, and as a result, one year later, after the Nazi assumption of power, Wiligut met Reichsführer SS Himmler at an esoteric conference. A short time later, Wiligut joined the SS in September of 1933 under the pseudonym Karl Maria Weisthor, and thereby his career as "Himmler's Rasputin" began.[13]

Two months later "Wiligut was officially appointed head of the Department for Pre- and Early History, within the *Rasse- und Siedlungshauptamt* (Main Office for Race and Settlement)" in Munich, and it was in this capacity that a "closer relationship developed between Himmler and his new adviser on ancient traditions."[14] By 1935, continuing to publish in *Hagal*, Wiligut-Weisthor had become such a close adviser to Himmler that he was moved to Berlin and given his own villa.

While Wiligut's work to some extent paralleled that of the notorious SS *Ahnenerbedienst*, "his work was essentially separate from that office. Wiligut worked *for Himmler personally*, whereas the *Ahnenerbe* was part of a much larger structure subject to more objective academic standards."[15] So important

---

10   Ibid.
11   Ibid., p. 47.
12   Ibid.
13   Flowers and Moynihan, *The Secret King*, p. 48.
14   Ibid.
15   Ibid., p. 49, emphasis added.

and influential was Wiligut during this period that he was instrumental in the selection and design of Himmler's infamous SS "Order Castle" at Wewelsburg, in the actual design of the SS ring, and the creation of SS rituals, that he issued a "steady stream of reports on esoteric matters of theology, history and cosmology...for the most part *directly to Himmler.*"[16] This will become quite the crucial point once we start examining the particulars of Wiligut's esoteric, and very alchemical, beliefs.

By 1938, however, Wiligut had also garnered powerful enemies within the SS, among them the powerful chief of Himmler's personal staff, SS General Wolff,[17] who discovered the facts of Wiligut's mental commitment, and eventually forced him to retire from the SS in 1939. Wiligut spent the rest of the war moving, or being moved, from one location or SS retirement house to another. Being interred at the end of the war for a short time by the British, he eventually was released, eventually wound up in Arlosen in 1945, where he suffered a severe stroke, and died on Jan 3, 1946.[18]

Karl Maria Wiligut, "Himmler's Rasputin"

b. And the SS Ahnenerbedienst

As was seen, Wiligut had, as far as is known, no direct contact with the SS' "occult and esoteric research" bureau, the *Ahnenerbediest*. But as both had a direct influence upon the formation of the esoteric culture of the SS and Himmler himself, a short review of some of the *Ahnenerbe's* principal beliefs,

---

16 Ibid., emphasis added.
17 This is the same Wolff that entered secret negotiations with OSS station chief in Zurich, Allen Dulles, toward the end of the war, to arrange a surrender of German forces in northern Italy.
18 Flowers and Moynihan, *The Secret King*, pp. 59–60.

set in the wider context of the "occult revival" that took place in Weimar Germany between the wars is in order.

(1) The Ahnenerbe, "Atlantis," and Esoteric Geopolitics

When Reichsführer SS Heinrich Himmler decreed the creation of the SS *Ahnenerbedienst* in 1935, one of its department objectives was the scientific and archaeological study of esoteric and ancient lore, legends, and myths, with the goal of the application of any military potential these may have held. In a certain sense, Himmler had willed into existence an entire government bureaucracy to do nothing but military studies of the esoteric, all under his personal control. This created an unusual if not unique first in modern history, because for the first time in modern history a technologically and scientifically sophisticated great power was acknowledging, even if covertly, the existence of a very ancient Very High Civilization whose science it was intent upon recovering. Himmler had decreed, in effect, that the Third Reich was not only going to look for a "paleoancient Very High Civilization," for "Atlantis," but more importantly, for its *science*.

Thus, while Wiligut's connection to the *Ahnenerbe* was such that he was never a member of it, as one of Himmler's closest prewar advisers, he could hardly have been unaware of it. Indeed, "he played an important role in conceptualizing and designing certain esoteric aims and practices of Himmler's elite circle within the SS, and this factor alone makes Wiligut a fascinating and unique study."[19] Wiligut may thus have possibly been an influencing factor on Himmler's decision to create the *Ahnenerbedienst*, for his own personal esoteric views were certainly in line with those held and developed by SS *Ahnenerbe* leaders themselves. In fact, the first director of the bureau, Herman Wirth, "deeply sympathized with pagan traditions."[20] Wirth, who became the *Ahnenerbe* director after Himmler's decree of 1935 that created the bureau, not only "held a longstanding antipathy for Christianity" but made it a point to promote his idea of a "pre-Christian Nordic-Atlantian ... culture" within the bureau.[21]

Another esotericist with whom Wiligut had close contact was Günther Kirchhoff. While Kirchhoff's views were rejected by the *Ahnenerbe* itself, Wiligut's endorsement to Himmler was enough to win Kirchhoff Himmler's ear, and as late as 1944 he was still preparing reports on occult ideas for Himmler.[22] Kirchhoff's chief concept that seems to have sparked Himmler's

---

19 Flowers and Moynihan, *The Secret King*, p. 70.
20 Flowers and Moynihan, *The Secret King*, p. 31.
21 Ibid., p. 32.
22 Ibid.

interest was the idea of an esoteric geomantic geopolitics based on his own unique conception of the world grid, i.e., the idea that certain ancient sites were laid out over "power points" designed to draw energy from the earth itself. On his view, this made Vienna the key to controlling Asia, via certain power points in Asia connected via a pattern of hexagons to the European city![23]

In yet another such geomantic geopolitics promoted during this time, one must take note of another idea promoted directly by Wiligut as a component of his "meta-history" of mankind. In Wiligut's version, the god Teut, a Germanization of the Egyptian god Thoth, made the Harz mountains of Germany a center of such activity. This may or may not be significant for our purposes, since the Harz region, as is now widely known, became a center for the SS' most secret and highly classified secret weapons projects, being honeycombed with a variety of underground installations, laboratories, and factories. Indeed, the Harz region, along with lower Silesia, might justifiably be called the most crucial component of SS General Hans Kammler's secret weapons think tank headquartered at Pilsen.

(2) The Ahnenerbe and the Scientific Decoding of Esoteric Lore

One of the most often commented upon, and yet paradoxically overlooked, aspects of *Ahnenerbe* research was precisely its interest in ancient folklore:

> As the example of Herman Wirth demonstrates, the interests of some National Socialist and SS-Ahnenerbe academics ran strongly in the direction of folklore and research into folk-symbols. This included medieval decorative arts, architectural styles…and runes. In literary fields this included folktales. In a certain sense such studies were sympathetic to paganism and also to a kind of "occult" knowledge — *it required a trained mind and eye to rediscover and unlock the ancient symbolic code which had been embedded in folk art, architecture, and lore.…* But the obsession with symbolic motifs — and even their "occult" meanings — aside, the aim of this research was to bring such things back out into the *open*, not to use them for arcane magical purposes. The ultimate goal was to rediscover the repressed ancient national character and to instill in the people a sense of nationalistic pride and identity. The same trend and method had been used in the nineteenth century to aid in the original forging of the German state in 1871.[24]

---

23   Ibid.
24   Flowers and Moynihan, *The Secret King*, p. 33, emphasis added. Flowers and Moynihan hint here at the deep secret society and esoteric connections that played such a significant role in the European nationalist revolutions of 1848 and later, in the formation of the German Empire in the wake of the

Flowers and Moynihan, whose remarks these are, get it only partly right, namely, one purpose of the *Ahnenerbe* was precisely to find and popularly promote the pre-Christian and pre-Classical roots of the German *Volk* and Culture. But they overlook the ultimately military and covert purpose of the *Ahnenerbe,* which Reichsführer SS Himmler's original decree discloses: the real, ultimate, and ultimately covert purpose of the organization was to recover the ancient lost science of that pre-existing "Atlantean" civilization, and the first step in the recovery of it is to decode its traces and remains in ancient texts, monuments, and symbols.

As Wiligut himself would state it in a short article entitled "Zodiacal Signs and Constellations" that he authored for the esoteric journal *Hagal* under his pen name Jarl Widar:

> In Atlantean times, then, these original Aryan Runes had a deep meaning by virtue of their logical ordering in the zodiacal signs, which was apparently lost as soon as the meaning was changed in an astrologically illogical way as a result of the variation brought on by the precession of the equinoxes..... So it was only after the demise of the Atlantean cultural epoch that the zodiacal signs and their written signs were transferred to the constellations of the ecliptic. This caused that confusion of knowledge between the processes in the great solar year and those of a terrestrial year. This confusion is still fundamentally in operation today. But with this knowledge the oldest cultural documents known to us can be measured and — viewed not from the perspective of terrestrial years, but rather solar ones — these can lead us to correct knowledge of the ages.[25]

One must pause and carefully observe what sort of program Wiligut is actually engaged in here.

First, there is the reference to a pre-classical "Atlantean" civilization, which, in Wiligut's hands, comes to an abrupt end via some sort of cosmic catastrophe. Second, this catastrophe he implies is "cosmic" in scale because the *physics* is not that of the solar system, measured in terrestrial years, but is truly a galactically-scaled physics, because the "true year" of the zodiac is the *solar* year, measured in terms of the precession of the equinoxes and its course through the signs of the

---

Franco-Prussian War. That the German Empire was the crowning achievement of Prussian statecraft and generalship cannot be gainsaid, but little is known of the possible role that the "Society of Lizards" and other lineal descendants of the Teutonic Knights may have played in this, beyond their obvious connection to the Von Hohenzollerns themselves. The best one-volume scholarly history of the period and the role of secret societies remains the former Librarian of Congress James Billington's study *Fire in the Minds of Men.*

25   Karl Maria Wiligut, a.k.a. Jarl Widar, "Zodiacal Signs and Constellations," *Hagal* 12 (1935), Heft 4, pp. 56–58, cited and translated in Stephens and Moynihan, *The Secret King,* p. 113.

zodiac, with each such year thus being approximately 25,000 terrestrial years in length! Finally, the runes, laid out in correspondence with this plan, evidence a logical development, an encoded physics, of this phenomenon.

In this, as any reader of my previous books beginning with *The Giza Death Star* up through *The Cosmic War* will notice, Wiligut is hardly doing anything unusual. But he is *one of this first to view the significance of astrological lore from a purely scientific and astrophysics point of view, and to maintain, on that basis, that there is an encoded science and an encoded "meta-history" of the solar system and of mankind within such lore.* Small wonder, then, that he caught the attention of Himmler and gained such an important advisory position to the Reichsführer!

Wiligut lays out this "meta-history" and cosmic catastrophe physics at the very beginning of another of his papers presented to Himmler — and marked as actually having been *read* by Himmler himself on 17 June 1936. In it, Wiligut states the following:

> Each of these evolutionary epochs which have occurred up to now were, according to the oral secret doctrine, brought about by an enormous world-wide catastrophe culminated by *unifications of our earth with one of the heavenly bodies attracted into its orbit.*
> ....In the process, the *remnants of humanity which remained on the earth assimilated with those who came "from heaven" (stars) to the "earth." This assimilation brought about similar intelligences and thus established a new humanity which instituted new racial types.*[26]

Once again, with his reference to planetary collisions, Wiligut has anticipated the postwar catastrophists such as Velikovsky and Alford by several decades. And with his reference to the interbreeding of humanity with off-world, possibly human-like, intelligent life, Wiligut has anticipated the views of some postwar researchers such as Zechariah Sitchin by several decades.

Such views almost compel further speculation.

Wiligut's views on the origins of human life, obviously, would have fit Nazi racial ideology quite well. But what is of interest here is that Wiligut's actual inspiration for such views seems, at best, obscure, permitting one to speculate. Wiligut claimed to derive them by a secret, and long-held, family tradition that was recorded on wooden runic carvings possessed and maintained by his family for millennia. These, he maintained, were destroyed in the 1848 revolution.[27] But the textual basis for such views of human origins are well-

---

[26] Karl Maria Wiligut, "Description of the Evolution of Humanity from the Secret Tradition of our Asa-Uana-Clan of Uiligotis," SS Document, 17 Hune 136, Marked read by H.H. (Heinrich Himmer), Archival File for Wiligut/Weisthor, *Bundesarchiv Potsdam* NS 19/3671, cited in Flowers and Stephens, *The Secret King*, p. 126, emphasis added.

[27] Flowers and Moynihan, *The Secret King*, pp. 126–127.

known to anyone who has read certain ancient Sumerian texts, and such texts were in abundance in pre-World War I Europe, the Germans having translated and published many of them.[28] One need not appeal to a secret family tradition to have knowledge of such ideas.

Similarly, Wiligut ties — almost in Alan Alfordesque fashion[29] — the relationship of "cosmic collisions" of planets and meteors with the earth to the origins and subsequent "corruptions" of humanity by interbreeding with off-world intelligent life. In this, he is almost the first on record to work out a complete meta-history of mankind and civilization on the basis of such recurrent collision-based catastrophist astrophysics.

But as my book *The Cosmic War: Interplanetary Warfare, Modern Physics, and Ancient Texts* argued,[29] there are massive problems with such a "paleophysics" model as a tool for decoding ancient texts. Using celestial collisions as the paradigm, the texts fade into transparency, and reveal nothing. The cosmically-scaled physics *is present* in such texts, thus hinting that the ancient legends of the wars of the gods were not *metaphors*, but the *literal truth*, as someone actually weaponized a cosmically-scaled physical phenomenon, not of collisions, but of *rotating systems.*

And Wiligut is one step away, in this memorandum, from making the same conclusion.

And Himmler himself has *read* this memorandum.

As we shall shortly see, Wiligut himself even avoids the *collisional catastrophist physics* of Alan Alford and so many moderns, but moreover, he even couches it in terms of rotating systems as his principle paleophysics paradigm. In short, of all the esotericists and occultists within the milieu of Himmler's SS, it is Wiligut himself who represents the best possible esoteric influence and basis for some of the SS' subsequent projects to *reconstruct the technology of that physics.* And note, in Wiligut's hands, the "bloodline myth," the holy Grail, has taken on a new aspect: for Wiligut rationalizes the SS quest for the Holy Grail, for the "cup of God", in an entirely new way: the cup — the blood — may indeed contain the blood, not of Christ, but of "the ancient gods," the off-world human "cousins" of humanity.

d. Wiligut and the Bloodline: Otto Rahn

In this respect, and before embarking on an examination of Wiligut's esoteric and "paleophysics" views, however, mention must be made of yet another significant episode in esoteric and SS history that Wiligut brokered.

---

28 Many of these texts are examined in my book *The Cosmic War: Interplanetary Warfare, Modern Physics, and Ancient Texts,* (Kempton, Illinois: Adventures Unlimited Press, 2007), pp. 100–233.
29 Q.v. my *The Cosmic War,* pp. 67–83.

This was, of course, Himmler's personal recruitment of an esoteric scholar, historian and archeologist into the SS, Otto Wilhelm Rahn,[30] author of *The Crusade Against the Grail*, and source of ongoing contemporary speculations of the nature and degree of SS research into the Holy Grail and the various bloodline myths of Europe.

Indeed, Wiligut maintained, as was seen, that his secret family tradition was destroyed in 1848 when ancient records were lost. As a part of this tradition, however, Wiligut also maintained that he himself was personally descended from the ancient "god" Wotan, and thus was the real "Secret King" of Germany. And it was Wiligut with whom Rahn had been in close personal contact prior to joining the SS and becoming part of Himmler's personal staff. In this capacity, Rahn would inevitably have worked closely with Wiligut.[31]

Seen in this light, then, the SS obsession with the recovery of the Holy Grail is less about the recovery of the cup of the Last Supper, or even with the recovery of any putative and alleged bloodline of Christ's descendants that so many Gnostic traditions of Europe maintained, as it is about the recovery of the lost bloodlines of the ancient gods — the very gods of the ancient Sumerian pantheon — who descended from heaven and sired offspring with humanity, for Wiligut maintained adamantly that he was descended from them. We shall have more to say about Rahn in the Epilogue.

2. Wiligut's Alchemical Views: The "Primeval Twist"

So what exactly *were* Wiligut's "esoteric physics" views? How might one characterize them?

In a word, they were alchemical.

And in being alchemical, they were also to that extent, Egyptian.

Consider only the following short stanza from a poem called "Number," submitted by Wiligut under his pen name of Jarl Widar once again, to *Hagal* magazine:

> "Spirit in Matter!" out of Aithar's form
>   concealed in the dual-idea —
> There rests *the "twist" as the primal beginning,*
>   The ring of "life's woes."[32]

This one short stanza aptly summarizes almost all of Wiligut's "alchemical physics," which was, like the Philosophers' Stone, to imprison or "em-body"

---

30 Flowers and Moynihan, *The Secret King*, p. 34.
31 Ibid., p. 57.
32 Karl Maria Wiligut, a.k.a. Jarl Widar, "Number," *Hagal* 11 (1934) Heft 8, pp. 104, cited in Flowers and Moynihan, *The Secret King*, p. 100, emphasis added.

spirit, the "spirit" of the form of the aether, the transmutative physical medium itself, in matter. This is the very quintessence of the alchemical operation.

But note Wiligut gives a further clue as to *how* this might be done, for that "spirit in matter" rests in "the 'twist' as the primal beginning." The primeval twist, the primeval *torsion* — if one may so interpret his words — is the method, for it is the means by which the diversity of creation itself came about in Wiligut's view.

3. Wiligut's Version of the Augustinized Trinity and the Tripartite Stone

As if this were not enough, Wiligut also reproduces faithfully and almost exactly the Augustinized Trinitarian formulation we encountered back in part one with the alchemists themselves and their formulations of the "tripartite Stone." Indeed, the whole key to deciphering the hidden physics of runes was based upon "the flow of Matter-Energy-Spirit" which was accomplished "by the *Drehauge* ('Rotating Eye')."[33] Even the mention of a "Rotating *Eye*" is suggestive of its own roots in the themes of ancient and esoteric lore, recalling the "All-Seeing Eye" so common to Egyptian mythology.[34]

The Triune Stone and the Augustinized formulation of the doctrine of the Trinity are exactly, and very concisely and subtly, wedded by Wiligut in the following very significant passage:

1. Gôt is Al-unity!
2. Gôt is "Spirit and Matter," the dyad. He brings duality, and is nevertheless, unity and purity.
3. *Gôt is a triad: Spirit, Energy and Matter.* Gôt-Spirit, Gôt-Ur, Gôt-Being, or Sun-Light and Waker (*Wekr*), the dyad;
4. Gôt is eternal — *as Time, Space, Energy and Matter in his circulating current.*

...

7. Gôt — beyond the concepts of good and evil — is that which carries the seven epochs of human history.

...

9. ...*He closes the circle at N-yule, at Nothingness,* out of the conscious into the unconscious, so that this may again become conscious.[35]

---

33 Flowers and Moynihan, *The Secret King*, p. 70.
34 Cf. my *The Cosmic War: Interplanetary Warfare, Modern Physics, and Ancient Texts*, pp. 285–294, and *The Giza Death Star Destroyed: The Ancient War for Future Science*, pp. 8–11.
35 Karl Maria Wiligut, "The Nine Commandments of Gôt," signed manuscript, cited in Flowers and Moynihan, *The Secret King*, p. 79, emphasis added.

There are several things of interest in this short passage. First, note the progression of Wiligut's "trinity" from unity, through a "middle step of two," the dyad, arriving finally at the now-familiar triad of the Triune Stone of "Spirit, Energy, and Matter."

If one were to place each of these three terms in the three vertices of the now-familiar "Augustinian Trinitarian Shield" and place "Gôt" in the center circle, one immediately sees the connection both to the Augustinian Trinitarian formulation as well as to the alchemists' triune stone. First, let us recall the original Augustinian Trinitarian shield:

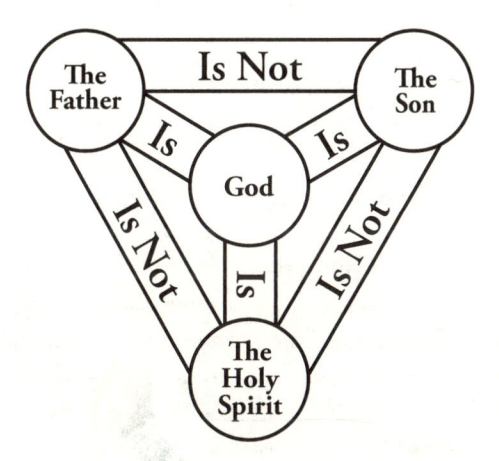

Original Augustinian Trinitarian Shield

And now, the Triune Stone version of the alchemists:

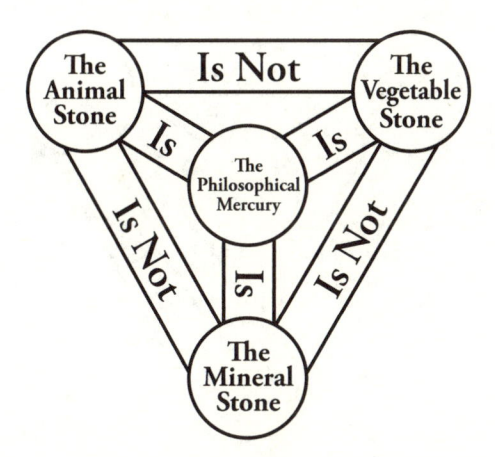

The Alchemical Triune Stone Version of the Augustinian Trinitarian Shield

THE PHILOSOPHERS' STONE | 261

And now, Wiligut's version of it:

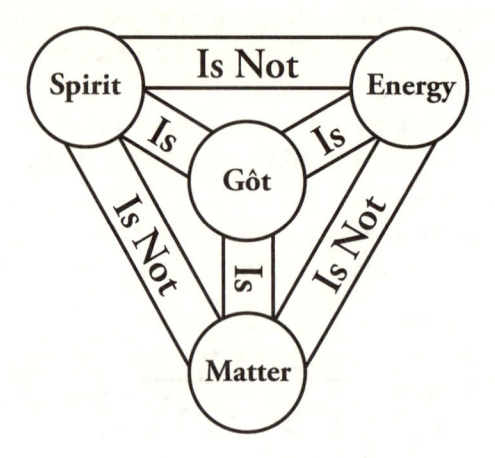

Wiligut's Version of the Triune Stone

But at this juncture, an important question occurs: What does Wiligut mean when he states "Gôt is eternal — *as Time, Space, Energy and Matter* in his *circulating current?*" The physics jargon employed — "circulating current" — suggests that he has in mind precisely a physics metaphor or interpretation for the whole scheme, for a circulating current suggests precisely *time* as the fundamental component of his "primal twist." In other words, the whole pictogram might represent — had he known of the terminology – "primeval torsion," time, as the underlying "hyper-space" unfolding into this world's manifestations of energy, matter, and "spirit," as the following diagram suggests:

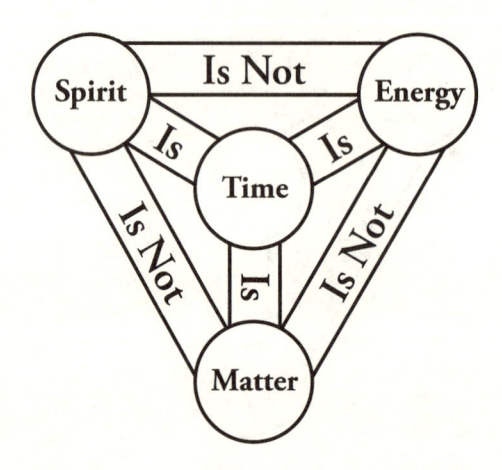

Implied Physics of Wiligut's Alchemical Trinity

262 | Joseph P. Farrell

That this implied physics is what Wiligut ultimately had in view is further suggested by the last comment in the cited quotation above: "He closes the circle...at Nothingness..." suggesting that "the cycle is concluded when it returns to the very same substance," nothingness, the undifferentiated physical medium, "from which it began."[36] By invoking a physics terminology with his reference to "circulating current," Wiligut is implying that this whole alchemical-theological theme is at root based in physics and not metaphysics. Thus, as he avers, this "Gôt" is "beyond the concepts of good and evil." In a Europe already concussed with loss of faith and a confidence in Christian institutions and morality, such a statement could only have disastrous moral consequences, for the whole mistaken thematic identity of this physical process with the Christian Trinity and therefore with the whole edifice of Christian morality inevitably led to a similar mistake in ethics: if they were based only in a physical process neither good nor evil, then there could be no ultimately good or evil action, only actions now more, now less, in harmony with the process itself. With Wiligut's "triune Stone" the way was open for the genocidal alchemical transmutation of man himself.

A final point should be noted about the immediately previous pictogram. By placing "Time" in the center, Wiligut is suggesting the very same type of torsion physics, with time as an *actual participating physical process* in physical mechanics as was suggested in the pioneering experimentation of Dr. Nikolai Kozyrev! But are there more explicit indicators that this type of thinking was actually a basis for physics experiments inside of Germany?

4. Wiligut's Opposing Spirals: The "Swastika Tensor" Revisited

Indeed, there *are* other statements that connect Wiligut's views much more directly to the "alchemical physics" embodied in Nazi Germany's most highly classified secret weapons research project, The Bell, statements that indicate that, in part, these esoteric views might have formed part of the rationalization for the project, at least, as far as the unscientific leadership of the SS was concerned.

For example, Wiligut, expounding on the nature of the "hooked cross" or swastika as an ideogram of that physics, states the following from an article entitled "Ancient Family Crest of the House of Wiligut," published again in the magazine *Hagal* in 1933:

> TWO HOOK-CROSSES (*Hakenkreuz*) in both directions of rotation: to the left turning to the right, absorbing inward; to the right turning to the

---

36 The phrase "The cycle is concluded when it returns to the very same substance from which it began" is actually that of Thomas Aquinas, used in reference precisely to the Augustinian formulation of the doctrine of the Trinity. *Summa Contra Gentiles, Book Four: Salvation* (Notre Dame), p. 145.

left, radiating outward. Korschelt was already aware of this differentiation, which is *confirmed by the latest investigations by Heermann Kassel, M.D. In his apparatus he separates inhibiting radiation (with rightward rotation) from the growing radiation (with leftward rotation). An experimental confirmation of this heraldic symbolism!*[37]

In other words, for Wiligut, the imagery of the "circulating current" of the cosmos was indeed not only a very deliberately chosen physics metaphor, but also very deliberately coupled by him to the view of the swastika or "Hakenkreuz" - in both its orientations – as an ideogram of that physics. Moreover, as he avers, some type of medical experimentation had been underway during or prior to 1933 involving precisely the *separation* of two rotating fields. This, as is now well-known and as we shall see in the next chapter, is the very *basis* of the rationalization behind the Nazi's Bell project![38] These were the ideograms of the "twist as the primal beginning," the very ideograms of the underlying hyper-dimensional transmutative medium itself.[39] Wiligut is implying, by means of his reference to the swastika and the torsion-based physics it symbolized, that the fabric of space-time has a spin orientation, just as was implied in the actual theorizing and experimentation of Dr. Nikolai Kozyrev. Moreover, he is also implying that as early as 1933 someone in the Third Reich is already performing medical experiments based on this conception, and since that person is a medical doctor, apparently the connection had *already been made* between that physics and life processes themselves.[40]

Wiligut makes even clearer his connection of this rotating physics view to that of alchemy by clearly stated references to the most alchemical of all alchemical maxims:

> Primal law: "Above as below, below as above!"
> Therefore in the middle there is a neutral force- (i.e., generational-) field!

---

37   Karl Maria Wiligut, "Ancient Family Crest of the House of Wiligut," *Hag All All Hag* 10 (1933), Heft 2/3, pp. 290–293, cited in Flowers and Moynihan, *The Secret King*, p. 81, emphasis added.

38   See my *SS Brotherhood of the Bell*, pp. 162–191 and my *Secrets of the Unified Field: The Philadelphia Experiment, The Nazi Bell, and the Discarded Theory*, pp. 262–288.

39   Karl Maria Wiligut, "Number," *Hagal 11* (1934) Heft 8, pp. 1–4, cited in Flowers and Moynihan, *The Secret King,* p. 100.

40   For those who have been following my argumentation as it began to be developed beginning in my book *The Giza Death Star Deployed* and continued in my books *The SS Brotherhood of the Bell* and *Secrets of the Unified Field,* this means that Lt. Col. Tom Bearden's arguments that this type of physics began to be fully investigated in the former Soviet Union is not, as I argued there, in fact the case. The presence of Wiligut's ideas in the SS, and the fact that he is a close personal advisor to Himmler, the fact that his ideas at least *conceptually* bear such a strong resemblance to the actual operative parameters of the Bell, and finally the fact that he himself states that some experimentation was already underway in separating these rotating fields is yet another argument in favor of my contention that the actual *modern* historical investigations and applications of this physics began in Nazi Germany.

> Spirals are apparently "contrary" to each other in their rotations and despite this, form — connected at their longitudinal ends — a "unity" from an oppositional "dyad," duality..."

Wiligut then reproduces the following diagram:

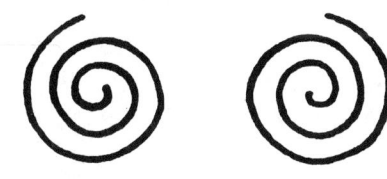

Wiligut's Double Opposing Spirals

He continues:

> From the two spirals, each triply wound and connected by the generational plane (the middle field), is developed the form of the two-ended pointed egg — the concept of the "World Egg."[41]

Once again, one is reminded of his fellow Austrian Viktor Schauberger's similar observations about the spiraling forms of energy which he called "implosion" and "explosion."[42]

Wiligut elaborates on the above conceptions a little further on in the article in a passage that, if one did not know it was by an Austro-German SS alchemist on the personal staff of Reichsführer SS Heinrich Himmler, one might be induced to conclude it had been written in some private diary of Dr. Nikolai Kozyrev!

> We observe the same process, which is demonstrated here by the snail whose shells are to be found in massive deposits in all primeval limestone formations of our earth, and we see this same process in the Al (cosmos) in the form of "spiral galaxies"....therefore:
> CREATIVE PRINCIPLE:
> Spiritual direction in Matter through Energy!
> CREATIVE INFERENCE:
> Eternal generation, and thus eternal Life and through this an eternal

---

41 Karl Maria Wiligut, a.k.a. Jarl Widar, "The Creative Spiral of the 'World-Egg'!" *Hagal* 11 (1934) Heft 9, pp. 4–7, cited in Flowers and Moynihan, *The Secret King*, p. 106.
42 See my *Reich of the Black Sun: Nazi Secret Weapons and the Cold War Allied Legend*, pp. 206–221.

circulation, through constant "turning" of Life-forms in Matter is implied — from cause comes effect and from this the new cause is formed. From this ensues the law of the conservation of Energy, which again implies the "material circulatory transformation" — that is, the concept of "eternal life."[43]

This cosmically scaled physics we have already found Wiligut identifying with the precession of the equinoxes, the solar year, and galactic physics.[44]

Wiligut makes one final comment which should be cited here for its possible direct bearing on the possible esoteric basis for the rationalization of the Bell project:

> FURTHER INFERENCES:
> Both of the spirals of the "World-Egg" therefore possess two poles. In turning (rotation) these poles are connected as an axis. These are the poles — "Above-Below," "Below-Above" — which are the World-Axis.[45]

So what does one make of Karl Maria Wiligut, a.k.a. Jarl Widar, a.k.a. Weisthor, "Himmler's Rasputin"?

Clearly, his basic conceptions and reconstructions of ancient "paleo-physics" is alchemical in nature. Moreover, many of his specific conceptions eerily parallel not only the thinking behind Kozyrev's later torsion experiments, but indicate that some similar sort of experimentation was underway in Nazi Germany as early as 1933 using *separated* and *counter-rotating systems of fields.* Indeed, as was seen, some sort of connection had already apparently been observed between such systems and the life-processes themselves. His views, therefore, since they had the ear of Reichsführer Himmler himself, would have sown fertile ground for anyone approaching the SS — or whom the SS approached! — with a more serious, and dangerous, project in mind.

A project like the Nazi Bell.

And someone like Nobel Physics Laureate Professor Dr. Walther Gerlach, the Bell's eventual project head…

---

43 Karl Maria Wiligut, a.k.a. Jarl Widar, "The Creative Spiral of the 'World-Egg'!" *Hagal* 11 (1934) Heft 9, pp. 4–7, cited in Flowers and Moynihan, *The Secret King,* p. 107.

44 Karl Maria Wiligut, a.k.a. Jarl Widar, "Zodiacal Signs and Constellations," *Hagal* 12 (1935) Heft 4, pp. 56–58, p. 113, cited in Flowers and Moynihan, *The Secret King,* p. 113. Previously cited here on p. 269.

45 Karl Maria Wiligut, a.k.a. Jarl Widar, "The Creative Spiral of the 'World-Egg'!" *Hagal* 11 (1934) Heft 9, pp. 4–7, cited in Flowers and Moynihan, *The Secret King,* p. 108.

B. The Tausend Affair and Prof. Dr. Walther Gerlach's Article

In my book *The SS Brotherhood of the Bell: NASA's Nazis, JFK, and MAJIC-12* I noted that Prof. Dr. Walther Gerlach, a Nobel Laureate physicist who eventually came to be the project head of the Nazi Bell, wrote a short article in a Frankfurt newspaper in 1924. Here, then, is what I wrote in that book about that article, in a chapter entitled "Gerlach's New Alchemy," about this article. I cite only the relevant portions here.

1. Prof. Dr. Walther Gerlach Takes Note of an Alchemical Paper: Excerpt from The SS Brotherhood of the Bell

"On Friday, July 8, 1924, Nobel Laureate physicist Prof. Dr. Walther Gerlach, published an interesting article in the evening edition of the *Frankfurter Zeitung* newspaper entitled 'The Transmutation of Mercury into Gold' (see picture page 285). The article, though brief, opens a Pandora's box of possibilities and speculations dark with abysmal promise. The editorial introduction begins innocently enough:

> To yesterday's telegraphically transmitted note on "Modern Alchemy," we present the following report as a welcome supplement and commentary.

And with that short notice, Gerlach's main text follows in all its significant concision:

> In no. 29 of the weekly magazine "Physical Science," Prof. A. Miethe from the Technical Higher School in Charlottenburg together with Dr. Staumreich *apprises us (of the fact) that it might be possible to induce the disintegration of mercury with relatively simple physical methods* and to clearly identify gold as one of the chemical and physical products of the disintegration. Until now we have known of two types of atomic bonding: spontaneous decay that occurs through no sort of external influence, and that through such external radioactive disintegration process such as was first carried out by the English physicist Rutherford, and as was shortly thereafter repeated in the Vienna Radium Institute by disintegrating lighter atoms (for example, lithium, boron, silicon, aluminum and so on) through bombardment with "rays." The general possibility of an "alchemy" is no longer in need of proof, since atomic research has demonstrated that all atoms are constructed from very simple building blocks, from hydrogen and helium, the lightest things...; various chemical elements may hereafter be distinguished by the number of components and perhaps through the types of atoms they (most readily) bind to.

Gerlach's 1924 Frankfurter Zeitung *Newspaper Article on the Transmutation of Mercury into Gold*

> The point of origin of Miethe's (investigation) was the observation that mercury lamps ... by means of very powerful and fast bombardment formed a dark incrustation on their interior winding. The research obtained greater amounts of such incrustation in the mercury lamps — it was made certain that such mercury lamps were free of any gold material, in the course of 70 – 200 hours with 70 volts potential and loaded with 400–2000 watts — yielding a measurable quantity of gold!....The amounts are small, but they are always between 1/10 and 1/100 milligrams, but tangibly weighable and analyzable. ...*A larger scientific exploration is to be hoped for, because there is every indication these authors have made a singular and thoroughgoing breakthrough, especially when one considers the astonishingly low stability of mercury when it is stressed.*[46]

On first reading, the article appears like so many other popular science articles of that era when nuclear physics was on the verge of the discovery of fission (by Otto Hahn fourteen years later in 1938): full of promise, yet naively unaware of the complexities of quantum mechanics that lurked just around the corner. A modern-day physicist would most likely dismiss it as the type of 'wishful thinking' that so often accompanies scientific discovery, before the 'reality' of subsequent observation sets in.

"But there are problems with that viewpoint too, not the least of which is that the article is by Gerlach, and that it *does* occur before the discovery of nuclear fission. Simply put, the transmutation being observed by professors Miethe and Straumreich [sic] appears to be induced precisely by 'electrolytical' methods, since the 'bombardment' being referred to is expressed in conventional terms of 'volts' and 'watts.' Yet, the context also indicates that radioactive bombardment — most likely by electrically generated X-rays — is also in mind.

"And there are other peculiarities about the article not to be missed, and that is the name of the professor supposedly who made the discovery: A. Miethe. Readers familiar with the Nazi Legend of the UFO will recognize the name of Miethe as being one of those allegedly involved with Schriever, Habermol, Epp, Schauberger, and Bellonzo in flying saucer research for the Nazis. Could this be the same Miethe, or even a relation?

"...Whatever the answer to these questions may turn out to be, one thing is clear from Gerlach's article, and that is that he is not thinking in terms of the standard models of transmutation via *neutron* bombardment that would obtain *after* the discovery of nuclear fission simply because those models do not yet exist, yet, he *is* thinking in terms of *some* form of transmutation via radioactive and electromagnetic bombardment, or rather, *stress*. Indeed, for Gerlach, whose 1921 experiment in magnetic resonance and electron spin won him the Nobel prize in physics, these results must have set his own mind buzzing with the possibilities of

---

46 Prof. Dr. Walther Gerlach, "The Transmutation of Mercury into Gold," *Frankfurter Zeitung*, evening edition, Friday, July 18, 1924. My translation from the German. Emphasis added.

what spin and resonance, under extreme conditions, might be able to achieve, given enough funding, interest, and research. This, I believe, is what must have actually been in his mind.

"These insights make his final two comments darkly revealing. 'A larger scientific exploration is to be hoped for,' Gerlach urges, 'because there is every indication these authors have made a singular and thoroughgoing breakthrough, especially when one considers the astonishingly low stability of mercury when it is stressed.' It is the familiar cry of a scientist making a plea for research funding. And given his notoriety and that of the newspaper in which he presents his plea, there can be only one intended target of his remarks: the German government and its many corporate financial backers.

"But notice the final remark concerning *'the astonishingly low stability of mercury when it is stressed.'* Gerlach here enunciates in a few words the line of investigation he wishes such research to take: subject a high-density, viscous, low-stability metal such as mercury to *stress*. And given the context, we know what kind of stress he intends to subject it to: *radioactive and electrical.* Gerlach has deduced the obvious: if such simple means can cause minute amounts of mercury to transmute into gold, then this must mean that these conditions induce a kind of instability in mercury. And with that, a whole new world of possibilities opened up before him, a world he himself calls 'alchemical.'"[47]

And there I let matters lie, at least, insofar as alchemy itself was concerned.

But what was, in fact, the wider context of Gerlach's remarks? What was that paper by professors Miethe and Staumreich [sic] to which Gerlach had referred?

This is where it gets…*interesting…*

2. Alchemical Papers by One Japanese and Two German Physicists

The story begins when a Japanese physicist named Nagaoka and his associates performed an experiment in March, 1924, the same year as the appearance of Gerlach's article some four months later. Studying various spectral lines of isotopes of mercury and bismuth, and particularly the spectral lines of gold, Nagaoka and his associates at the University of Tokyo

> Discharged about 15 x 10⁴ volts/cm for 4 hours between tungsten and mercury (terminals) under a dielectric layer of paraffin oil. They used the Purple of Cassius test to detect (gold) in the viscous residue… The black mass was purified *in vacuo*, then by combustion with oxygen and extraction with

---

47 Joseph P. Farrell, *The SS Brotherhood of the Bell: NASA's Nazis, JFK, and MAJIC-12,* (Kempton, Illinois: Adventures Unlimited Press, 2006) pp. 272–276.

(hydrochloric acid) to yield (gold), either in aqua regia solution or as ruby-red spots in the glassware. Microscopic films of (gold) were found on occasion.[48]

Shades of David Hudson!

Nagaoka also observed that "In order to be sure of transmutation, repeated purification of (mercury) by distilling in vacuum *at temperatures below 200 is essential.*"[49] Bear that very low temperature in mind, for it will become a crucial part of the story in the next chapter.

But what Nagaoka then proposed was breathtaking, for at that time, no high-spin state nucleus model had then been observed or proposed — at least and as far as is known from the extant literature, not *publicly*:

> If the above assumption as to the (mercury) nucleus is valid, we can perhaps realize the dream of alchemists by striking out a hydrogen-proton from the nucleus by (alpha-)rays or by some other powerful methods of disruption (to produce (gold) from (mercury)).[50]

In short, Nagaoka was suggesting that a proton could be 'slightly detached' from the nucleus" of mercury and removed.[51] Professor Nagaoka's method of removal is obvious from the quotation above: subject the nucleus to bombardment and see if a proton is kicked loose. Such ideas, circulating at the time, would have surely caught Gerlach's attention, and indeed one of Nagaoka's articles is written in May of 1924, precisely in the German scientific journal *Naturwissenschaften*. This *might* have suggested to Gerlach other methods of such transmutation, such as high spin, his specialty.

But at approximately the same time at the Photochemical Department of Belin's Technical High School, Professor Adolf Miethe "found that mercury vapor lamps used as a source for ultra-violet rays ceased to work after a time because of a sooty deposit which formed in the quartz tubes."[52] Testing these deposits he discovered that they contained gold. On May 8, 1924, he and his associate Dr. Hans Stammreich were issued a German patent, number 233,715, entitled "Improvements in or Relating to the Extraction of Precious

---

48 Robert A. Nelson, "Adept Alchemy," www.levity.com/alchemy/nelson2_ html, p. 1, citing Nagaoka, H., "Transmutation of Hg into Au," *Naturwissenschaften,* 13:682-684 (1925); *Naturwissenschaften,* 14: 85 (1926); *Nature* 114 (August 9, 1924), 197; *Nature* 117 (#2952, May 29, 1926), pp. 758–760.

49 Robert A. Nelson, "Adept Alchemy," www.levity.com/alchemy/nelson2_ html, pp. 1-2, citing Nagaoka, H., "Transmutation of Hg into Au," *Naturwissenschaften,* 13:682–684 (1925); *Naturwissenschaften,* 14: 85 (1926); *Nature* 114 (August 9, 1924), 197; *Nature* 117 (#2952, May 29, 1926), pp. 758–760.

50 Robert A. Nelson, "Adept Alchemy," www.levity.com/alchemy/nelson2_ html, p. 2, citing Nagaoka, H., *Journal de Physique et la Radium* 6:209 (1925).

51 Nelson, "Adept Alchemy," p. 2.

52 Ibid.

Metals."[53] The amount of gold depended, in part, on the difference in electrical potential, the quantity of current, and the pressure of the mercury vapor.[54]

Such considerations as these would also, let it be noted, have provided a *further* stimulus to the possible Nazi development of laser isotope separation, as seen in the previous chapter, for it takes little imagination to see that a more efficient means of separating and even transmuting elements or their isotopes would be precisely via bombardment by *coherent* electromagnetic radiation — such as laser light — that was precisely tuned to the resonant frequencies of the elements or isotopes to be separated.

But with Meithe and Stammreich's patent, the alchemical genie was let out of the bottle, and scientists around the world began arguing for, and against, the possibility and the results of their experiments. And as has been seen, Gerlach himself had already weighed in, with his article in the *Frankfurter Zeitung*, in favor of the possibility. For example,

> O. Honigschmid and E. Zintl determined the atomic weight of Miethe's mercuric (gold), using potentiometric titration or auric salt with $TiCl_2$. It was found to be 197.26, which is heavier than ordinary (gold) (197.2). They emphasized the need for a mass spectrographic analysis.[55]

Other scientists weighed in against the whole scenario, arguing principally that the methods being proposed for the observed transmutations were simply inadequate to account for the phenomenon.

Then in December of 1924 the magazine *Scientific American* "announced that it would arrange for a comprehensive and exact test of the Miethe experiment."[56] Tests were performed and a replica of Miethe's equipment actually procured from Germany, but in no case could any traces of gold be discovered. Some suggested therefore that Miethe had only detected it because the original starting mercury contained gold.[57] But perhaps another mechanism was in play, one that we know from Kozyrev's experiments and from the observations of the alchemists themselves: *timing of an operation was crucial to its success.*

The Germans remained undaunted. In 1926, the giant German electrical firm of Siemens got into the act, announcing in April 1926 that "a 10,000

---

53   Nelson, "Adept Alchemy," p. 2.
54   Ibid., citing "attempts at Artificial Gold" in *Literary Digest* (14 March 1925 and 12 December 1925). No page references given.
55   Ibid., citing Honigschmid, O. & Zintl, E., "The Atomic Weight of Au..." in *Naturwissenschaften* 13:644 (1925).
56   Robert Nelson, "Adept Alchemy," p. 9.
57   Ibid., citing *Scientific American* (Dec. 1924); ibid. (November 1925), p. 256; ibid., (April 17 1926) p. 90; ibid., 138 (128) p. 208.

fold increase in yield had been obtained in the production of mercuric-gold process." The Siemens firm had bombarded mercury with "electrons in extremely high vacuum" and obtained a hundred milligrams "from 1 (kilogram) of (mercury)."[58] *Siemens und Halske Aktions Gesellschaft* subsequently obtained a German patent, number 243,760, for the process.[59]

The entire affair raged on, with the original performers of the various experiments, Professors Nagaoka, Miethe, and Stammreich, quietly handling the criticism. But "the entire issue was never resolved."[60]

What emerges from the whole affair are two things. One has already been seen, namely, that someone in Germany may have connected the mechanisms at work either to *spin resonance phenomena,* on the one hand, or to the possibility of coherent optical separation via a "cascade" or lasing principle on the other. After all, what had started the whole affair was pulsing mercury vapors electrically. Why not pulse them with precisely tuned resonant frequencies of coherent laser light? Had the Germans asked that question — and as was seen from the previous chapter a circumstantial case can be made that they did — then it would have inevitably led them, as Richard Hoagland proposed and I argued in the previous chapter, to cohered light — lasers — as the only truly mechanism to accomplish it.

But the other thing that emerges from this affair is…

3… *The Franz Tausend Affair*

With Franz Tausend, alchemy itself enters the picture quite openly, not hiding behind scientific papers in obscure journals, for Tausend was self-admittedly a practicing alchemist. And, if one is to believe his claims, he began producing gold from mercury in the 1920s.[61]

But this is not his significance. His significance lies rather in the fact that by 1925 Tausend was a close associate of none other than General Erich Ludendorff, the famous general who led Imperial Germany's last offensive in the spring and summer of 1918 against the Western Allies, an offensive that came uncomfortably close to splitting the British and French armies and driving the former into the sea.

And Erich Ludendorff was, by 1925, closely associated with none other than Adolf Hitler and a strong supporter of the Nazi party. Tausend managed to obtain approximately 100,000 gold Reichsmarks from Ludendorff for his experiments, and the financial backing of other German industrialists such as

---

58 Ibid., p. 5., citing *Scientific American* (April 17 1926), p. 90.
59 Ibid.
60 Ibid., p. 7.
61 Nelson, "Adopt Alchemy," p. 7.

Alfred Mannesmann.[62] This led to the formation of "alchemical corporations" with Tausend naturally at the center. Its other prominent backers included, of course, Himmler, and later high-ranking SS general Oswald Pohl.

There even exists, as the following photograph attests, a postwar letter from Pohl attesting the existence of these "alchemical companies" for Himmler:

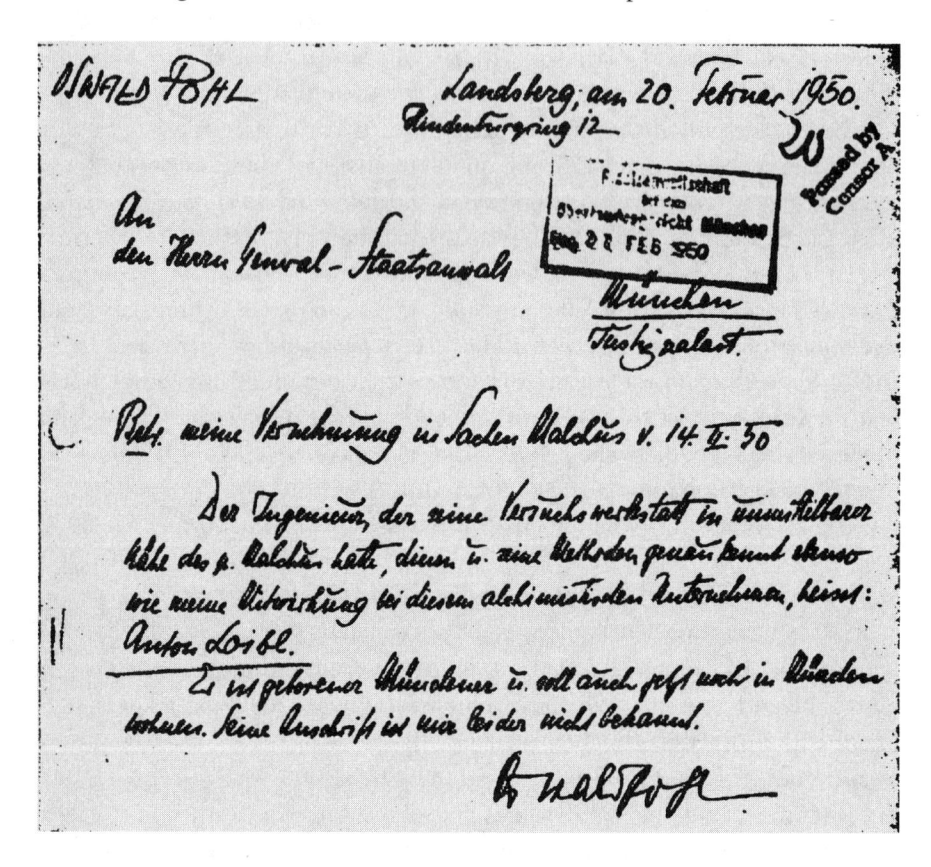

Oswald Pohl's Postwar Letter Concerning the German and Nazi "Alchemical Corporations"

The Tausend affair erupted into a something of a national and international scandal for the Nazis and their associates, however, for much of the money went to supporting the scheme's luxurious tastes, and the process cost more than the value of what it produced, eventually leading Tausend to a popularly publicized trial for fraud in Germany, a trial that received some international attention, even making, in the first instance, the front page of *The New York Times*.[63]

---

62 See Franz Wegener, *Der Alchemist Franz Tausend: Alchemie und Nationalsozialismus* (KFVR: 2006), ISBN 3-931300-18-8.

63 See, for example, "Alchemy's Deceitful Trail: 'Gold' by Transmutation Still Deludes Victims,

C. Conclusions

Whatever one makes of the Tausend Affair, one thing is clear if one places it within the wider context of the peer-reviewed "alchemical" science of the times in general, and of the alchemical philosophy of Wiligut in particular, and that is that there is ample enough conceptual soil for the SS to have pursued, quietly and off the books, an "alchemical" physics. And with the appearance scarcely a decade later of Otto Hahn's actual discovery of nuclear fission by neutron bombardment, at least one clear method was available, one that the German scientists — like scientists elsewhere — readily appreciated could lead to a horrible new weapon of mass destruction, the atom bomb.

And there was enough conceptual ferment both within the scientific culture of quantum mechanics — Germany's "home-grown" and thus "Aryan" science — that would have also suggested to them a radical and very advanced form of isotope separation: laser isotope separation depending on the principle of the photon cascade. As was seen in the previous chapter, there is a circumstantial case that can be made that this was precisely what they did, fully four decades in advance of its public disclosure in the scientific literature. Nor in the final analysis need one balk at the possibility of such advanced rationalizations of technology in that time frame, for a similar process was also very much in evidence with their other secret weapons projects in general, and with the Bell in particular.

Finally, with Karl Maria Wiligut, for a brief period prior to the war Reichsführer SS Heinrich Himmler's close personal advisor in occult matters, one has an alchemical philosopher of the physical medium itself, a medium moreover that he viewed as having *spin orientation characteristics,* a concept that could not have been lost on someone like Prof. Dr. Walther Gerlach, project head of the Nazi Bell.

---

but Early Experimenters Aided Science," *The New York Times,* Sunday, November 17, 1929, p. SM 9; "German's Alchemy Called Huge Hoax: Leading Munich Chemists and Newspapers So Brand Test — Leaks in Supervision Disclosed," *The New York Times,* Friday, October 11, 1929, Section, Radio, p. 37; "German Produces Gold in Synthetic Test; Denies Swindling Ludendorff and Others," *The New York Times*, Thursday, October 10, 1929, p. 1.

The Nazi Serum

"IRR Xerum 525," The Nazi Bell, and The Recipe

•••

"...how did it happen that scientists from the 1940s understood exactly where they were heading? They had applied after all ideas from XXI century physics.... What arguments did they lay down (before the launch of work) that caused them to win the race for funds...? ...The unusualness of all this is summed up by the fact, that descriptions of mercuric propulsion had appeared as long ago as in ancient times — in alchemy and old Hindu books...It may prove that an explanation of all the technical questions related to work from the time of the war, will reveal a far greater mystery..."
Igor Witkowski[1]

The Bell was Nazi Germany's most highly classified secret weapons project, classified even higher than its atom bomb project, given the classification of *Kriegsendscheidend*, or "War Decisive." That designation says it all, for if their atom bomb project — a project which was possibly far more successful than the postwar Allied Legend would have one believe[2] — then by comparison the Bell's potential was truly awesome and horrific.

The story of the Bell was first disclosed by Polish military history researcher Igor Witkowski in works published in his native Poland, and subsequently compiled into the last chapter of his magisterially authoritative work on Nazi secret weapons, *The Truth About the Wunderwaffe*.[3] The actual story of the Bell

---

1 Igor Witkowski, *The Truth About the Wunderwaffe,* translated from the Polish by Bruce Wenham (Farnborough, England: Books International and European History Press, 2003), p. 284.
2 See my *Reich of the Black Sun: Nazi Secret Weapons and the Cold War Allied Legend*, pp. 3–158.
3 See Igor Witkowski, *The Truth About the Wunderwaffe* (Farnborough, Hampshire, England: European History Press, 2003), pp. 231–288.

was first outlined to the English-speaking world, however, by British defense journalist and *Jane's Defence Weekly* staffer Nick Cook in his bestselling book *The Hunt for Zero Point*.[4] British author Geoffrey Brooks also contributed additional information concerning the possible postwar whereabouts of the project in his book *Hitler's Terror Weapons: From V1 to Vimana*, tracing the project to Argentina.[5] These books all made me aware that with the Bell, one was dealing with a profoundly different paradigm of physics theory, experimentation, and technological development, a theory and development that appeared to have some connection to the type of physics I believed to be in evidence in the Great Pyramid of Egypt, and accordingly, I reprised Nick Cook's research at great length in my book *The Giza Death Star Deployed*.[6] While the connection to Egypt may seem outlandish or even contrived, there would appear to be at least conceptual connections, as will be seen a little later on.

Subsequently obtaining a copy of Witkowski's vitally important book, I reviewed his research and extrapolated on its possible physics basis and implications in my books *Reich of the Black Sun, The SS Brotherhood of the Bell* and *Secrets of the Unified Field*.[7] While coming to somewhat different conclusions about many aspects of the underlying physics and construction of the Bell than Witkowski or Cook, their work remains vital to any approach to the topic. But one component of the Bell was, and remains, a mystery: its "fuel," the enigmatic "IRR Xerum 525," the Nazi Serum. In order to see how, and what this mysterious compound might have been, a review of the Bell project is in order. In doing so, many possible "alchemical" connections will become evident.

A. The Black Projects Reich within the Reich: The Kammlerstab Think Tank

The story of the coordinating body for Nazi secret weapons, the SS *Sonderkommando Kammler* (Special Command Kammler), also known as the *Kammlerstab* (Kammler Staff), has been told in a variety of books by various competent researchers, but the story first broke to a general public when

---

4 See Nick Cook, *The Hunt for Zero Point*, pp. 181–190.
5 Geoffrey Brooks, *Hitler's Terror Weapons: From V1 to Vimana* (London: Leo Cooper, 2002), p. 204, n. 147.
6 See my *Giza Death Star Deployed* (Kempton, Illinois: Adventures Unlimited Press, 2003), pp. 121–127.
7 See my books *Reich of the Black Sun: Nazi Secret Weapons and the Cold War Allied Legend* (2004), pp. 331–344; *The SS Brotherhood of the Bell: NASA's Nazis, JFK, and Majic-12* (2006), pp. 141–308; *Secrets of the Unified Field: The Philadelphia Experiment, The Nazi Bell, and the Discarded Theory* (2008); pp. 227–313. All these books were published by Adventures Unlimited Press.

one of its civilian members, Dr. Wilhelm Voss, shared his story with British journalist Tom Agoston. Voss swore Agoston to secrecy until his death, at which point Agoston told his story in a short book that disclosed the existence of what, for all intents and purposes, was the world's first "secret weapons think tank," a deeply classified covert SS think tank whose mission brief far exceeded, or rather dwarfed, that of the American "Manhattan District" atom bomb project.

Headed by SS *Obergruppenführer* — equivalent to a four-star general — Dr. Ing. Hans Kammler, this group recruited scientists from all over Germany and throughout Europe, and essentially told them to map out the technology trees to second-, third-, and fourth-generation weapons technologies and platforms. General Kammler himself was not only the head of this group, but significantly, also the very architect who had designed the notorious Auschwitz concentration camps, and who also was head of the entire SS Buildings and Works Department. Thus, Kammler is probably connected to the vast "Buna" factory of I.G. Farben at Auschwitz — a plant that we have previously argued was actually a large uranium separation and enrichment facility that employed a variety of exotic separation technologies, including perhaps a rudimentary laser isotope separation technique — and as head of the Buildings and Works Department of the SS, he could tap into a potential slave labor pool of over ten million people.

1. Agoston and Dr. Wilhelm Voss

It is this probable connection of Kammler to the Auschwitz "Buna" factory that is significant, for as Dr. Wilhelm Voss told Agoston, the Kammler Group's purpose

> was to pave the way for building nuclear-powered aircraft, working on the application of nuclear energy for propelling missiles and aircraft; laser beams, then still referred to as "death rays"; a variety of homing rockets and to seek other potential areas for high-technology breakthrough. In modern jargon, the operation would probably be referred to as an "SS research think tank." *Some work on second-generation secret weapons, including the application of nuclear propulsion for aircraft and missiles, was already well advanced.*[8]

---

8 Tom Agoston, *Blunder! How the U.S. Gave Away Nazi Supersecrets to Russia* (New York: Dodd, Mead, and Company: 1985), p. 12, emphasis added, cited in my book, *Reich of the Black Sun*, pp. 104–105, without emphasis. For the whole story of the Kammlerstab, see *Reich of the Black Sun*, pp. 99–116, and my *Secrets of the Unified Field: The Philadelphia Experiment, the Nazi Bell, and the Discarded Theory*, pp. 227–238.

The significance of the remark concerning lasers has already been explored in chapter ten in connection with the possible development of a rudimentary tunable laser isotope separation process by the Nazis and the SS.

Here a further clue is provided, for such an advanced technology of isotope separation would be valuable not only in the separation and enrichment of needed isotopes for an atom bomb — uranium-235 and plutonium-239[9] — but also in the separation of elements and isotopes needed as the fuel for any nuclear propulsion as well.

However, there is one more significant point to bear in mind. The term "nuclear propulsion" as used here is rather vague and all-encompassing; it *could* mean, and it will be argued subsequently that it *does* mean, any form of *field* propulsion or anti-gravity technology that relies on some form of radioactive isotopes in the creation of those field effects.

In this context Igor Witkowski mentions a significant fact, and that is, by war's end, General Kammler had also been bestowed the title of *Reichsbevollmächtigter für den Aufbau des Forschungszentrums in Pilsen zur Frage atomaren Technologien für den Antrieb von Lenkwaffen und Flugzeugen,* or "Reich Plenipotentiary for the Construction of Research Centers in Pilsen concerning Atomic Technologies for Propulsion of Guided Weapons and Aircraft."[10] Such a title, and the considerable authority that went with it, bestowed by Hitler himself, on a practical engineering projects manager such as Kammler, is a strong indicator that Nazi Germany had progressed things to the point of practical operational deployment of at least *some* of these technologies, and that other more advanced ones were showing signs of imminent success. Success beyond anyone's wildest dreams.

Something like the Bell.

2. The Reichprotektorat *of Bohemia and Moravia*

Dr. Voss was assigned to the engineering division of the famous Skoda Munitions works in Pilsen, Czechoslovakia, after the German annexation of that country in March 1939. It was in this department that Kammler set up and headquartered his secret weapons think tank. When Germany made all of Bohemian and Moravian Czechoslovakia a *"Reichprotektorat,"* all of Bohemia — including the Skoda Works and Kammler's think tank — fell under the jurisdiction of the SS under the command of none other than Reinhard Heydrich, Himmler's virtual "second in command" of the entire SS.

---

9 Remembering, of course, that the manufacture of *any* plutonium required a functioning reactor technology inside of Nazi Germany and/or Fascist Italy!

10 Igow Witkowski, "Supplement 2, to be Added at the end of the chapter on nuclear weapons, page 219 of the English edition" of *The Truth About the Wunderwaffe,* personal communication to the author concerning the forthcoming German edition of Mr. Witkowski's book.

This fact, plus the presence of Kammler's think tank at Pilsen, effectively made the *Reichprotektorat* — the entire province — Nazi Germany's version of the Nevada Test Ranges, a vast secret projects preserve under SS jurisdiction. And within this "secure zone," with its Gestapo agents crawling the countryside ferreting out sedition, Kammler's think tank was surrounded with a triple belt of security of political, economic, and technological counterintelligence personally headed by none other than close Martin Bormann associate, the head of the actual Gestapo, Heinrich "Gestapo" Müller![11]

Dr. Wilhelm Voss and Nazi bigwigs at the Skoda Works in Pilsen. Dr. Voss is the civilian in the hat on the right side of the front row

3. *The* Forschungen, Entwicklungen, und Patente

Another group connected to Kammler's "think tank" and tied directly to the Bell project was the F.E.P., or *Forschungen, Entwicklungen, und Patente* (Researches, Developments, and Patents) group of the SS, headed by an obscure but apparently extremely important SS *Obergruppenführer* named Emil Mazuw.[12] The mission brief of this group was to scour German and other European scientific journals, personnel, and patents for anything that could be

---

11   For the implications of this relationship between Kammler, Bormann, and Müller, see my *Secrets of the Unified Field*, pp. 230–236.
12   See my *SS Brotherhood of the Bell*, pp. 144–148.

of value for a potential technological breakthrough and military application.[13] Another mission brief of this group concerned the development precisely of theories and technologies that could lead to German energy independence.

Consequently, with the F.E.P. one has suggestive clues as to some of the purposes of Nazi research with the Bell project, namely, that in addition to its possible purpose in the development of field propulsion, yet another purpose may have been to access new sources of energy that would free Germany from dependence on "fossil" fuels and other non-renewable energy sources, which were scarce in Europe.

In any case, one of the things that surely would have come to the attention of this group were precisely the patents of Miethe, Stammreich, and the Siemens corporation, and the article of Prof. Dr. Walther Gerlach concerning mercury to gold transmutations and the scientific possibility of an "alchemy." This, coupled with the probable SS knowledge of esoteric Sanskrit and other texts concerning mercury propulsion, and the general "alchemical" climate within the SS that was close to Himmler — namely Wiligut et al. — would have provided fertile group culture for the SS to commit to the study and research of the actual scientific possibilities that such texts could have held. If there *was* a group within the SS that would have made such connections, and been capable of fulfilling the *Ahnenerbe's* brief of the military applications of such texts and traditions, the F.E.P. was it.

4. The SS Entwicklungstelle IV

Yet another shadowy group connected to all this highly classified SS secret black projects relating to the Bell was the *SS Entwicklungstelle IV*, or "Development Area 4." While the term "area" may refer to a group investigating a certain "area" of research, there is another possibility, and that is that "area" might indeed refer to an actual geographical location of underground and secret installations conducting the research. Many such SS research groups were indeed designated by the actual area where the research was conducted.

If this "SS Development Area IV" sounds vaguely familiar, that's because it is. Well-known UFOlogist Bob Lazar claimed to have worked at a secret area *within* Area 51 in Nevada called "S-4", hidden underground in a mountain near Papoose dry lake, south of Groom Lake, or Area 51 itself. While this author has many difficulties with aspects of Mr. Lazar's story and the claims of his associate John Lear, two aspects of the story stand out as possibly significant in the context of Nazi research into advanced field propulsion concepts.

---

13 Thus, in this respect, with the F.E.P., one finds the Nazi equivalent, and perhaps the historical root, of the subsequent Soviet "research bureaus" that Lt. Col. Tom Bearden believes were established by Stalin in postwar Russia to seek out in the scientific literature any overlooked potential area for such a breakthrough development leading to a super weapon.

The first is, of course, the designation "S-4," a dim resonance of "SS Development Area IV." The second is that both groups appear to have been working on more or less the same thing: advanced field propulsion or "UFO" technology. But the third is perhaps the most unusual and interesting of them all.

Lazar claimed that the "captured alien flying saucers" he was allowed to work on at "Area S-4" utilized element 115 as the fuel for the matter-antimatter reactions that powered the craft. At the time that Lazar first made these claims, element 115, now known as unumpentium, had not yet been synthesized. When it was, however, it was done at the high energy physics center in Darmstadt, *Germany.*

These claims constitute, for this author, some of those problematical areas of Lazar's story, for the underlying physics assumptions implied by these claims is that the matter-antimatter relationship in the universe is more or less one of equilibrium, i.e., that the same amounts of both exist. This author does not subscribe to this view. But the second physics problem posed by this assertion is that Lazar claimed he and Lear actually obtained an amount of it. However, the element's half-life is far too short for a substantial amount to be accumulated, much less obtained and held for a period any longer than a few fractions of a second.

However, might the real intention of the episode have been ultimately not to disclose "crashed and recovered UFOs" and a secret black project designed to reverse-engineer the technology, but something else? Might the whole episode have been contrived by someone — not necessarily Lazar — to point a very clear, though very subtle, finger toward Germany as the source of some of these "extra-terrestrial" technologies, with its references to a secret "Area S-4" dimly evocative of "SS Development Area IV" and to "element 115," subsequently actually synthesized in Germany? Certainly neither Lazar nor his associate John Lear have ever — to this author's knowledge — made such claims. But it is interesting that Lazar subsequently indicated, after going public with his story, that he had been contacted by the group in Germany that had synthesized unumpentium to come to Germany and assist in their research, for they had heard of his story and its mention of element 115! Thus, yet another subtle clue may have been released, pointing yet another finger to that country as a "player" in the development of such technologies.

B. The Nazi Bell, its Operation, Effects, and Scientific Rationalizations: A Review

Before proceeding to outline a speculative case of what the mysterious IRR Xerum 525 might have been composed of, and what its use in the Bell was designed to do, a brief review of the Bell itself — its construction, operation, effects, and scientific rationalizations for them — is in order. Briefly stated, its construction, operation, and effects, are these:

1) While the size of the Bell's dimensions vary according to which sources one consults, the standard size may safely be said to be from 12–15 feet tall, and 9–12 feet wide;[14]
2) It was cylindrical and bell-shaped in appearance, hence its codename among the Germans, *die Glocke,* the Bell;[15]
3) Its outer shell was composed of some type of ceramic or metallic material;[16]
4) It was cryogenically cooled by liquid nitrogen or oxygen;[17]
5) Around its outside there were several ports for high-voltage electrical cabling, a fact that indicates that the Bell could be transported to various sites for testing and "plugged in." Additionally, the Bell made a buzzing-hissing sound when in operation, a fact that led to its nickname among the Germans, *die Bienenstock,* the Beehive;[18]
6) Inside the device were two counter-rotating drums capable of being spun at extremely high velocities;[19]
   a) Witkowski has argued that these drums were nested one inside the other, as in a typical plasma focus device;[20]
   b) I have previously argued that another possible arrangement was one on top of the other;[21]
7) The top of the device had some sort of hook-like attachment, indicating that it could perhaps be suspended from a balance scale or similar device;[22]
8) In the central core of the Bell was a stationary shaft, into which a "thermos bottle" containing the Serum 525 was placed;[23]
9) Mercury and possibly the Serum 525 itself were placed on the surfaces of the rotating cylinders;[24]
10) The Bell gave off extremely toxic, and in some cases lethal, field effects to a distance of some 150-200 meters from the device when

---

14 See Nick Cook, *The Hunt for Zero Point,* pp. 192-193.
15 Ibid., see also my *Reich of the Black Sun,* p. 331 and *SS Brotherhood of the Bell,* pp. 171-179.
16 Ibid., see also my *SS Brotherhood of the Bell,* pp. 174-175.
17 Ibid.
18 See my *Secrets of the Unified Field,* pp. 273-274; and Witkowski, *The Truth About the Wunderwaffe,* p. 234.
19 Nick Cook, *The Hunt for Zero Point,* pp., 192-193.
20 Igor Witkowski, *The Truth About the Wunderwaffe,* pp. 249-250; see also my *SS Brotherhood of the Bell,* pp. 175-179, and my *Secrets of the Unified Field,* p. 270-280.
21 See my *Secrets of the Unified Field,* pp. 268-280.
22 Witkowski, *The Truth About the Wunderwaffe,* pp. 232-233.
23 Ibid., see also Nick Cook, *The Hunt for Zero Point,* pp. 192-193.
24 Ibid.

in operation.²⁵ Accordingly, a variety of organic materials and life forms were tested in its presence. Among its many effects, the following two are of noteworthy significance:

a) In the case of plants, exposure to its field resulted in the cellular breakdown of plants within hours to weeks, with the plants decaying suddenly and finally to a state of a brown-grayish goo;[26]

b) In the case of humans, its first tests resulted in the deaths of five of the seven scientists involved. While subsequent tests appear to have mitigated these effects, people exposed to its field reported the sensation of "pins and needles" on their skin, of sleeplessness, and a persisting metallic taste in the mouth;[27]

c) the Bell's above-ground test-rig, the so-called Henge structure at Ludwigsdorff, shows signs of exposure to heavy ionized and neutron radiation. Additionally, the structure's steel reinforcement bolts running around the top of the structure appear to have been sheared or broken off by some abrupt and powerful force;[28]

d) the Bell also exploded light bulbs and created damage in electrical equipment in the surrounding environs during its tests above ground;[29]

11) In the case of outdoor tests conducted at night, some concentration camp workers associated with the project report seeing "barrel-like" devices ascending above the tree-lines with a pale blue glow;[30]

12) The Bell was tested under the most extreme security around the environs of Ludwigsdorff and Castle Fürstenstein in the mountains of Lower Silesia;[31]

13) Approximately sixty of the scientists and technicians connected with the project were murdered by the SS prior to the end of the war, and the device, and all its project documentation, along with General Kammler and a massive six-engine Junkers 390 heavy lift ultra-long-range aircraft went missing at the end of the war;[32]

14) Its scientific head, Prof. Dr. Walther Gerlach, was one of the

---

25 Ibid. Witkowski notes that the test crews had to be issued special rubber clothing and polarized viewing glasses even at these distances from the device.
26 Witkowski, *The Truth About the Wunderwaffe*, p. 234.
27 Ibid., pp. 234–235, see also my *SS Brotherhood of the Bell*, p. 177ff.
28 See the recent works of Rainer Karlsch in Germany.
29 Witkowski, op. cit., p. 234.
30 Witkowski, op. cit., p. 263; see also my *SS Brotherhood of the Bell*, pp. 185–188.
31 Witkowski, op. cit., pp. 260–261; see also my *SS Brotherhood of the Bell*, pp. 185–188.
32 Cook, *The Hunt for Zero Point*, pp. 192-193; see also my *Reich of the Black Sun*, pp. 107–110 and *SS Brotherhood of the Bell*, pp. 167–171.

interred scientists at Farm Hall, who was heard talking mysteriously of magnetic fields separation, vortex compression, the gravitation of the earth's magnetic field, and so on. Additionally, Dr. Gerlach was the only one of the Farm Hall scientists subsequently taken to the United States for further interrogation. Upon his return to Germany, he never wrote again of gravitation or spin resonance and other subjects that had been his specialty before the war.[33]

15) One of the Bell's project scientists, and the designer of its power plant, the electrical engineer Dr. Kurt Debus, subsequently ended up as a senior flight director during the NASA Mercury, Gemini, and Apollo space projects at Cape Canaveral;[34]

16) A mathematician from the University of Königsberg, Dr. Elizabeth Adler, was brought in as a consultant during the project, for reasons never disclosed in the currently extant evidence concerning the project;[35]

17) The Bell's code names, *Das Laternenträgerprojekt,* The Lantern-bearer Project, and *Projekt Kronos,* Project Time, or alternatively, Project Saturn, are evocative of an esoteric and occult influence possibly at work in the background of the project, since "lantern-bearer" or "light-bearer" is a loose conceptual relationship to Lucifer, the light-bearer, and "Kronos" is the Greek word both for time and for the planet Saturn. Thus, the code name *Projekt Kronos* could suggest a project having to do both with the manipulation of time, and with some field propulsion experiment designed to give lead to access to distant planets, symbolized by Saturn.[36] Finally, there is also a reference to the Bell project also being code-named *Tor* or "Gate" at some point in the literature.[37]

---

[33] Witkowski, *The Truth About the Wunderwaffe,* pp. 237, 255; Nick Cook, *The Hunt for Zero Point,* pp. 182–190; see also my *SS Brotherhood of the Bell,* pp. 148–153.

[34] See my *SS Brotherhood of the Bell,* pp. 155–157.

[35] Witkowski, *The Truth About the Wunderwaffe,* p. 235; see also the crucial remarks in my *Secrets of the Unified Field,* pp. 282–286 and my *SS Brotherhood of the Bell,* pp. 152–153, 159–161.

[36] The reference to Saturn may have yet even stranger associations and implications. See my book *The Cosmic War: Interplanetary Warfare, Modern Physics, and Ancient Texts,* pp. 385–398.

[37] Witkowski, *The Truth About the Wunderwaffe,* p. 235; see also my *SS Brotherhood of the Bell,* pp. 166–167. The reference to the code name "Gate" has led many to contact me concerning the possibilities that the intention of the project was really to open a kind of "hyper-dimensional gateway" to contact "extra-dimensional intelligences," and other extreme forms of speculation. While the physics implied in the Bell certainly indicates that one of its direct project goals was precisely the manipulation of the fabric of space-time, such extreme speculations cannot be supported from the evidence. At best, they remain just that: extreme speculations. That said, there is another more mundane, but equally far-reaching series of implications. If indeed the Bell was conceived as some kind of "hyper-dimensional gate" to *allow energy to pour down into our lower-dimensional world,* that would certainly fit in with the stated aims of the F.E.P. to render Germany energy-independent. Such a technology and physics also inevitably implies a weaponization potential far beyond that of mere nuclear and thermonuclear bombs. Hence, such an interpretation

The scientific rationalizations of some of these points will now be briefly explored, while other points relating to its project environs and history will be left for later.

1. Kaluza-Klein, Torsion, and the Layout of the Counter-rotating Drums

In my book *Secrets of the Unified Field,* I explored at some length the cogent arguments of Bell researcher Igor Witkowski that the counter-rotating drums within the device had been laid out in the form of a typical plasma focus, one inside the other. Indeed, Witkowski's arguments are compelling, and rationalize the device quite well.[38] As outlined in *Secrets of the Unified Field,* however, a case can be argued that the Nazis may have rationalized the device along the lines of the incorporation of some form of torsion tensor in the unified field theories that began to be published in the Weimar Republic between the wars and before the Nazi assumption of power. There I argued that the presence of a mathematician from the University of Königsberg — the same university that saw the development of the *first* such hyper-dimensional unified field theory, that of mathematician Dr. Theodor Kaluza — within the Bell project pointed to a modified form of Kaluza's theory, that included torsion, as the basis of the project. Additionally, that theory's "geometry" was cylindrical.[39]

Since one of the effects of torsion is similar to wringing an empty soda can like a dishrag, producing spiraling folds and pleats in the fabric of space-time similar to the spiraling folds and pleats in the can that result from its "wringing," I speculated that in order to maximize this torsion effect within the Bell, that the counter-rotating drums were analogous to this image: the drums, in other words, could equally be rationalized to have been placed one on top of the other, *in order to maximize the torsion effect produced by the device.*

a. Wiligut

However, as has now been seen in our previous examination of the esoteric influences at work within the Nazi Party, the SS, and Germany itself between the wars, there is yet another possible conceptual influence hovering in the background: the alchemical views of Karl Maria Wiligut, and his very influential position in the immediate circle of advisors around Reichsführer

---

provides a basis for why the Nazis would have classified the project as "War Decisive," and why some sources maintain that the Nazis were indeed working in areas of "doomsday physics." (See *SS Brotherhood of the Bell*, pp. 192–241).

38   See the summary of his interpretations and extrapolations in my *SS Brotherhood of the Bell,* pp. 179–185.

39   See my *Secrets of the Unified Field,* pp. 262–288.

SS Heinrich Himmler. As noted, Wiligut himself promulgated views that the physical medium, in his understanding of esotericism, was the result of "the primal twist," an image that immediately conjures the notion of torsion. Moreover, Wiligut also viewed the image of the Nazi Party, the swastika itself, as an ideogram of that physics. Thus, in addition to the *scientific* culture of Germany itself, there is within the SS an additional impetus to the consideration of such views, namely, the alchemical influence of Wiligut.

b. Rudolf Hess and Nazi Hermeticism:
The Egyptian, Alchemical and Unified Field Theory Connections

The alchemical views of Wiligut gain an additional significance in this respect when little-known connections between alchemy and the various unified field theories then in vogue in Germany are considered. The British researcher Geoffrey Brooks mentions a little-known, seldom-discussed fact about the British interrogations of Rudolf Hess in which Hess makes some disturbing, or — depending on one's lights — insane statements:

> Rudolf Hess, interrogated in the Tower of London under the effect of a truth serum, stated that National Socialists valued the occult sciences highly and might even be, through Hitler, the puppets of a clandestine Directorate in the Orient.[40]

Such an insane assertion would surely be dismissed, had it come from anyone else other than Hess, at the time the actual Nazi Party head as its *Reichsleiter,* a general in the SS as well, and himself highly interested in occult and esoteric matters, having imbibed them from his youth growing up in... *Alexandria, Egypt.*

And the assertion, disturbing as it is, permits perhaps other high speculations. Might Allied knowledge of the existence of such a relationship, or the awareness that the Nazis themselves, at the highest levels, believed that such a relationship existed, have been a factor in the Allied declaration that nothing but unconditional surrender of Germany would do, a declaration that virtually assured that the war would be a fight to the literal death of one or the other of the sides, and of their leaders? Whatever the answer to that purely speculative question may be, it is Hess' and the SS' interests in the occult, and his connection to Egypt, that is of crucial importance here.

As Brooks observes, even though

---

[40] Geoffrey Brooks, *Hitler's Terror Weapons: From V1 to Vimana* (2002), pp. 26–27, citing OSS Interrogation Archive document #12678 *Nazi Occult Organizations.*

The underlying philosophy of Nazi science is undocumented, it is likely to have been the Hermetic tradition. Hitler was a disciple of the Buddhistic thinker Schopenhauer, and his success stemmed from a profound knowledge of magical causes occasioned by reading Schopenhauer's treatment of Hermeticism. By this is meant the ancient *Trismegistic literature of the Hermetic tradition of which uncontested Egyptian treatises survive and thus for the second time...we (find) ourselves confronted by the spectre of Ancient Egypt in connection with National Socialism.*[41]

But what has this alchemical connection — for Hermeticism is but another name for alchemy — to do with physics in general and the various unified field theories of the 1920s and 1930s that were published in Germany?

Brooks states the connection as ably as any:

Hermetic science states that each element in matter has as its crystal a unique geometric form.... All matter, mineral or organic, is merely a molecular structure held together by a keynote, from which one can infer that everything in the material universe is the result of vibration, a fortuitous concourse of atoms.... This was the secret behind the philosophy of Schopenhauer, the guru of Hitler, that the material world is only an illusion and no physical object has any permanent reality. The only reality is the vibration.[42]

On this view, the lattice structure of the elements — its "pure crystalline form" — is thus a representation of those vibrations. As a consequence of this view, the physical forces can in turn be understood to be defects or distortions in that structure. Bear these views in mind, for we shall encounter them again in the next chapter.

Thus arose the idea of "Aryan Physics," a term coined by two Nobel Laureate German physicists, both members and ardent supporters of the Nazi party: Dr. Philipp Lenard and Dr. Johannes Stark.[43] The direct targets of this artificial paradigm were, of course, the Jewish physics of Einstein represented by his relativity theories. But another lesser known target were the quantum mechanical theories of Heisenberg with its apparent lack of structure, and its uncertainty principle.[44] In short, Stark and Lenard were, like Dr. Kozyrev much later in Russia, very uncomfortable with the counterintuitive nature of relativistic and quantum mechanical physics. While the vast errors and contradictions of their "Aryan physics" are well-known and almost tiresomely

---

41   Brooks, *Hitler's Terror Weapons,* p. 27, emphasis added.
42   Ibid., emphasis added.
43   Brooks, *Hitler's Terror Weapons,* p. 27.
44   Ibid., p. 28.

rehearsed as examples of the persistent breakdown of physics inside of Nazi Germany, the truth in fact may not be so simple, for if projects such as the Bell existed in Nazi Germany, and if they achieved the sensational results that are alleged for them — and they most assuredly did — then what "Aryan physics" really represents is an attempt, howsoever clumsy and contradictory, to lay out the outlines and philosophy for a wholly different paradigm of physics.

In this light, Brooks makes a rather interesting series of observations:

> Much earlier....Aryan Physics *suspected that when Einstein was working on the unified field theory, he realized that relativity could not be accommodated in it,* but by then it was impossible for him to admit his earlier error. While dismissing relativity altogether, *Aryan Physics also refuted quantum physics on the grounds that all unified field theories continued to view space-time in Einstein's terms. Since Hermetic science accepts other dimensions beyond our own dimensions beyond our own continuum, presumably this, together with the assertion that the sub-atomic world has no independent structure, was what Aryan Physics considered was lacking in quantum theory.*[45]

In short, Aryan physics is a clumsy attempt to develop the same thing that Kozyrev developed, namely, a *science of time and "causal mechanics" based on the spiraling motions and distortions in space-time represented by torsion and crystal defects,* for note what "Aryan Physics" is actually saying:

1) The geometry of general relativity is inadequate to the unification of gravity and electromagnetism;
2) when considering certain effects, it would appear there is a sub-structure to quantum mechanics not yet known, and deserving of investigation;
3) higher-dimensional theories are acceptable since the indeterminancy of quantum mechanics could only be overcome by means of such higher-dimensional theories.

As I have already argued in my book *Secrets of the Unified Field,* the work of Gabriel Kron, crucial to any practical applications of unified field theories, no doubt came to the attention of the Nazis and impelled them to the development of both the physics and technologies, not of *Einstein's* unified field theory, but of Theodore *Kaluza's* unified field theory. Moreover, this theory, with its *four* dimensions of space and one of time, was a fully fledged hyper-dimensional theory untainted by having been originated from a Jewish

---

45 Brooks, *Hitler's Terror Weapons,* p. 28, emphasis added.

physicist. Additionally, Kaluza remained in Germany at various teaching posts throughout the Nazi era.

If this was indeed what was actually going on behind the scenes in Nazi Germany, as it has every appearance by now of being, then the German physicists could not have avoided coming to the same conclusion that Kozyrev came to, and with the same result of having *their* work highly classified as was Kozyrev's much later: time was *not a scalar,* but an active *force.*

That force was torsion.

And it was evident in the SS via alchemical influences of Wiligut, the scientific milieu of "Aryan Physics," the various unified field theory papers of the 1920s and 1930s.

And it could be engineered.

2. Two Types of Electrical Potentials and Ultra-High Speed Mechanical Rotation

Yet another area of speculative difference between Witkowski's interpretation of the Bell and my own is that Witkowski maintains that the Bell's electrical operation was continuous and relied upon normal alternating current. My own speculations, developed in *Secrets of the Unified Field*, were that the Bell's nickname, "the Beehive," and its characteristic hissing buzzing sound was perhaps indicative of the rapid opening and closing of a very high voltage *direct* current switch, delivering several thousands of high-voltage DC pulses in the device. Furthermore, the close proximity of a power plant to Bell's test installations may have been further testimony to this possibility, delivering enormous amounts of power that might not have been capable with immense amounts of batteries.

But what would the purpose of the utilization of two types of electrical potentials be?

To answer this question, we must again return to Witkowski's very credible rationalization of the Bell's scientific principles:

> Yes, plasma sometimes creates a kind of vortex, but this is usually a side effect. Nobody yet, nobody after the war — has built a "plasma focus" device *chiefly* for the fast spinning of heavy ions…the internal construction of every plasma is purely static. The conception of rotating or counterrotating cylinders remains unknown. Nobody has struck upon the idea of doing this![46]
>
> ….
>
> I imagined a large, metal drum, in which a small amount of mercury

---

46 Witkowski, *The Truth About the Wunderwaffe*, p. 250

was present. *The drum would then be accelerated to a speed of say tens of thousands of revolutions per minute. Under the influence of the centrifugal force the mercury, as a liquid, would cover the walls of the drum creating a thin layer. After achieving the target speed a high voltage electrical discharge would be created between the circumference of the drum (the mercury layer) — and its axis — the core. Theoretically this would accelerate the ions of mercury towards the core, with a speed of many kilometers per second. But since the mercury would already possess a certain torque, in due measure of approaching the core its angular velocity would increase ... thus developing an increase in rotational speed. In the case of the drum with mercury this would lead to an overlapping of the two speeds — created by a preservation of the torque and a result of the flow of electric current. From my approximate calculations it followed that by this means it would be possible to achieve a speed of the ultimate "compressed" vortex of the order of even hundreds of thousands of revolutions per second.*[47]

So far, so good, but as I noted in my book *Secrets of the Unified Field*, this rationalizes the *rotation* but not *the counter-rotation* of the drums.[48]

If such effects were achieved with counter-rotating drums nested one inside the other, then this would have created a kind of sheer effect within the device itself, a kind of "temporal neutral zone" between the two counter-rotating drums and the stationary core of the device. Thus, the whole device would not have been within this zone, a rather self-defeating exercise if the goal were both the manipulation of space-time and the creation of a field propulsion technology.

Hence I argued that the possible configuration of the drums was one on top of the other. Here the effect of the discharge is different, while the principles underlying Witkowski's analysis remain the same, for the "neutral zone" is created along *two* axes: between the drums and the stationary core, parallel to the rotational axis of the drums, or "vertically" so to speak, and a *horizontal* zone between the drums, extending beyond the device in the plane perpendicular to the rotational axis. Again, the analogy of wringing the soda pop can comes to mind, for the area of maximum stress will be precisely on the rotational plane perpendicular to the axis of rotation.

One may extend the speculative analysis further, in a much more speculative direction. With a high-viscosity, low-stability metal such as mercury, the high spin of the drums will have two possible effects: first, it will squish the nuclei of atoms — or better, the mercuric ions and plasma — into *one of the nuclear deformities of high-spin-state atoms*, similar to those discussed in the series of papers referenced by David Hudson. This will in turn induce the "new

---

47 Witkowski, *The Truth About the Wunderwaffe*, p. 251. Emphasis added.
48 Farrell, *Secrets of the Unified Field*, pp. 271–280.

radioactivities" and models of fission that recent physics papers — decades *after* the war! — have observed and recounted. Secondly, this high-spin-state of an ionized plasma will in turn produce an electrical charge according to the spin orientation of the nuclei. Note now what happens to the relationship of the plasmas and compounds being spun in the two drums: a slight charge differential is built up between the two, with the intervening "neutral zone" between them![49] If the top drum is spun to create a positive charge, and the bottom one a negative one, the difference will possibly produce a levitation effect, exactly as witnesses described seeing in the Bell's test environs, when they described seeing pale blue barrel-like objects floating in the sky.

With sudden and repeated discharges of high-voltage DC current from the drums to the stationary core, all other principles remain more or less as Witkowski described, with one additional possible effect. Such discharges would amount in effect to a "sudden stoppage" of the rotation of the ions, releasing enormous amounts of energy as, for a brief moment, they lost their angular momentum. This energy would be released most likely in the form of very high frequency radiation: X-rays and gamma rays, in a fashion similar to Hudson's observed gamma bursts from *his* high-spin-state platinum group metals, in a kind of "torsion bomb."

In other words, the principle rationalization in evidence once again is that of *maximizing the sheer effect of torsion in three dimensions — and not the one dimension implied by Witkowski's analysis — by any practical means available with 1940s technology.* With all of the massive and repeated "short-circuiting" that such a system deliberately designed into it, it is little wonder that the Bell gave off the characteristic buzzing and hissing sound when in operation that earned it its nickname of the Beehive.

If we now assume that the central stationary core was the one utilizing an alternating and a direct current on the drums, with the central core given a positive and the drums a negative charge relative to each other, even more sheer effect is created.

Of course, the electrical engineering of all this would have been complicated, given the charge differentials, two types of electrical potentials, and the

---

49 Let it be noted that this neutral zone would be similar to the Bloch wall in an ordinary magnet, where the magnetic polarities cancel out, hence the phrase "magnetic fields separation" in reference to the Bell. Compare this principle to the so-called Schappeller device unearthed by researcher Henry Stevens, *Hitler's Flying Saucers* (Kempton, Illinois: Adventures Unlimited Press), pp. 177–179. See also my discussion in *Reich of the Black Sun*, pp. 228–231.

The principle being utilized is this the creation of *three* electrical charge differentials between 1) the two counter-rotating drums themselves relative to each other in the vertical axis of rotation, as between positive and negative charge, producing a horizontal "neutral zone," and 2) between the two counter-rotating drums as one system, and the central stationary core, as another system. If this rationalization be true, then it is the closest anyone has come to conceiving of an electrical "tri-pole," and is an indicator that the Nazis were rationalizing their "electro-gravity" to an extraordinary degree.

high voltages involved, and would have required someone with an expertise in such things, and with handling tricky high-voltage direct current and its accurate measurement and control.

Someone like Dr. Kurt Debus, the Bell's powerplant designer, and subsequent NASA Apollo flight director.

a. A Curious Parallel: the Varo Annotated Edition

The idea of a sudden high voltage direct current pulse of a magnetic vortex — one of the principles possibly in operation in the Bell to maximize torsional sheer — is one that is found in another very unusual place: in the Varo Annotated edition of early UFOlogist Dr. Morris K. Jessup's *The Case for the UFOs*, the annotated edition of the book being the ultimate source for the origin of the Philadelphia Experiment. While an extensive review of that book and its relationship to the physics explored here in connection with the Nazi Bell is not possible,[50] there is one reference that the anonymous annotators make in direct regard to such concepts, though little hint is given whence they — clearly not educated as scientists — could have gained it.

> What happens when a bolt of lightning hits at a point where there is a "node" such as a "swirl" in the magnetic sea or where a mag. [sic, et passim] "dead spot" caused by the neutralization of mag. Sea contra gravity especially, what, when the node & bolt both act over bronze inlay.[51]

As I remarked there, "interestingly, this annotation seems to hint at some sort of effect achieved by pulsing the center of a magnetic vortex with extremely high-voltage electrostatics....While it is very unlikely that the annotators of the Varo Edition, whoever they were, knew of the Nazi Bell project, it is an intriguing comment for them to make."[52]

b. Another Curious Parallel: Lyne and Tesla

According to researcher William Lyne, similar ideas — using two types of electrical potential in charge differentials — were also at the heart of Nikola Tesla's idea for an electrical airplane:

---

50 See my *Secrets of the Unified Field,* pp. 45–85.
51 Varo Annotated Edition of Jessup's *The Case for the UFOs,* cited in *Secrets of the Unified Field,* pp. 281–282, cited here without the annotators' peculiar capitalizations, italicizations, underlinings, and other emphases.
52 Farrell, op. cit., p. 282.

Lyne, summarizing Tesla's views on the aether and electrical circuits as *open* systems, notes that electrical motive force is not due merely to "varying currents" but to "rarefaction and compression of the ether" brought about by rapidly varying and different *kinds* of currents.[53] The implication was that, while as early as 1884 Tesla had done calculations for his flying machine when he first immigrated to the United States,[54] he later claimed that his "dirigible torpedo" could achieve speeds of 300 miles per second, or an incredible 1,080,000 miles per hour![55]

The idea of the compression and rarefaction of the aether via electrical stress caused by two types of electrical potential, coupled with the idea of the *structure* of matter being nothing but vorticular motion in that aether, makes Tesla's views far ahead of their time.[56]

This is strange company indeed — the Varo annotators, Nikola Tesla, and the Nazis — but nonetheless, it is the same principle that unites them all: utilization of two types of electrical potential and charge differential at high voltages. And notably, Tesla's utilization of it was allegedly for the express purpose of an "electrical" airplane.

3. Cryogenic Cooling and Superconductors

A third principle in evidence in the Bell is that of its cryogenic cooling, either by liquid nitrogen or oxygen. This property, as is well known, is a component of superconductors, for the extremely low temperatures lower electrical resistance to the point that electrical current circulates "freely" and for all intents and purposes "forever." Another property of superconductors recently observed is also their tendency — as Hudson observed — to lose weight. This is another indicator of how thoroughly the Nazis had scientifically rationalized the project, and is also another indication that the principle in evidence again is to maximize a torsion sheer effect when the whirling, super-cooled Xerum 525 was electrically pulsed.

a. Professor Nagaoka's Observation

But there is another fact that should be recalled.

In our previous examination of the "alchemical" experiments of transmutations of minute quantities mercury into gold that were part of the science

---

53   Ibid., p. 71.
54   John J. O'Neill, *Prodigal Genius,* p. 66.
55   Lyne, *Occult Ether Physics,* p. 72.
56   Farrell, *The SS Brotherhood of the Bell,* pp. 98–99, citing Willian Lyne, *Occult Aether Physics,* pp. 71–72, and John J. O'Neill, *Prodigal Genius,* p. 66. For a more extended discussion of the development of this principle, see my *SS Brotherhood of the Bell,* pp. 97–100.

of the 1920s, it will be remembered that Japanese physicist Nagaoka observed that "In order to be sure of transmutation, repeated purification of (mercury) by distilling in vacuum *at temperatures below 200° essential.*"[57] Since the paper regarding his experiment was published in the prestigious German scientific journal *Naturwissenschaften* the reference could hardly have been lost on the Nazi scientists, especially since Prof. Dr. Walther Gerlach had himself written an article concerning this "alchemy," since his specialty was spin, gravity, and magnetic resonance, and even more especially since he was the Bell project's scientific head!

b. A Brief Note on The Bell and Nazi Centrifuge Isotope Separation Technologies

Some have proposed that the Bell was actually nothing other than an advanced form of the Nazi centrifuge isotope separation technologies. While not subscribing to this view as the Bell's ultimate purpose or design, it does raise interesting questions. What exactly happened to its mysterious fuel, Serum 525, while being spun and electrically pulsed inside the Bell? Given all that has been learned about high-spin states and the actual construction and possible internal arrangement of the Bell's counter-rotating cylinders, it is just remotely feasible that some of its most classified results might indeed have been unusual transmutations of the Serum's constituent elements and their radioactive properties.

*4. The Operative Principle in Evidence Thus Far:
Maximizing Torsion Sheer Effects in the Medium*

So, putting all of this together, what do we have?
We have:

1) The super-cooling of the device, to greatly lower electrical resistance and increase the spin state of the Serum being rotated;
2) The mechanical high spin itself;
3) The counter-rotation of the substance in two drums, possibly arranged one on top of the other;
4) The possible rationalization of the device in a modified form of Kaluza's unified field theory, one which incorporated torsion in a cylindrical geometry;

---

57 Robert A. Nelson, "Adept Alchemy," www.levity.com/alchemy/nelson2_html, pp. 1–2, citing Nagaoka, H., "Transmutation of Hg into Au," *Naturwissenschaften,* 13:682–684 (1925); *Naturwissenschaften,* 14: 85 (1926); *Nature* 114 (August 9, 1924), 197; *Nature* 117 (#2952, May 29, 1926), pp. 758–760.

5) The possible use of two types of electrical potential — alternating and direct current — in the device under conditions of high-voltage stress of the Serum.

Throughout all of these things, one principle stands out as the probable basis for all these things, and that is that every conceivable effect and artifice was employed in order to maximize a torsion sheer effect in the physical medium of space-time itself.

C. The Alchemical Serum: The Provisional Role of "IRR Xerum 525" in Light of the Previous Analysis

With this principle in hand, it is an easy and short logical deduction to the conclusion that the mysterious compound, the "fuel" of the Bell, was as essential a component in the actual operation of the device and the achievement of the sought-after effects as all the others, namely, that the substance had been specifically designed and confected *to maximize and possibly also to* **measure** *this torsion sheer effect at one and the same time.*

1. The Thorium and Radium Mysteries

In order to argue this case, one must return to a little-noticed comment of Igor Witkowski in his magisterial *The Truth About the Wunderwaffe* concerning the actual chemical composition of the compound.

> Before each trial some kind of ceramic, oblong container was placed in the core (it was defined as a "vacuum flask"...), surrounded by a layer of lead approx. 3 cm thick. It was approx. 1–1.5 m long and filled with a strange, metallic substance, with a violet-gold hue and *preserving at room temperature the consistency of "slightly coagulated jelly."* From the produced information it followed that *this substance was code-named "IRR XERUM-525" or "IRR SERUM-525" and contained among other constituents the thorium oxide and beryllium oxide (beryllia). The name "Xeron" also appeared in the documentation. It was some kind of amalgam of mercury, probably containing various heavy isotopes.*
>
> *Mercury, this time already in pure form, was also present inside the spinning cylinders. Before the start of each experiment, and perhaps also for its duration, the mercury was intensively cooled.* Since *information appeared about the use of large quantities of liquid gas — nitrogen and oxygen, it appeared that it was precisely these that were the cooling medium. The entire device, i.e., the cylinders and core was covered with the aforementioned ceramic housing, of a bell-like shape —*

a cylinder rounded at the top crowned with some kind of hook, or fastening. The entire device was about 1.5 m in diameter and about 2.5 m high...[58].

With respect to beryllium oxide and *thorium* oxide, it is worth recalling that the Nazi obsession with the elements was something of a mystery to the Allied "Alsos" intelligence teams entering the Reich and scouring it for every trace of evidence concerning Nazi nuclear projects. Indeed, this obsession with thorium (and radium) is one of the enduring mysteries of the war, as Witkowski relates: "...the U.S. Alsos mission was unable to explain the high role of thorium in the German research."[59]

But why thorium and radium amalgams with mercury?

a. One Explanation: Thorium Isomers

The answer lies, once again, with the particular form of high-spin states of nuclei known as isomers. As was seen in the examination of Hudson's anomalous platinum group metals in part two, such high-spin superdeformed nuclei have a variety of odd, asymmetrical shapes. Interestingly enough, elements in the radium to thorium range take what are known as "octupole" shapes, *and have low excitation thresholds to reach their isomeric, high-spin, high-energy state.* [60]

Even more significant is a recent discovery emanating from Lawrence Livermore Laboratory. Researcher Anne Stark, in an article entitled "Researchers Move Closer to Switching Nuclear Isomer Decay On and Off," mentioned some unusual properties of a thorium isomer. It is best to cite her actual words, in order to allow the full significance of the discovery to be exhibited in the light of the previous chapters:

> Researchers at Livermore studied an isomer of Thorium-229. *This isomer is unique in that its excitation energy is near optical energies, implying that one day scientists may be able to transition Th229 nuclei between the ground and isomeric states using a table-top laser.*
>
> ...
>
> For years, researchers have been fascinated with this isomer because it could lead to new science and technology breakthroughs. Among them

---

58   Igor Witkowski, *The Truth About the Wunderwaffe*, pp. 232–233, emphasis mine, cited in my *SS Brotherhood of the Bell*, pp. 172–173.

59   Witkowski, "Supplement 2, to be added at the end of the chapter on nuclear weapons, page 219 of the English edition) for the forthcoming German edition of *The Truth About the Wunderwaffe*, personal communication to the author.

60   See Renée Lucas, "Nuclear Shapes," *Europhysics News*, (2001) Vol. 31 No. 7, p. 5, www.europhysicsnews.com/full/07/article1/article1.html.

are: a quantum many-body study; *a clock with unparalleled precision… a superb qubit (a quantum bit) for quantum computing; testing the effects of the chemical environment on nuclear decay rates. Isomers also may serve as a battery for storing large amounts of energy.*[61]

Could the Nazis — if they had developed not only solid state optical lasing cavity lasers such as a ruby laser, but rudimentary tunable gas lasers for isotope separation — have noticed this property of thorium-229's unusual isomer? The evidence arguing in favor of the possible existence of a rudimentary version of such a laser isotope separation technology inside the Third Reich's black projects has already been presented. This fact, coupled with the unusually high importance that thorium had in the Nazi nuclear projects, would seem to argue in favor of the answer to that question being a tentative "yes," for the presence of thorium oxide within the recipe of the "Nazi Serum" is best rationalized by the Nazis having discovered its peculiarly low excitation states.

Moreover, its isomeric, high-spin state, in an amalgam with mercury, would also be another example of the Serum's deliberate confection as a substance to maximize the torsion sheer effects evident in the other operating and design parameters of the Bell. Moreover, it is conceivable, given Serum 525's reddish-violet color and high density, that it was perhaps a precursor to the whole "Red Mercury" story and perhaps in its own right a powerful "ballotechnic explosive."

But there is another clue in the article just cited, and that is its use not only as a maximizer of torsion, but as a *measurer* of it.

b. Another Highly Speculative Explanation: Super-dense Matter

But there is another highly speculative explanation, one that deserves mention in spite of its extremely low probability. Here again, one is confronted with the supreme mystery of just how radical the Nazi quest for superweapons really was.

The mystery begins with something I discovered when writing my book on the Nazi atom bomb project, *Reich of the Black Sun*. The mystery was found in a Japanese communiqué from the Imperial military attaché to the Japanese Embassy in Stockholm to Tokyo, a communiqué intercepted and decoded by the Allies in 1944 just prior to the beginning of the Battle

---

61 Anne Stark, "Researchers Move Closer to Switching Nuclear Isomer Decay On and Off," Department of Energy, Lawrence Livermore National Laboratory, April 6, 2007, p. 1, emphasis added, www.eurekalert.org/pub_releases/2007-04/dlnl-rmc040507.php.

of the Bulge (Dec. 16, 1944). The communiqué outlined what Japanese intelligence had learned about German atom bomb research, research that the communiqué leaves no doubt is in a high state of development, and not the eviscerated lackadaisical effort that the postwar Allied Legend of Nazi nuclear incompetence would have one believe.

But then, toward the end of the communiqué, a stunning passage occurs, indicating that the Nazis were after more than just mere atom bombs:

> Naturally, there have been plenty of examples even before this of successful attempts at smashing individual atoms. However,
> Part 5.
> as far as the demonstration of any practical results is concerned, they seem not to have been able to split large numbers of atoms in a single group. That is, they require for the splitting of each single atom a force that will disintegrate the electron orbit.
> *On the other hand, the stuff that the Germans are using has, apparently, a very much greater specific gravity than anything heretofore used. In this connection, allusions have been made to SIRIUS and stars of the "White Dwarf" group. (Their specific gravity is (?6>) 1 thousand, and the weight of one cubic inch is 1 ton.)* [62]

While I did not comment at length on this passage in *Reich of the Black Sun*, it is prudent to do so here.

Note what the Japanese communiqué actually *says:*

(1) The Germans were actually *using* some sort of material with a great "specific gravity." Given that the Japanese military attaché probably did not have more than a passing knowledge of physics, he is using the most accurate term he could think of to describe the substance. But an important, though radical, clue is afforded in his next statement:

(2) That the material has some connection to white dwarf stars and to *super-dense matter.*

So, putting it together, what the communiqué is saying is that the Germans were trying to come up with a super-explosive, not via the "conventional" routes of nuclear fission or thermonuclear fusion, but via *super-dense matter.*

The idea for a super-dense explosive as the fuel for a bomb was certainly not foreign to the Nazi way of rationalizing physics, for as I noted in *Reich of the Black Sun,*

---

[62] Farrell, *Reich of the Black Sun* (Kempton, Illinois: Adventures Unlimited Press, 2004), p. 45, emphasis added, citing Inter 12 Dec 44 (1.2) Japanese; rec'd 12 Dec 44; Trans Dec 44 (3020-B).

Was first patented prior to World War II in Austria, and a modification of the idea was patented in Germany in 1943.[63] Its inventor, Dr. Karl Nowak, explained the reason for his invention as being to create a superbomb without the radioactive fallout effects that were evident from atomic and thermonuclear explosions![64] In other words, the Nazis were already looking *past* the thermonuclear age toward the creation of second and third generation weapons systems that would give the same offensive and strategic "punch" but without the side effects! In theory, the bomb is workable, but was way beyond the technological capabilities of Germany, or any other power, in that time period.

These ideas, of course, recall the alleged "ballotechnic" properties of Red Mercury. But here there is an important difference, for the Japanese communiqué makes it clear that, whatever the material's density was that the Germans were using, it was of a density of matter not normally encountered on Earth. Granted that it is highly unlikely if not downright implausible that the Germans laid their hands on super-dense "white dwarf star" matter during the war much less a material that would weigh a ton per cubic inch, the Japanese attaché is nonetheless very clear that it is of a density *far greater than any then known or discovered element*. So, if they did in fact possess some sort of super-dense matter, *what was it?*

The answer, once again, lies with thorium and with some highly unusual claims for it that a group of scientists has only recently made.

On April 29, 2008, Fox News reported that scientists may have found traces of the super-heavy element 122 (called Unbibium, with the chemical symbol Ubb), and moreover, the element was *extremely stable*, unlike most super-heavy elements which decay in a matter of a few parts of a second, and moreover, that they found traces of this element occurring *naturally, and in thorium*.[65] Of course, almost immediately "mainstream science" stepped in and denounced the results, calling into question the scientists' methods, alleging either faulty equipment, faulty sampling, faulty methodology, and/or erroneous conclusions.[66]

But the authors of the scientific paper making the claim refuse to back off from their assertions and findings, notwithstanding the rejection of their

---

63 German patent 906.847, March 16, 1943, cited in Thomas Mayer and Edgar Mehner, *Hitler und die "Bombe"*, p. 159.

64 Ibid.

65 "Possible New Element Could Rewrite Textbooks," *Fox News*, Tuesday, April 29, 2008, www.foxnews.com/story/0,2933,35298,00.html.

66 See "Addressing Marinov's Element 122 Claim," www.chemistry-blog.com/2008/04/29/adressing-marinovs-element-122-claim/, and also "Heaviest element claim criticized," 2 May 2008, www.rsc.org/chemistryworld/News/2008, May/02050802.asp

paper by such prestigious scientific journals as *Nature*. Indeed, the scientists making it form a broad international team of eight people, representing Israel, Sweden, France, and two Germans (of course), and two Americans. The paper itself is entitled "Evidence for a long-lived superheavy nucleus with atomic mass number A=292 and atomic number Z ≅ in natural Th." In other words, these scientists claim to have found very minute amounts of element 122, Unbibium, occurring in natural thorium!

It is what the abstract and the paper itself says, however, that is even more breathtaking:

> Evidence for the existence of a superheavy nucleus with atomic mass number A=292 and abundance (1-10) x 10-12 relative to ^{232}Th has been found in a study of natural Th using inductively coupled plasma-sector field mass spectrometry. The measured mass matches the predictions for the mass of an isotope with atomic number Z=122 or a nearby element. Its estimated half-life of t$_{1/2}$ >108 (years) suggests that a long-lived isomeric state exists in this isotope. The possibility that it might belong to a new class of long-lived high-spin super- and hyperdeformed isomeric states is discussed.
>
> The question "how heavy can a nucleus be" is a fundamental problem in nuclear physics. Experimentally, elements up to Z=118 have been produced synthetically by heavy-ion reactions, with the half-lives of the Z=106 to 118 nuclei ranging from a few minutes to about a millisecond. However, in a recent study of natural (thorium) substances, *long-lived isomeric states with estimated half-lives (of 100,000,000 years), 16 to 22 orders of magnitude longer than their corresponding ground states have been observed in the neutron-deficient 211,213,217,218 (thorium nuclei).*[67]

Note carefully what the authors of this paper are asserting:

(1) A super-dense element — Unbibium or element 122 — may exist in natural thorium in extremely minute quantities (from 1 part in a quadrillion to 1 part in 100 trillion);

(2) The longevity of that element, the feature that makes it *stable* over millions of years, is its isomeric state, that is, its stability is gained by its *shape* and high-speed *rotation; and,*

(3) The element was isolated by more or less standard mass spectrography methods.

---

67 A. Marinov, I. Rodushkin, D. Kolb, A Paper, Y. Kashiv, R. Brandt, R.V. Gentry & H.W. Miller, "Evidence for a long-lived superheavy nucleus with atomic mass number A=292 and atomic number Z≅122 in natural Th," unpublished internet paper, p. 1, emphasis added.

And note the implication:

> (4) If the element is stable for millions of years and separable by "standard" methods of mass spectrography, then theoretically one might accumulate quantities of the element by stockpiling vast amounts of thorium and separating the unbibium from it.

So, what if mainstream science is wrong, and the authors of the paper are right? Moreover, what if the Nazis, utilizing some prototypical technology of laser isotope separation and mass spectrography, the latter technology which they certainly possessed, had also noticed something very similar? After all, their own scientist and one of Gerlach's most brilliant students, Dr. Ing. Ott Christian Hilgenberg, had predicted the occurrence of superheavy elements.[68]

However one chooses to view the Nazi obsession with stockpiling massive amounts of thorium, it is clear that their obsession with the element cannot be rationalized on the basis of standard nuclear physics. Indeed, as was already seen, this was what most mystified the Allied intelligence and scientific teams entering Nazi Germany and searching for clues into its nuclear program. In fact, *there is no good way to rationalize it other than one of the two previously mentioned alternatives,* and notably, *both* of these alternatives involve the high rotational state of isomers. And let it also be noted that the extremely high-spin state of isomers is also the best candidate to explain the cryptic references to Nazi *use* of super-dense matter in their experiments.

There is likewise another possibility that Unbibium also opens up. As was mentioned in part one of this work, one attribute claimed for the Philosophers' Stone—and indeed for the ancient Sumerian "Tablets of Destinies" that formed such a technological centerpiece of an ancient "Cosmic War"[69]—was precisely its indestructibility. A super-dense matter on the order of unbibium or any other isomeric stable state matter would also qualify for "indestructibility."

2. Hoagland's Measurement Model: Torsion Effects, Radioactive Decay Rates and Shielding, and a Proposal

During the course of our email discussions on the Farm Hall Transcripts and laser isotope separation, Mr. Hoagland also mentioned a possible rationalization of the Bell's Xerum 525. On this view, he argued that it was less a *"fuel"* for the device, and more of an exact *measuring device* for the amount

---

68 See my *Reich of the Black Sun,* p. 199, and my *SS Brotherhood of the Bell,* chapter six.
69 See my book, *The Cosmic War: Interplanetary Warfare, Modern Physics, and Ancient Texts,* chapters seven through nine.

of torsion sheer and temporal displacements the device actually produced. Again, this view is fully accounted for by the Lawrence Livermore discoveries outlined above.

Under the normal conditions of radioactive or isomeric decay back into an element's ordinary ground state, such decays are precisely measurable in terms of time. Thus, if extreme torsion and temporal displacements occur, the best tool for measuring them would be via changes in rates of radioactive decay. Under the conditions of the extreme torsion sheer effect being argued as the basic principle of the Bell's operation, however, such decay rates would dramatically change, and under certain conditions, radioactive *shielding* properties could even result.

Consequently, there is much to commend Mr. Hoagland's hypothesis. Indeed, if one follows Witkowski's analysis carefully, this "Serum" was placed within the "thermos bottle" at the Bell's stationary core and axis of rotation, *an eminently logical position for a "measuring compound" to occupy.* In *that* position, its high-spin properties are *not* being employed to further maximize the torsion sheer effect operative in the device.

However, it is this very property that also argues that the Serum was also the *fuel* of the Bell, with its high-spin mechanical rotation in conjunction with the possible presence of thorium-229 isomer, the principle of maximizing torsion sheer would again be in evidence. Thus, the *best* way for the substance to function as a measurer is if it were present both in the core *and* in the rotating drums, where, under the extreme conditions of Kozyrev-like "alchemical" stress that it was subjected to in the Bell, any number of strange transmutations and effects may have been observed by the Nazi scientists, among them possibly bursts of gamma ray radiation and asymmetrical nuclear fission transmutations associated with high-spin state elements.

3. A Prototypical Technology and Three Potential Applications

Viewed in all this context, it is *together* that the Bell *and* its "fuel," the alchemical Nazi Serum 525, represent a prototypical technology of a "unified" physics, a technology that like the physics itself was a "unified" technology capable of a variety of potential applications: first, for "free energy" drawn from the medium of space-time itself via a hyper-dimensional "torsion gate" that the two constituted.

But it is equally the case, if the above speculations and arguments be true, that the two technologies of the Bell and Serum 525 also represent the first prototypical steps toward a field propulsion technology, via the approach of maximizing a torsion sheer effect to manipulate the fabric of space-time. As Witkowski so concisely and brilliantly observes, such a new *physics* was not only necessary for any *really* practical manned exploration of deep space, but it was also signally important for a much more immediate reason: only "a propulsion based on a new physics...was the only thing that could change the course of the war!"[70] The need for a new, hyper-dimensional physics had been foreseen by the Nazis, not only as the only sound basis for deep space manned missions, but ultimately, for their plans for global domination as well.

But there is a final potential application of the Bell and its Serum 525 "fuel," and that is as not only a weapons *platform*, but as a *weapon*, for a technology capable of engineering the fabric of space-time for energy independence and field propulsion is also a technology with a horrific weaponization potential, one that would dwarf the destructive power of the largest thermonuclear bombs.[71] Indeed, if either of the speculations concerning thorium above were at work in the Nazi obsession with the element, then its role as a possible constituent — given its high-spin isomeric state — in the Serum 525 is virtually assured, since the Bell itself was a device reliant upon high-speed rotation.

Indeed, if the enigmatic Serum 525 was a compound of mercury and thorium-229 isomer oxide, then its alchemical reddish-purple color, its heavy goo-like liquidity under ordinary room temperature, would make it the ideal candidate for the type of "ballotechnic" explosive in its own right later claimed for the Soviet "Red Mercury" scare of the 1990s. And given that the Soviets made off with their own fair share of the Nazi technological spoils, the possibility that it may even have formed the basis for the subsequent Red Mercury legend and possible Soviet investigations and synthesis of similar compounds cannot be discounted. Moreover, as will be seen in the next chapter, there is a tenuous circumstantial case that one of postwar Germany's most brilliant physicists may have been involved in wartime research of precisely that nature. Serum 525 is testament to the fact that the alchemical Reich may very well have succeeded in all these aims for the Bell project.

D. The Mystery of Two More Scientists

But there is more to the mysteries of the strange compound than just its chemical composition and uses inside the Bell, and this concerns the case of

---

70 Igor Witkowski, "The Third Reich — A Key to Secret Technology," article in preparation for the German edition of *Nexus* magazine, p. 4, personal communication to the author. Given the extraordinary degree to which the Bell project was scientifically rationalized down to the smallest detail, Witkowski's words regarding a kind of two-track space program hypothesis under way in the Third Reich gain additional importance, given the recent revelations contained in Mr. Richard C. Hoagland's and Mr. Michael Bara's recent book on the same subject inside of NASA, *Dark Mission*. Witkowski states "It seems some kind of alternative program had existed, being carried out for a long time, and quite a serious one at that." (*The Truth About the Wunderwaffe*, p. 259).

71 See my *Secrets of the Unified Field*, pp. 282–286.

two more little-known, but very significant scientists. They are the medical doctor Hubertus Strughold, subsequently an Operation Paperclip scientist who ended up in an influential position within NASA's space medicine community, and the little-known but highly important Dr. Pascual Jordan, one of the key developers of quantum mechanics. Both were strong supporters of the Nazi regime, and the work of both in the secret projects of the Third Reich are virtually unknown to this day. And once again, it is thanks to the penetrating and brilliant research and insights of Igor Witkowski that the connection of either man to the Bell project has even been surmised or suggested.

1. *The Strange Case of Dr. Hubertus Strughold*

Within the rats' run of secret underground SS installations in Lower Silesia around the test environs of the Bell, there was one in particular that appeared to be connected to the medical arm of the Bell project, that arm which sought to investigate its effects and means of limiting them on humans. There was, Witkowski writes, "an experimental test station to test the control system of the new flying craft and its influence on humans – mentally disabled children, directed by Hubertus Strughold *(all in or near Fürstenstein bei Waldenburg)*..."[72]

Castle Fürstenstein was one of the centers connected to the Bell project, and indeed, there are large underground bunkers located beneath the castle and its environs. Interestingly enough, Witkowski also referred to an interview that Dr. Strughold gave to the Polish author Kakolewski in the 1960s, in which he stated that he had indeed carried out experiments during the first part of 1945 "in Fürstenstein" and with "a simulator of space flight," presumably a centrifuge or pressure chamber.[73] The significance of Strughold's involvement in medical experimentation within one of the installations and centers directly connected with the Bell project will be examined at the end of this chapter.

For now, however, it is important to note that it is in this context that Witkowski summarizes the actual *size* of the Bell project. It was not merely a laboratory project encompassing only one installation, and one device. Rather, if one combines it and all the other super secret classified SS black projects being undertaken by the Kammler Group within and around the *Reichprotektorat* of Bohemia and Moravia and Lower Silesia, then "we will get an undertaking that employed (with the Gross-Rosen [concentration camp] prisoners) well over 60,000 people — in fact, something that seems to be the real equivalent of the American Manhattan Project. From my perspective, however, it's much more important for quite another reason — it may, potentially, enable us

---

[72] Igor Witkowski, "The Third Reich — A Key to Secret Technology," article in preparation for the German edition of *Nexus* magazine, p. 10, personal communication to the author.

[73] Witkowski, email to the author, 15 April 2008.

to understand the technology crucial for the future, a completely different approach to gravity, the 'gateway to the stars.'"[74] And with the presence of Dr. Hubertus Strughold at Fürstenstein Castle, and later within NASA's space medicine program, one has a further clue into the mystery of its strangely influential Nazis.

And let us take note of one implication of Strughold in connection with the Bell: *extreme torsion can alter the normal decay rates of radioactive materials.* Thus, if the Bell was doing so, then one significant technology was opening itself to the Nazis for the possibility of radioactive shielding.[75]

2. The Even Stranger Case of Pascual Jordan

There is, however, an even stranger case of a scientist from Nazi Germany, and that is Dr. Pascual Jordan, who, along with Heisenberg, Schrödinger and a few others, was one of the seminal contributors to the early development of quantum mechanics. And with Jordan, we are truly confronted with a *significant* enigma. Once again, it is Witkowski who ferreted out his possible connection to the Bell project:

> It seems likely that yet another scientist played some role in this project, although his name did not appear in the original interrogation protocols; his unique works could constitute part of the theoretical basis of the experiments. It's about professor Pascual Jordan, one of the most outstanding physicists of the Third Reich (whose career was as enigmatic as Gerlach's!). During the war he worked on a theory which described the phenomenon of "separation of magnetic fields," linking the isolation of magnetic fields with gravitational effects. Shortly after the war it was "perfected" and combined with the works of another scientist — presently it is known as the Jordan-Thiry theory. It is considered as one of the most fundamental achievements of the XXth century physics and Jordan was an almost certain candidate to receive the Nobel prize in 1954. Eventually, however, he was disqualified after his role during the war was revealed... the Jordan-Thiry theory forms the basis to analyze relativistic plasma vortices...[76]

But this does not constitute the real mystery of Jordan's work during the war.

---

[74] Igor Witkowski, "The Third Reich — A Key to Secret Technology," article in preparation for the German edition of *Nexus* magazine, p. 10, personal communication to the author.

[75] Apollo hoaxers take note!

[76] Igor Witkowski, "Supplement 3, to be inserted on page 260 of the English edition," of *The Truth About the Wunderwaffe* for pending German edition, personal communication to the author. Witkowski cites Jordan specifically in reference to "magnetic fields separation" that played such a prominent conceptual role in the Bell project.

Part of the mystery consists of the fact that, in spite of being one of the fundamental founders of quantum mechanics and having authored a number of important papers and even textbooks in the subject, he abandoned the subject prior to the war, and switched to cosmological physics, beginning to elaborate concepts that would eventually lead to his theory that all the fundamental constants of physics were variable over time — shades of Kozyrev!

a. His Mysterious Absence from Wartime Nazi Secret Weapons Research

But the biggest mystery of Jordan was the fact that he was a Nazi, and that, being one of Germany's most brilliant theoretical physicists, he was — so the public legend goes — never a major participant in *any* of the Third Reich's advanced black projects. Joining the Nazi party in May of 1933 he subsequently enlisted in the notorious Brown Shirts, the *Sturmabteilung* or S.A of Ernst Röhm. By 1939 he had enlisted in the Luftwaffe and did some minor work at the Peenemünde rocket development center.

However, Jordan was not merely a "joiner," but a true believer. His "conviction that the 'Bolshevist peril' had to be eradicated drove him into the arms of the Nazis."[77] In spite of having offered the Nazi government a number of proposals for advanced weapons projects, Jordan was ignored, or at least, that is the story:

> This ideological support remained one-sided since (the Nazis) never rewarded him with a leading position in their weapons research program as they did many other cases. During the 30s, after the Nazis took power, he became increasingly isolated even within the German physics community.... This explains why several important contributions by Jordan which were ahead of times went unnoticed.[78]

There are several ways to construe Jordan's conspicuous absence from Nazis secret weapons projects.

The first, and most obvious, is that he was ignored not only for his seminal role in the development of quantum mechanics but for his close association with Jewish physicists during the pre-Nazi era. Such ideas and associations would have made him anathema to the "Aryan Physics' ideology being promoted by Lenard and Stark, and a potential "security risk" to the SS which oversaw the secret weapons projects.

Such an explanation, however, ignores two salient facts: Jordan's brilliance, and the SS Kammler Group's willingness to suspend normal required adherence

---

77 Bert Schroer, "Physicists in Time of War," December 2005, p. 7.
78 Ibid.

to Nazi ideological proscriptions, for the potential reward that "out-of-the-box thinking," of which Jordan was certainly a master, could bring. Simply put, it does not stand to reason that the Nazis, or rather, the SS, would not have made use of him, and probably in some highly classified project.

The second explanation is that Jordan was not made privy to the highest secrets for much the same reason that Heisenberg and other high-profile scientists were not: being such a well-known scientist would have made Jordan a high target priority for capture or assassination by the Allies. To have entrusted sensitive secrets to him would have placed those secrets at risk of compromise in the event he was captured by the Allies and tortured for information.

But again, the explanation fails, not only for the previously cited reasons, but also for the fact Jordan, after joining the Nazi party, appeared only too willing to abandon quantum mechanics and to pursue the cosmological ideas that would eventually lead to his formulation of the idea that the constants of physics were variable over time, a conception quite in keeping with the physics implied by the Bell.

And time variability, as Dr. Kozyrev demonstrated so amply, was an effect of torsion.

Moreover, Dr. Pascual Jordan could read the paper of Kaluza, and do the mathematics of the torsion tensor, as well as anyone of his day.

This leads us inevitably to the third, and I believe the most plausible, explanation, namely, that the public version of his non-involvement with sensitive Nazi secret weapons projects was just that, a public version promoted to deflect attention away from the fact that his theories dovetailed perfectly with the conceptual parameters of the Bell project, and that he was, as Witkowski avers, not only possibly involved with it, but *probably* involved with it. And as will be seen in the next chapter, Jordan surfaces in a most unusual connection in the postwar period, along with, of all people, Dr. Wernher Von Braun!

b. A Connection to Hilgenberg and Gerlach?

A final speculative connection must be mentioned. As noted, Jordan began to shift to cosmology and to develop views that would eventually lead him to conclude that the constants of physics, including that of gravity itself, were variable over time. In terms of planetary physics, then, this implies the view that a planet's gravity changes over time *because its mass changes* as torsional energy enters the planet via the now familiar mechanism of its rotation.

The linkage between changing mass, gravity and such rotating systems had, in fact, also been made by one of Dr. Walther Gerlach's most brilliant and gifted students, Dr. Ing. Ott Christian Hilgenberg, in a series of prewar papers. Central to Hilgenberg's ideas was even the idea that the earth itself was gradually growing and expanding as a result of these factors. And Gerlach himself was, as we now know, not only the Bell's project head, but also a close friend to the Austrian physicist Lense, who had predicted *dragging* effects of (the then theoretical idea of) objects in earth's orbit, effects which resulted from its rotation. With such rich conceptual soil on which to grow projects like the Bell, and with the close-knit nature of the German physics community prior to the war, it stands to reason that Jordan would have been aware of his colleague's thoughts, and they of his.

If Walther Gerlach, ardent supporter of the Nazi regime though he was, was never a member of the Nazi party, and if, with his own close association with "Jewish physics" and Jewish physicists before the Nazi assumption of power — after all, his partner in the famous Stern-Gerlach experiment that earned him his Nobel Prize was Jewish! — and if, finally, his own spirited defense of relativity and quantum mechanics against "Aryan physics" was not enough to disqualify him from heading the Bell project and eventually even the Nazi atom bomb project, then it hardly stands to reason that Jordan, not only a Nazi Party member but a member of the SA and the Luftwaffe, would be excluded from such projects.

E. Conclusions and Speculations

It has been argued that Dr. Hubertus Strughold was possibly a member of the Bell's medical research team. If so, then the presence both of Dr. Hubertus Strughold and of Dr. Kurt Debus within NASA is a strong indicator that a secret and rudimentary anti-gravity or "electro-gravity" technology was possibly in play during the Apollo program. Moreover, the fact that Strughold may have been involved in medical experimentation to limit the Bell's and Xerum 525's toxic and lethal field effects on humans — using luckless concentration camp victims in his experiments — and the fact that these experiments appear to have had some limited success, further buttresses the Two Space Programs hypothesis, an overt one for public consumption, and a covert one based on advanced secret technology, a two-track space program that began during World War II, if not before.

One area of crucial applicability of this speculation is the so-called "Apollo Hoax" hypothesis, namely, that we never went to the Moon at all. Many have erroneously concluded that the program was hoaxed because of the alleged lethality of passing through the Van Allen radiation belts to any humans so weakly shielded from its effects as in the Apollo Command Module. Granting this proposition for the sake of argument, the presence of a torsion- or scalar-based rudimentary anti-gravity technology within NASA, a technology based on that of the Nazis, then one effect of such technology would be to radically alter what for public consumption physics are the unalterable decay rates of radioactivity, and a technology that additionally can shield against it. If the Bell and its "fuel," Serum 525, could indeed create a kind of "space-time bubble" around the device, then it follows that one and the same technology as got us *off* the Moon might also have got us *through* the Van Allen Belts and *to* the Moon.

Hubertus Strughold, like Kurt Debus, is another potential clue that something *else* is going on within NASA and other American programs besides mere rockets, and *that* "something else" has been going on since the end of the war. Their presence also indicates that the Nazis were, to some extent, exercising influence and control over its development, even in postwar America.

For the technological, historical, and political implications of the Bell, of Serum 525, and of their implied physics are as clear now as they would have been to the rocket-ridden Nazis then: if there ever is to be a *practical* human presence and penetration into deep space, then another technology and physics from mere rockets is altogether *necessary*. It is a *sine qua non* for such a presence. This would have been as clear to the Nazis then, as it is to us now, and as we shall see in the next chapter, as it is finally publicly, though quietly, admitted by NASA.

Thus the implication is as disturbing as it is speculative: it would appear that the postwar Two Space Programs hypothesis had its beginning in World War II itself, with the Nazis rockets representing, as they did for the postwar American and Soviet programs, the "public consumption" program, and the exotic black projects in anti-gravity, torsion, and the manipulation of the fabric of space and time representing the black one.

But is there any *further* proof that the Germans would have scientifically rationalized things to this extreme extent during the course of the war?

While there is no *direct* proof, there is a very intriguing case of a little-known German physicist who began thinking along similar lines shortly after the war, and who did unusual work for the Nazis *during* the war, work that no one seems to know anything about, but which cost him his hands, his sight, and his hearing, but that left him with his genius mind…

…and he has only in the last decade come to the attention of NASA…

…or at least, that is what the public record would have us believe…

13

NASA Shows an Interest
(With a Little Help from the Germans)

•••

"A Heim space is a quantized space comprising elemental surfaces with orientation (spin)...A Heim space may comprise several subspaces..."
Walter Dröscher and Jochem Häuser[1]

It is appropriate to end this survey of the alchemical wedding of physics and ancient hermeticism where it began: in the void, but the void not of the transmutative alchemical medium, but of space. Even here, however, there are connections between the two voids, and it took a little-known German theoretical physicist to make them with formal mathematical precision. This physicist was, moreover, one of those genuine intellectual giants, for his ideas were formed entirely — like Einstein — in his head, but unlike Einstein, forged under much more personally limiting circumstances, for he was almost blind and deaf, and had no hands, all the results of injuries he suffered during a mysterious wartime accident doing research for the Nazis.

The German physicist's name was Burkhart Heim.

And his theory, Heim theory, is a fully geometrized and quantized hyper-dimensional unified field theory, a modern-day heir to the line of classical unified field theories going all the way back to the first such hyper-dimensional theory of Theodor Kaluza, himself a German mathematician, in 1921, and

---

[1] Walter Dröscher and Jochem Häuser, "Guidelines for a Space Propulsion Device Based on Heim's Quantum Theory," *40th AIAA/ASME/SAE/ASEE Joint Propulsion Conference and Exhibit* (11–12 July, 2004), p. 6.

heir as well of the many versions of the unified field theory put forward by Albert Einstein in the 1920s and early 1930s.

But as will be seen, with Burkhart Heim's theory one is dealing with a conceptual framework that in its extraordinary depth and predictive power is not only an engineerable and testable theory, but a theory that moreover unites the threads suggested throughout the previous pages. It is a theory with hyper-dimensions, sub-spaces, hyper-spaces, and most important of all, a latticework of a fully quantized *spin-oriented* space-time itself, able to predict masses of fundamental particles from first principles. All in all, it is a towering monument of a towering intellect, a monument and an intellect made even more mysterious because the theory is almost totally neglected outside of Germany, save in certain quarters.

And what of the man and the intellect behind it? The answer to *that* question contains an even greater mystery.

German Theoretical Physicist Burkhart Heim, 1925–2001

A. Heim Theory
1. Getting Acquainted: The Book

I first became acquainted with the name of Burkhart Heim years ago, when reading a book that is now considered to be something of a classic on UFOs, Leonard Cramp's *UFOs and Anti-Gravity: Piece for a Jigsaw.* Cramp's book was unusual in that it was not the typical UFOlogy book, even in its treatment of the possible science of UFOs, for it stuck with analysis based on mainstream, though little-known, physics. And Heim's name figured prominently in the text:

Burkhard Heim, extremely handicapped by war wounds (he is blind, armless and nearly deaf), has evolved a six-dimensional theory containing the general relativity theory, and the quantum theory of fields as special cases. Which means that the formalism of these two theories is derived by neglecting certain members in Heim's field equations. Therefore Heim's theoretical approach has a more universal character than any other presently treated theoretical approach. *In addition it bridges some fundamental contradictions between general relativity and quantum theory, which tends to restrict them to the macro and microcosmic domain. The reader will not be surprised therefore to learn that B. Heim's six-dimensional theory yields results which predict physical phenomena mentioned elsewhere in this book.*[2]

This utterly captivated me, for indeed, if Heim had done what Cramp says he did, and did it in only six dimensions — after all some versions of string theory had up to twenty-six dimensions and were completely *un*testable — then he had indeed found the Philosophers' Stone, the Holy Grail, of modern theoretical physics: the unification of relativity and quantum mechanics. And the clincher that "sold" the theory were the "physical phenomena mentioned elsewhere in this book," namely, the theory predicted a coupling of electromagnetism, and more importantly, of *magnetism itself,* with gravity, with specific anti-gravity implications.

I was "hooked" on the possibility, and read on, wanting to hear more of this little-known German physicist. A short while later into Cramp's book, I found it:

>...(It) becomes obvious that we must look to the very core of gravitation if we are ever to accomplish real space flight. No doubt *it is such considerations as these which led Burkhard Heim to develop his six-dimensional field theory. His results concern in particular "mutual relations between the gravitational force and the matter which generates it."* Heim postulates that since electro-magnetic waves are special cases of material fields, then electro-magnetic [sic et passim] fields must be accompanied by gravitational ones. Heim's theoretical investigations have shown that the *"meso-field"* may exist in two states, *"contrabaric"* and *"dynabaric."*
>
>When *"contrabaric"* it is able to transform electro-magnetic **waves directly** into gravitational ones, it could induce the **acceleration of mass** from a direct conversion of **electro-magnetic waves.** Heim believes the energy required for this phenomenon could be derived directly from nuclear processes.

---

2 Leonard Cramp: *UFOs and Anti-Gravity: Piece for a Jigsaw* (Kempton, Illinois: Adventures Unlimited Press, 1996, reprint of the 1966 edition), p. 94, emphasis added.

Although hitherto Heim focused attention chiefly on the consequences for space flight, it becomes obvious that such an identical conversion process could be employed to generate kinetic energy from electricity without any wasteful intermediate thermodynamic process, that is, practically free from loss.

In the "dynabaric" state, the "intermediary field" is inverted so that electro-magnetic energy is directly liberated from matter, without accompanying heat or other waste. Fantastic prospects would result if engineering techniques were available to harness the dynabaric state.

In addition, it must be stressed, Heim's approach does not conflict with known laws of nature, in fact it strictly agrees with the quantum theory. UFO researchers will note the parallel with electro-magnetic saucer effects here.

Heim's theory goes on to predict mutual interactions between inertia forces and electro-magnetic radiations. Einstein said they were equivalent to gravitational forces....

As an open-minded scientist, A.R. Weyl's following comments will be encouraging to UFO researchers. Writing on Heim's work he said:

"In Heim's theory the member which represents electro-magnetic radiation is related, by way of an 'operator' (that is, an instruction for carrying out definite manipulations of computations in an ordered manner), to a '*space density* **variable** *in time.*' It supposes an inertia force of such a kind that the operator's treatment of the electro-magnetic radiation produces effects upon the inertia, from the radiation.

"If it were possible to realize this 'operator' physically, that is to devise means which actually carry out the formalistic manipulation of the theory, the direct transfer of electro-magnetic waves (for example, light) into mechanical force (gravitational waves) would become reality, subject to some efficiency factor, of course. Also, mass could be directly converted into radiation without the production of heat.

....

"Heim's intermediary field (represented by the spirit which motivates his 'operator') would also show the possibility to neutralize or to reverse gravitational acceleration...at **direct** expense of mass-energy, by way of an electro-magnetic/gravitational conversion which would impose practically no loss. A dynabaric state of the intermediary field should also be capable of producing the direct conversion of mass into electro-magnetic energy, without heat of the formation of waste products.

"If Heim's conclusions from his comprehensive theory should prove realizable, certain fantastic consequences somewhat of the kind usually ascribed to 'flying saucers' (including their immunity from the effects of rapid accelerations) would become attainable.

"It would also appear possible to propel space vehicles from external sources of natural energy, that is, **from the conversion of light or other electro-magnetic radiation.**"[3]

This was a very heady wine to imbibe, for note what Cramp has implied with his remarks:

1) Heim appeared to have developed his theory *precisely as a response to the needs for a genuine human presence in deep space, a presence requiring the development of a wholly new physics paradigm;*
2) Heim's hyper-dimensional theory appeared to predict the gravitational shielding and even anti-gravity effects, which Heim calls by its more technically correct term of "contrabary," effects, moreover, of some states or types of matter, shades of the alchemists and of David Hudson's American "gold"!;
3) Space itself had a *"density"* that was "variable in time," shades of the alchemists and Nikolai Kozyrev!

Beyond these vague generalized assertions, however, Cramp said nothing more about Heim theory.

So I purposed to find out more about the man and his remarkable theory....

...and ran into a stony brick wall of utter silence.

Search as I might and certainly did over the years — and I read of Heim's work over twelve years ago! — I simply could find nothing else out about the man other than that mentioned by Cramp in his book. I even began to doubt that he, or his theory, ever even existed.

2. Getting Acquainted: The Documentary

However, one day many years after reading of Heim in Cramp's book, and quite by chance, I purchased an old used copy of a documentary on UFOs from a local bookstore that also specialized in used movies, videos, and unusual documentaries. The documentary was called *Above Top Secret, Part Two.*[4] As I began watching the documentary it became clear that I was *not* watching the typical UFO featurette, full of wild-eyed hysteria about "coming disclosure" and unsubstantiated tales of abductions and gruesome medical experiments. Certainly the video contained such elements, but these were

---

3 Cramp, *UFOs and Anti-Gravity: Piece for a Jigsaw,* pp. 110–111, italicized and boldface italics emphasis added, boldface emphasis only is in the original. Cramp does not cite a source for Weyl's remarks.

4 *Above Top Secret: Part Two,* Questar Video, Inc (Chicago: 1994).

sandwiched between a series of otherwise very staid interviews with some very unusual people. I would have written the documentary off as more nonsense, had it not been for the presence of one man who made me pull up from my reclining and dozing slouch, and pay *very* close attention.

The man was Baron Jesco Von Puttkamer, a tall, distuinguished-looking silver-haired man in a well-tailored suit, calmly explaining in his heavy German accent what NASA had to do in order to look beyond chemical rockets toward a new physics and new technology. But that is not what caught my attention. What caught my attention was the video's description of him as simply an "advisor" to NASA and to Gene Roddenberry, creator of the famous *Star Trek* and *Star Trek: The Next Generation* science fiction television series.

But I knew what most viewers of the documentary probably did *not* know: Jesco Von Puttkamer was no ordinary German who just happened to work for NASA or as a consultant for Gene Roddenberry. Jesco Von Puttkamer was none other than Adolf Hitler's personal Naval Adjutant to Staff, the *Kriegsmarine's* representative to the *OberKommando der Werhmacht* (O.K.W.) before and during the Second World War! There is even a picture of a very young *Kapitän zur See*[5] Von Puttkamer standing with Reichsmarschal Göring and Adolf Hitler, who is bent over a map, in the map room of the Reich Chancellery in Berlin, planning the invasion of France in John Toland's celebrated biography of Hitler! Von Puttkamer began the war as a Captain in the German navy, but by war's end had been promoted to Admiral.[6]

What was Adolf Hitler's personal naval adjutant to staff doing working for NASA!?!? But before the question had even been formulated in my mind, the documentary then cut away for another interlude of hysteria, before returning to another elderly silver-haired German, talking quite slowly and with measured deliberation, being interpreted by a commentator.

That man — so the caption told me — was Burkhart Heim.

As I sat and listened to Heim's brief remarks, however, it became quite apparent that the documentary's producers either hadn't the faintest clue of the significance of his theories, or that they *did,* and were deliberately going out of their way to avoid talking about them, for Heim himself was merely speaking out against nuclear energy and its environmental consequences and connections to the military-industrial complex (of West Germany, of course)!

The reason why I suspected a deliberate obfuscation of Heim's theory was the presence of Puttkamer himself, Hitler's naval adjutant, for as I was also well aware, the German Navy, beginning during the Weimar era and continuing

---

5 Essentially a Commodore.

6 For other interesting things about Von Puttkamer and his peculiar interest in physics and the esoteric, see my *Reich of the Black Sun: Nazi Secret Weapons and the Cold War Allied Legend,* p. 50, and my *Giza Death Star Deployed,* p. 81.

into the Third Reich itself, had shown an interest in the development of alternative energy technologies such as the Coler Coil,[7] and moreover, a German admiral was somehow involved in the Bell project itself.[8] Nor was the connection of naval services to such alternative technologies limited to Germany, for as I also pointed out in my previous book *Secrets of the Unified Field: The Philadelphia Experiment, the Nazi Bell, and the Discarded Theory*, the U.S. Navy had taken such an interest as well, in the form of the Office of Naval Research and its Project Orbiter-Vanguard committee, which included, among others, Dr. Wernher Von Braun as a possible recipient of the Varo Annotated Edition of Morris Jessup's *The Case for the UFOs*, a book outlining the Philadelphia Experiment![9]

I concluded that the inclusion of Jesco Von Puttkamer, and Heim, and the neglect of any mention of the extraordinary implications of Heim's theory, in the documentary was hardly an accident. It had been a deliberate attempt to point a subtle finger once more to Germany, to *Nazi* Germany, with the subtle message of "look closer, look here."

And that brings us to Heim's extraordinary theory, and to the even greater mystery of what he was doing for the Nazis that cost him his arms, sight, and hearing.

3. The Basics of Heim Theory
a. Motivations of the Theory

Unusually, it was when Heim died in 2001 at the age of seventy-six that his theory first gained notoriety outside of Germany, most notably in the United States, in a series of papers presented at various technical conferences by his associates Walter Dröscher and Jochem Häuser, who had revised and extended his theory along the lines Heim himself was outlining at the time of his death.

As was noted above, Heim himself, according to Leonard Cramp, began to work out the details of his theory precisely in reference to solving the problem of a permanent human presence in deep space. Such a presence, Heim reasoned, could only be established by an entirely new paradigm of physics, with all of its implied revolutionary technologies. In their paper on Heim theory presented to the 40th AIAA/ASME/SAE/ASEE Joint Propulsion Conference and Exhibit in July of 2004 in Fort Lauderdale, Dröscher and Häuser mention this very problem in connection with NASA's "Breakthrough Propulsion Program":

---

7   See my *Reich of the Black Sun,* pp. 226–231.
8   See my *SS Brotherhood of the Bell, NASA's Nazis, JFK, and MAJIC-12,* pp. 144–148, 160–161.
9   See my *Secrets of the Unified Field: The Philadelphia Experiment, The Nazi Bell, and the Discarded Theory,* pp. 291–296.

For effective and efficient lunar space transportation as well as interplanetary or even interstellar space flight a **revolution** in space propulsion technology is needed.

Regarding the requirements of NASA's *Breakthrough Propulsion Program (BPP)* a revolutionary space propulsion system should

- use **no or a very limited amount of fuel**,
- possibility for **superluminal speed,** and
- requirement for a **low energy budget.** This immediately rules out any device flying close to the speed of light, since its mass is going to infinity, according to (Special Relativity). A spacecraft having a mass of 105 kg, flying at a speed of 1% of the speed of light, carries an energy content of 4.5x1017 (joules). Even if the spacecraft can be provided with a 100 (megawatt) nuclear reactor, it would take some 143 years to produce this amount of energy.

It is understood that the laws of current physics do **not allow** *for such a revolutionary space propulsion system. Propulsion techniques of this type can only emerge from novel physics, i.e., physical theories that deliver a unification of physics that are consistent and founded on basic, generally accepted principles, either removing some of the limit, or giving rise to additional fundamental forces, and thus providing alternatives to current propulsion principles.*[10]

Since Heim's theory itself is only published and available in Germany — and from a very unusual source as we shall see — our examination of his theory will be via its revised form as presented by his associates.

In doing so, one must assume that their critiques of the contemporary state of theoretical physics are reflections of those of Heim himself. Having stated the need for a breakthrough physics and technology, Dröscher and Häuser point out some of the fundamental difficulties of quantum mechanics and relativity dogma:

> (Quantum) theory… in its current form does not lead to an explanation of the elementary structures of matter, and does not lead to a consistent cosmology either….In (quantum theory) the existence of matter is taken for granted, defining an elementary particle as a point-like structure. In classical physics, including the General Theory of Relativity…, science starts from the belief that space and time are *infinitely divisible, in other words, that spacetime is continuous* (a differentiable manifold in the mathematical sense). Both ideas contradict **Nature's** all-pervading **quantization principle** and

---

10 Walter Dröscher, Jochem Häuser, "Guidelines for a Space Propulsion Device Based on Heim's Quantum Theory," *40th AIAA/ASME/ASME/SAE/ASEE Joint Propulsion Conference & Exhibit,* AIAA 2004-3700, 11–14 July, p. 4, italicized emphasis added, boldface emphasis in the original.

immediately lead to contradictions in the form of infinite self-energies or self-accelerations.[11]

Moreover, the general problems of the contemporary standard models go much further, and lead to (intentional?) dead ends:

> ...(Quantum theory) has not been able to deliver the mass spectrum of elementary particles, nor is there a theoretical explanation for their lifetimes, neither can quantum numbers be derived. None of these theories is able to explain the **nature of matter and inertia**, topics that are **essential for the physics of a completely novel propulsion system.**[12]

This leads to the first and most fundamental principle of Heim Theory:

b. *The Geometrization and Quantization of Space-Time Itself*

Dröscher and Häuser point out that Heim did not reject one fundamental insight of Einstein, both in his General Theory of Relativity and more importantly, in his subsequent versions of the Unified Field Theory, namely, that physical forces could be described geometrically. Thus, forces not only were functions of the geometry, the geometry *was* the forces. But there is one crucial insight to Heim's version of the theory, borrowed from the standard model of quantum mechanics, and that is "the two important ingredients that Einstein did *not* use," namely "a discrete spacetime and a higher dimensional space, provided with special, additional features."[13] Thus, Heim's theory is a logical extension of the insight of quantum mechanics *and* the various Unified Field Theories of the 1920s and 1930s in that all physical interactions are described as geometries, and in the crucial insight that space-time *itself* is quantized, that is, that it is built not of ever smaller infinitely divisible units, but of discrete "smallest possible" units of space and time.[14]

Thus, Heim's theory is not only a "completely geometrized unified field theory" but it is one with profound implications, for it gives rise "to a

---

11 Ibid., p. 1, boldface emphasis in the original, italicized emphasis added.
12 Walter Dröscher, Jochem Häuser, "Guidelines for a Space Propulsion Device Based on Heim's Quantum Theory," p. 5, boldface emphasis in the original.
13 Ibid.
14 Dröscher and Häuser point out in a previous paper, also presented to the AIAA, that "Space-time itself is quantized. The current area of a Metron," which is Heim's term for the smallest possible unit of quantized space-time , "τ is $3Gh/8c^2$ where G is the gravitational constant, h denotes the Planck constant, and c is the speed of light in vacuum. The Metron size is *a derived quantity and is not postulated.*" (Walter Dröscher and Jochem Häuser, "Physical Principles of Advanced Space Propulsion Based on Heim's Field Theory," *38th AIAA/ASME/SAE/ASEE Joint Propulsion Conference and Exhibit,* AIAA 2002-4094, 7–10 July, 2002, pp. 6–7., boldface emphasis in the original italicized emphasis added).

novel concept for an advanced space transportation technology, permitting, in principle, *superluminal* travel."[15] This unification is accomplished, in its extended version, by an eight-dimensional quantized "and *spin-oriented space.*"[16] As a result, its predictive power is enormous, for it not only predicts fundamental particle properties such as masses and lifetimes,[17] but also that "a transformation of electromagnetic wave energy at specific frequencies into gravitational like energy is possible."[18]

In coming to this view, Heim followed the principle of General Relativity and universalized it. In General Relativity, matter bends space-time itself, or, to put it differently, matter *is* a "bend" in the fabric of space-time; "matter and spacetime curvature are equal."[19] Since the geometry and matter itself are equivalent, and to that extent, since matter is caused by the geometry of space-time distortion, then the essence of Heim's Theory is that every physical interaction or fields including gravity, electromagnetism, and the strong and weak forces of quantum mechanics are distortions in what otherwise would be a non-distorted undifferentiable space-time. Heim first presented this view "in 1952 at the International Congress on Aeronautics in Stuttgart, Germany,"[20] a date rather close to the end of World War II, and not without its own suggestive implications, as will be seen. Heim further formally developed these ideas in the first versions of his theory during the late 1950s and early 1960s, the precise period in which Leonard Cramp's book appeared, outlining his views for the first time to an English-speaking general public.

c. Spin-Orientation Potential of the Lattice Structure of Space-Time

This insight leads in turn to the next fundamental component of the extended eight- and twelve-dimensional versions of his theory. Dröscher and Häuser present this aspect of his theory in the following manner:

> According to Heim, the whole universe comprises a grid of *Metrons* or metronic lattice. Space that does not contain any information consists of a discrete Euclidean grid....However, empty space must be **isotropic** with regard to spin orientation. If all metronic spins of a 6D volume pointed outward or inward, such a world would not have a spin potentiality. Therefore, cells with all spins outward have to have neighboring cells with all spins inward and vice versa. This alternative spin structure satisfies the

---

15 Ibid., p. 3, emphasis in the original.
16 Ibid., emphasis added.
17 Ibid., p. 4. These predictions have, in part at least, been recently verified.
18 Ibid., p. 3.
19 Ibid., p. 5.
20 Ibid.

isotropy requirement, but provides empty space with spin potentiality....
Thus empty space is void of physical events, but has inherent potentiality for
physical events to happen.[21]

In other words, the normal void of space itself, absent any information distortions such as mass or matter, is a pure potential for "things to happen," it *is* the transmutative medium of the ancients. As a void it is a non-differentiated nothing, since there is a balance of spin orientations in all its cells.

It is thus when that balance or equilibrium is unbalanced and enters into non-equilibrium, where a preponderance of spin orientation of one or the other type prevails, that all physical interaction, and all matter, arises. Non-equilibrium and spin are the very *mechanisms* not only for the differentiations of creation, but therefore are also the very mechanisms of the unification of physics, or, to put it in alchemical terms, for the "em-bodiment" of the medium itself within normal matter. Thus, Heim's theory incorporates and corroborates the fundamental insights of Jordan and Kozyrev, for particles themselves for they are "geometrical entities that possess an internal structure which is changing cyclically in time...Elementary particles are not point entities, but do consist of **Metrons**."[22] Thus, in a sense peculiarly paralleling some ancient views, "space and time are not the container for things, but are, due to their dynamic (cyclic) nature, the things themselves."[23] In Heim's hands, this dynamic view leads to a truly extraordinary cosmology, because the higher dimensional space is composed of such a lattice of "metrons" there are no "singularities" or infinities that plague the standard models, and that have to be gotten rid of by the fancy mathematical accounting trick of renormalization.[24] The universe thus began as a single metron covering its entire surface (the undifferentiated medium), and as the number of metrons increased, their size decreased.[25]

In short, the spin-orientation and active force characteristics of time itself is the fundamental pillar of Heim's theory.

If all this is beginning to sound familiar, hold on, there's more.

In the twelve-dimensional extension of Heim's original theory, there are

five semantic units, namely, the subspaces $R3$ (space), $T1$ (time), $S2$ (organization), $I2$ (information) and $G4$ (steering of $I2$) where superscripts denote

---

21   Dröscher and Häuser, "Physical Principles of Advanced Space Propulsion Based on Heim's Field Theory," p. 6.
22   Dröscher and Häuser, "Physical Principles of Advanced Space Propulsion Based on Heim's Field Theory," p. 7.
23   Ibid., for the actual mathematical description of particles in Heim theory, see p. 12.
24   Ibid., p. 16.
25   Ibid.

dimension. Except for the 3 spatial dimensions, all other coordinates are imaginary. Several metric tensors can be constructed from these subspaces.... Analyzing the metric tensors acting in **R4**, the theory predicts *six fundamental interactions,* instead of the four experimentally known ones.[26]

These two additional interactions are the basis behind the claim that the theory holds out the promise of a breakthrough in physics and in propulsion and energy technologies.

The first of these reactions, a weak gravitationally *repulsive* action, which is analogous to the standard model's "dark energy," Heim calls by the peculiar term "quintessence," a term with its own alchemical associations, as has been seen, and thus perhaps suggestive of influences at work on Heim's thinking other than the purely physics-related, namely, the alchemical. But there is also a "gravitophoton" reaction that "enables the conversion of electromagnetic radiation into a gravitational like field."[27]

It will have been noted by the careful reader that Heim's hyper-dimensional space is actually comprised of five sub-spaces, one of which is our normal three-dimensional space. Put differently, our three-dimensional space is influenced by distortions in the higher nine-dimensional space composed of its sub-spaces, recalling esotericist Manly P. Hall's observation that the successful confection of the Philosophers' Stone must be realized by one operation occurring simultaneously in four separate worlds.

It is precisely Heim theory's prediction of a direct gravitational-electromagnetic coupling that Dröscher and Häuser propose as the basis "for the novel space propulsion concept" that will be presented momentarily.[28]

d. *The European Space Agency's and the U.S. Air Force's Anti-Gravity Experiment, Versus Heim Theory*

Taking as their point of departure a series of experiments sponsored jointly by the European Space Agency and the U.S. Air Force, Dröscher and Häuser summarize the experiments and its implications as follows:

> In a recent experiment (March 2006), funded by the European Space Agency and the Air Force Office of Scientific Research, Taimer et al. report on the generation of a toroidal (tangential, azimuthal) gravitational field in a rotating accelerated (time dependent angular velocity) superconducting

---

26 Ibid., p. 1, emphasis in the original.
27 Dröscher and Häuser, "Guidelines for a Space Propulsion Device Based on Heim's Quantum Theory," p. 5.
28 Ibid., p. 7.

Niobium ring. IN July 2006, in a presentation at Berkeley university [sic], Taimar showed improved experimental results that confirmed previous experimental findings. Very recently, October 2006 and February 2007 the same authors reported repeating their experiments employing both accelerometers as well as laser ring-gyros that very accurately measured the gravito-magnetic field. The acceleration field was clearly observed, and its rotational nature was determined by a set of four accelerometers *in the plane of the ring.*

Since the experiment generates an artificial gravitational field, which is in the plane of the rotating ring...it cannot be used as a propulsion principle. It is, however, of great importance, since it shows for the **first time that a gravitational field can be generated other than by the accumulation of mass.**[29]

Again, the mention of superconductors recalls Hudson's observations of mass loss anomalies in his strange material, and of physicist Hal Puthoff's explanations for them.

But note what Dröscher and Häuser conclude about the configuration of the experiment: since the artificial gravitational field was generated in the plane of the rotation of the superconductor, and not in the plane parallel to the axis of rotation, no practical propulsive benefit could accrue from the experiment. And this, precisely, is where Heim's unique theory steps in:

> ...(Theoretical) considerations obtained from (Extended Heim Theory) lead to the conclusion that an experiment should be possible to generate a *gravitational field acting parallel to the axis of rotation of (a) rotating ring...* and thus, if confirmed, could serve as a demonstrator for a **field propulsion principle.**[30]

The authors then propose an apparatus for testing this hypothesis consisting of two counter-rotating systems *arranged, as per the predictions of Heim's theory,* **one on top of the other,** *exactly as I have conjectured was the actual internal configuration of the Nazi Bell's counter-rotating drums!*[31]

---

29 Dröscher and Häuser, "Current Research in Gravito-Electromagnetic Space Propulsion," *Institut für Grenzgebiete der Wissenschft* (Innsbruck, Austria), p. 10, italicized emphasis added, boldface emphasis in the original.

30 Dröscher and Häuser, "Current Research in Gravito-Electromagnetic Space Propulsion," *Institut für Grenzgebiete der Wissenschft* (Innsbruck, Austria), p. 10, italicized emphasis added, boldface emphasis in the original.

31 See my *Secrets of the Unified Field,* pp. 268–282.

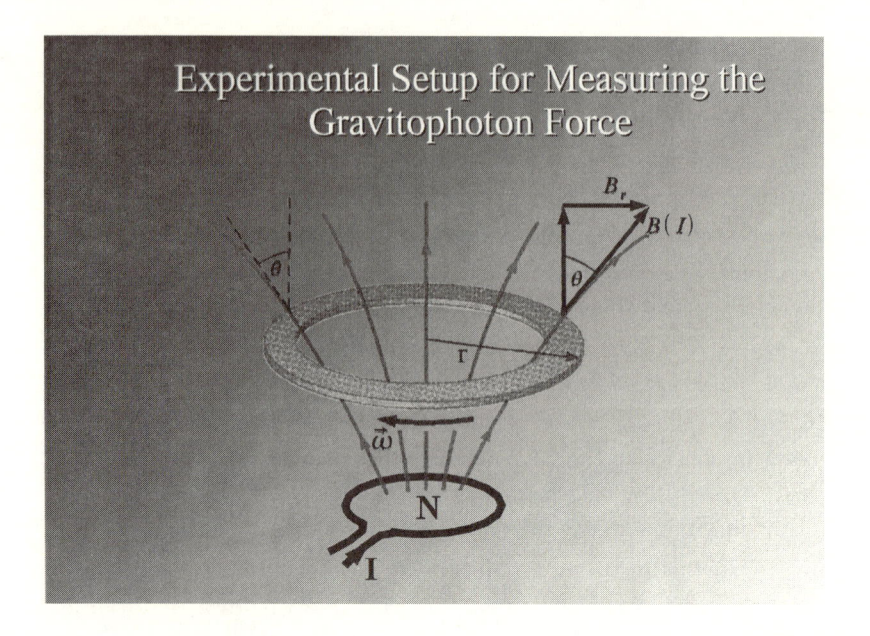

Dröscher and Häuser's Proposed Apparatus for the Testing of a Field Propulsion Principle in a Plane Parallel to the Axis of Two Stacked Counter-rotating Systems

That this theory would predict so specifically such effects, within such a short time after the end of World War II, with its suspicious resemblance to the actual proposed configuration of the Bell's internal counter-rotating drums, is suggestive that there is perhaps more in play in the formation of Heim's theory than meets the eye. And this is only confirmed by the mystery of what sort of research Heim was actually engaged in during the war.

B. The Mystery of Heim's World War II Research

As mentioned, Heim performed most of his theoretical work entirely in his head, having lost his hands, sight, and hearing in a wartime accident doing vague and never-specified research for the Nazis during the war. Try as one might, one simply cannot find any material or indications on what Heim might have been doing, so one is thrown back upon threadbare clues and a process of reasoned extrapolation and speculation from them.

1. The Accident

All that is known about the accident that left Heim handicapped for life was that sometime ca. 1942–1943, Heim was drafted to do service in the

Luftwaffe. At some point during this period he sent a letter to "the Chemical-Technical Reichsanstalt in Berlin, whereupon he was summoned to work there on the development of the proposed new explosives. It was here that he met with the accident..."[32] But what type of explosives might this have been that so caught the eye of the Luftwaffe?

2. Heim and Heisenberg on a "Clean" Hydrogen Bomb: Ballotechnic Explosive Work?

A clue, perhaps, is afforded by the fact that sometime during 1943, apparently in the same time frame as this research, Heim actually met the famous founder of quantum mechanics, Werner Heisenberg. Heim informed Heisenberg "of his plan to use chemical implosion to facilitate an atomic explosion. This design was based on his idea he developed for *a clean hydrogen bomb when he was 18.*"[33] But as has been seen in our examination of the Soviet "Mercury" and the whole "Red Mercury Legend," the idea of a clean hydrogen bomb, one *not* reliant upon the far dirtier atom bomb as the fuse, required precisely a breakthrough in conventional explosives technologies: the so-called ballotechnic explosive. And as was seen, one proposal for such an explosive was precisely the high-spin state of certain isotopes known as *isomers,* which, if induced to "suddenly stop" spinning, would release the energy of their angular momentum in a massive flash of electromagnetic energy in the form of high-frequency gamma rays which, if properly configured in an implosion detonator, *might* obtain sufficient energies to initiate nuclear fusion, and unleash the far greater destructive potentials of the hydrogen bomb. Given the prominent and pivotal role of spin orientation in Heim's subsequent theory, it seems that one must hold out the possibility, highly speculative though it be, that Heim may have been engaged in such research for the Nazis. And if his accident is any indicator, that research may have met with some measure of success. And there is one final factor to recall in this respect, and that is that Heisenberg, according to the latest research of Igor Witkowski, headed precisely some group engaged in nuclear research at the University of Dresden! Might Heim's contact with Heisenberg and his comments about a clean H-bomb in fact be related to the otherwise unknown purposes of that research?

It seems a possibility that one must entertain.

And if Heim's theory makes such specific predictions as to the nature of the gravity-electromagnetic coupling effect that an experiment can be formulated on its basis using stacked counter-rotating systems, then the connections to the

---

32 "Burkhard Heim, Academic Work and History," www.experiencefestival. com/a/Burkhard_ Heim_-_Academic_and_work_history?id/4, p. 2.

33 Ibid., emphasis added.

Bell and its mysterious Serum 525 — a compound that I have suggested may have employed high-spin-state thorium isomers — are even more stunning. It is beginning to look as if Heim may possibly have been involved in some *very* secret research for the Nazis, and that his subsequent postwar theory, which so closely approximates the details of the scientific rationalizations I have advanced for the Bell project, may have been precisely an attempt to work out in greater detail the results of that project!

3. His Curiously Alchemical Reference: Quintessence, and His Publication of His Theory with a New Age Publisher

A further mystery that surrounds Heim is his persistent mystical nature. As has been pointed out, Heim himself called his fifth physical interaction the "quintessence" force, a term with specifically alchemical roots. As if to hint at this influence, Heim curiously chose to publish two highly mathematical and technical books, in German, outlining his theory, not with any scientific publishing house, but with a New Age publisher! Given what we have learned about the alchemical influences of Wiligut at work in the SS, and *its* close resemblances to the actual parameters of the Bell project, once again the conceptual link to Heim's subsequent formalized theory is curious, and adds further weight to the speculation that he may have had some connection to the project.

4. Pascual Jordan and Wernher Von Braun

As if to make the strange connection of Burkhart Heim to Nazi secret research even more compelling, shortly after word of the nature of the first six-dimensional version of his theory circulated throughout Germany, Heim was visited by both Wernher Von Braun and none other than Pascual Jordan, whose own theories shared many similarities with Heim's. Both men praised his work as containing breathtaking implications![34]

So what does one have with the quiet German physicist, Burkhart Heim, a man who deliberately avoided learning English so that his astonishing theory would never be circulated outside of Germany?[35] One has, at the minimum, a brilliantly penetrating intellect with a possibly dubious past working for the Nazis in some very sensitive projects, a past that he perhaps deliberately tried

---

34 "Mars in Three Hours — Theoretically," Sandia Z Machine — the Hyperspace Forums, www.mkaku.org/forums/showthread.php?p=1159, p. 2. The article also connects the U.S. military's interest in Heim's concepts with apparent efforts underway at Sandia National Laboratories to use the enormous power of its "Z Machine" to power the massive ring magnets Dröscher and Häuser propose as the basis of a practical experiment to test a usable field propulsion technique.

35 Ibid.

to atone for with his sincerely-felt postwar advocacy for peace and for forms of energy production that were not based on nuclear power or non-renewable resources.

In this theory, a fully geometrized and quantized space and time were for Heim the very nature of the transmutative physical medium itself.

Space is, in some sense, the Philosophers' Stone, for in order for there to be a genuine human presence in deep space, it will require, as Heim — and the Nazis before him — saw, a breakthrough in physics, a breakthrough dependent upon the ability to embody the transmutative medium in the matter of human technology, an embodiment that will, in some sense and for good or ill, "alchemically transform" mankind himself. And that, in turn, might in fact require the confection of some new Serum 525, to be utilized in some new Bell.

Or, to put it differently, the old Serum and the old Bell have simply continued to be worked upon, somewhere, in utmost secrecy, and brought to a perfection only hinted at in the original Nazi experiments.

Epilogue Is Prologue
The Holy Grail of Physics and Alchemy

∴

"First and foremost, Otto Rahn was a modern-day alchemist…"
Nigel Graddon[1]

With Heim's hyper-dimensional theory, complete with its spin-oriented space, its view of space as a lattice of information, as a "crystal" whose defects mirror, and indeed *are* the physical forces, we have come full circle, and ended where we began: on the plateau of Giza, gazing up at the massive crystal of the greatest alchemical, hyper-dimensional machine of mankind, the Great Pyramid. We end where we began, with the view of space and of the physical medium itself as a transmutative medium whose differentiations — giving rise to the universe — arose from a "primary differentiation," from an ultimate asymmetry and non-equilibrium, from the physics of a "primal twist" (to employ Wiligut's suggestive phrase) whose ideogram is the swastika. If there is one thing that unites the three alchemical quests for the "triune Philosophers' Stone" examined in the previous pages — the American "Gold," the Soviet "Mercury," and the Nazi "Serum" — it is rotation, spin, *torsion*.

This latter point may best be illustrated by once again appealing to the alchemical version of the Trinitarian shield of the *Augustinian* formulation of the doctrine of the Trinity, a formulation that, it was suggested back in part one, was actually nothing more than a formulation of physics in the guise of a metaphysical and religious doctrine. If one places the "Philosophers' Stone"

---
1   Nigel Graddon, *Otto Rahn and the Quest for the Holy Grail: The Amazing Life of the Real Indiana Jones* (Kempton, Illinois: Adventures Unlimited Press, 2008), p. 215.

of the concepts of "spin and torsion" in the center circle, and each of the three exotic substances we have examined — the American "Gold," the Soviet "Mercury," and the Nazi "Serum" in each of the three vertices of the shield, one sees the link:

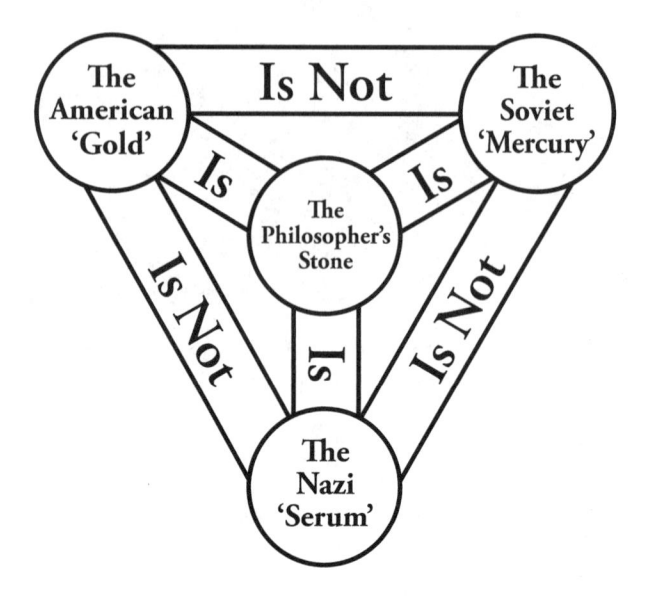

The Triune Stone: Spin and Torsion as the Unifying Concept

The truly paradoxical thing about this journey has not been its alchemical nature, however, but that so much not only appears to have been known (or at least, preserved) both by the alchemists, but known or at least rediscovered by the Soviets and the Nazis. With Hudson, we encountered a paradoxical, though largely accidental — and for that reason all the more compelling — rediscovery of the Stone, and by processes indeed very similar to those within the tenuous technological grasp of the alchemists. With Kozyrev and the Soviets, the principle of rotating open systems, of ever-changing torsion and time, confirmed yet another alchemical insight: that timing was everything to the performance of a successful alchemical confection of the Stone. With Kozyrev, accidental discovery had been abandoned for real science, for real testability, and for real *engineering* of the forces of time. One senses with the Soviets a closely rationalized development. Perhaps in the end the Red Mercury legend is really a testament to the explosive success of their research.

But it is with the Nazis that we find *all* the threads fully rationalized: the actual esoteric and occulted influence of alchemy in the SS via Wiligut, the profound understanding of the swastika as an ideogram of that "primal twist," the alchemical nature of space-time itself, the blend of accidental discovery and a well-conceived program to recover that science of the control of space-time. All the threads led to, and came together in, the Nazi Reich, the Bell and Serum 525, and again afterward in its Project Paperclip survival inside the USA in the form of some of the personnel — Debus and possibly Strughold — who worked on it.

The story, however, does not end there.

In chapter eleven, "The Greater German Alchemical Reich," the connection between Reichsführer SS Heinrich Himmler and SS archaeologist and esoteric scholar Otto Rahn via "Himmler's Rasputin," Karl Maria Wiligut, was mentioned. But there is more to Rahn than being merely some sort of Nazi "Indiana Jones" or Himmler's favorite SS esotericist, for the modern apocryphal rumors surrounding him are almost as strange as the man himself.

A. Otto Rahn, The Grail, and the Languedoc

Otto Rahn's claim to fame rests on a very few seminal works on alchemy and the quest for the Holy Grail. Indeed, it was Otto Rahn, on the basis of his scholarly exposition of the whole legend as first embodied in Wolfram Von Eschenbach's mediaeval Grail epic *Parzifal*, who first suggested that the characters and place names in the epic could be traced to specific individuals and places in the strange province of southern France known as the Languedoc. Rahn, following Eschebach rather closely, was not of the view that the Grail constituted so much the lost cup of Christ, but rather, a lost *stone* or *tablet of knowledge*. Eschenbach described it in rather vivid terms:

> Guyot, the master of high renown
> Found, in confused pagan writing,
> The legend which reaches back to the prime source of (all) legends...
> On an emerald green achmardi
> She bore the perfection of Paradise.
> It was the object called the Grail.[2]

The reference to an emerald-green achmardi recalls the esoteric and alchemical tradition that Thoth, the Egyptian wisdom god the Greeks would later call Hermes Trismegistus, inscribed all his lore and science on an Emerald Green tablet. It also recalls similar tales from Sumerian mythology regarding the lost Tablets of Destinies, over which many wars of the gods were fought.

---

2   Von Eschenbach, *Pazrzifal*, cited in Nigel Graddon, *Otto Rahn and the Quest for the Holy Grail*, p. 20.

In short, Rahn was *decoding* the alchemical tradition, and this, as has been seen, was a primary concern to the upper and innermost circles of Himmler's SS.

For Rahn, the Grail was not a lost *cup*, but a lost *stone:*

> The etymology of the word Grail was said by French mystic-philosopher René Guénon to derive from: Grail = Gor = precious stone and Al = stylus, combined giving precious engraved stone. This Grail reveals the Book of the Key or the final secret. The stones were engraved by a race of pre-Flood, Ayran superman, residents of Hyperborea (Thule, White Island)....[3]

It was not, therefore, hidden treasure or the lost cup of Christ that the Nazis, through Rahn, were searching for in the Languedoc of southern France, but lost *knowledge,* lost *science*, and lost *technology.* Indeed, during Rahn's trip there he wrote a letter in September 1935 directly to Wiligut, "telling him excitedly of a place he was visiting in his search for the Grail and asking for complete secrecy except for Himmler."[4]

Indeed, as Rahn researcher Nigel Graddon points out, it is this lost knowledge, this lost antediluvian *stone tablet*, and its possible location in the Languedoc that might have formed a completely *hidden* and occulted agenda behind both the genocidal Albigensian crusade in the region launched by the papacy and the king of France, and the later massacre of French peasants in the village of Ouradour by the Waffen SS in the same region centuries later in 1944.

1. The Grail, the Stone, the Fleece, and the Cosmic War

Rahn, moreover, like alchemists before him, believed on the basis of his research that the Golden Fleece of mythological fame was one and the same thing as the Philosophers' Stone.[5] But there was a further connection that Rahn had made:

> Raoul Lefèvre wrote his *Histoire de Jason* in 1456 in which the remarkable feature is the inclusion of both the Golden Fleece and the parchment of the Emerald Tablet of Hermes Trismegistus. A king called Apollo received a parchment from the god Mars containing "all the mysteries that should be preserved to achieve such a lofty thing." The lofty thing is the Fleece.[6]

---

3 Ibid., p. 21.
4 Ibid.
5 Nigel Graddon, *Otto Rahn and the Quest for the Holy Grail,* p. 230.
6 Ibid., p. 232.

The reference to the planet Mars, ancient god of war across so many ancient human cultures, makes the connection between the Philosophers' Stone, Thoth's-Hermes' Emerald Tablet, the ancient Sumerian Tablets of Destinies — a connection first mentioned in part one of this book — and the suggestive theme of Mars and a Cosmic War. The connection is disturbing, for if this was apparent to Rahn — and there is no *direct* evidence that it was — then it would mean that his presence and the Nazis' interest in the Languedoc was to find and militarily exploit this lost knowledge.

They were trying to *recover* it.

2. The First Rumor: The Vril Society and the Time Project

Yet another rumor associated with Rahn, or rather, with his SS masters, was that the SS, as far back as its pre-Nazi forebears in the Vril Society, was attempting to construct the technologies of time travel.[7] While there has as yet been no real evidence for this beyond a much-repeated telling of the story, Graddon does mention that the Nazi interest in Tibet may have some connection to this story, and that "the time-technology work of Dr. Karl Obermayer, Nikola Tesla, and Rudolf Steiner" on behalf of a so-called "Prometheus Foundation" was supposedly "carried out allegedly to thwart the worst excesses of Hitler's true agenda," an agenda he strongly implies was that of controlling time.[8]

3. The Second Rumor: Of Superconductors

A second rumor connecting Rahn to these mysterious goings-on in connection with time-control technology is that there are rumors to the effect that some object with superconductive properties was smuggled out of Germany to the U.S.A. sometime in 1938.[9] If indeed Rahn was closely associated to Himmler's Rasputin, Wiligut, as his letter to him suggests, then it is just possible that Rahn had figured out this aspect of the lost science and wisdom he found in his studies of esoteric texts.

4. Proto-Star-Clouds and Micro-Diamonds

In a final tantalizing bit of information, Rahn researcher Nigel Graddon also mentions the legend of Asteria, "who hung millions of crystals by silver

---

7   Graddon, *Otto Rahn and the Quest for the Holy Grail*, p. 215.
8   Ibid., p. 217.
9   Ibid., p. 204.

threads inside the heavenly dome and called hem her stars."[10] He also cites a passage in verse from a section of the Vedas called *Indra's Net:*

> There is an endless net of threads
> Throughout the universe.
> The horizontal threads are in space.
> The vertical threads are in time.
> At every crossing of the threads,
> There is an individual.
> And every individual is a crystal bead.
> The great light of absolute being
> Illuminates and penetrates
> Every crystal bead, and also,
> Every crystal bead reflects
> Not only the light
> From every other crystal in the net,
> But also every reflection
> Of every reflection
> Throughout the Universe.[11]

As was already seen in part one, it is part of the lore both of astrology and of alchemy that every star and planet has its own unique association with a specific type of gemstone, or crystal. And as was also seen, in Heim's theory space-time is not only a latticework of spin-oriented cells, but the very physical phenomena of objects and forces are themselves distortions and defects in that "crystalline" latticework.

Graddon then points out a peculiar recent discovery:

> In 1992 the annual Lunar and Planetary Science Conference held in Houston, Texas, reported that an analysis of studies carried out at the Infrared Telescope Facility at Mauna Kea in Hawaii indicated that molecular clouds contain a most unusual carbon atom bonded to a hydrogen atom and to three other carbon atoms. Further analysis showed that these blobs of gas and dust, which eventually give birth to stars, are full of floating micro-diamonds. Is it just accident that ancient peoples believed that the twinkling stars in the heavens were crystals, or did they have information from a time long lost in the distant past?[12]

---

10 Ibid., p. 271.
11 Graddon, *Otto Rahn and the Quest for the Holy Grail,* pp. 271–272.
12 Ibid., p. 271.

Given what we now know of Wiligut, Rahn, and the whole mission brief of the SS *Ahnenerbedienst,* one would have to speculate that the Nazis had already answered that question in favor of the passing down of information from a remote past.

B. World War II: An Alchemical War to Control the Technology of Time?

In a final piece of tantalizing information, Nigel Graddon states that

> Writer Elizabeth Van Buren regards (World War II) as a colossal attempt by the darkside [sic] of humanity to harness the secrets of Time. Similarly, Maurice Magré believed that the swastika symbolized the power of Time. Thor's hammer, Mjolnir, is associated with the alchemical characteristics of the swastika, symbolizing the transmutation and origins of matter.[13]

But if World War II was, in some sense, a war to create and control the technology of the manipulations of space-time, then in that sense, it was an alchemical war.

Close scrutiny and consideration of this radical and highly speculative proposition will reveal its profundity, for Nazi Germany had to free itself from the alchemically-based financial system of the West imposed on Weimar Germany in the Treaty of Versailles. It had, in short, not only to *find* but to *fund* the physics in order to break free of the dynastic cartels of international finance capital and the few supremely private and wealthy hands that controlled it. Modern fiat money and reserve banking is indeed a manifestation of the transmutative "nothingness" of the Philosophers' Stone, for from the creation of credit out of nothing, gold is produced. By nationalizing that institution and wresting it from private, secretive hands, and using it to fund the alchemical physics it was beginning to develop as the ultimate energy source, as the ultimate power to transport mankind, and as the ultimate power for destruction on a doomsday scale, the Nazis indicated that they had understood the nature of the Stone. They had seen, and fully understood, the connection between alchemical physics, and alchemical finance. And they were willing to put it to supremely evil uses.

But that connection between alchemical physics and alchemical finance is, perhaps, a relationship that requires its own exposition....

Epilogue is Prologue....

---

13 Graddon, *Otto Rahn and the Quest for the Holy Grail,* p. 261.

BIBLIOGRAPHY

(No Author) (Intelligence intercept) Inter 12 Dec 44 (1.2) Japanese; rec'd 12 Dec 44; Trans Dec 44 (3020-B).

(No Author) "Addressing Marinov's Element 122 Claim." www.chemistry-blog.com/2008/04/29/adressing-marinovs-element-122-claim/

(No Author) "Heaviest element claim criticized." 2 May 2008, www.rsc.org/chemistryworld/News/2008/May/02050802.asp

(No Author) "Possible New Element Could Rewrite Textbooks." Fox News, Tuesday, April 29, 2008, www.foxnews.com/story/0,2933,352980,00.html

(No Author). "Alchemy's Deceitful Trail: 'Gold' by Transmutation Still Deludes Victims, but Early Experimenters Aided Science." *The New York Times*, Sunday, November 17, 1929, p. SM 9.

(No Author). "Ballotechnic nuclear bomb." everything2.com/index.pl?node=ballotechnic%nuclear%20bomb

(No Author). "German Produces Gold in Synthetic Test; Denies Swindling Ludendorff and Others." *The New York Times*, Thursday, October 10, 1929, p. 1.

(No Author). "German's Alchemy Called Huge Hoax: Leading Munich Chemists and Newspapers so Brand Test — Leaks in Supervision Disclosed." *The New York Times,* Friday, October 11, 1929, Section Radio, p. 37.

(No Author). "Mars in Three Hours — Theoretically." Sandia Z Machine — the Hyperspace Forums, www.mkaku.org/forums/showthread.hp?p=1159

(No Author). "Micro-Nukes — Can and Do They Exist?" www.wtcnuke.com/micronukes.php

(No Author). "Red Mercury." *Wikipedia.* en.wikipedia.org/wiki/Red_mercury

(No Author). "The Secrets of Mercury." mmmgroup.altervista.org/e-mercur.html (quoting *The Secret Book of Artephius*).

(No Author). *Above Top Secret: Part Two.* Questar Video, Inc. Chicago, 1994.

(No Author). "Burkhard Heim, Academic Work and History." www.experiencefestival.com/a/Burkhard_Heim-Academic_and_work_history?id/4

(No Author). OSS Interrogation Archive document #12678 *Nazi Occult Organizations.*

(No Author). *Rosarium Philosophorum,* Tomus II, *De Alchemica Opuscula complura veterum philosophorum.* Frankfurt, 1550, 18th-century English translation in Ferguson 210.

A Chymicall Treatise of the Ancient and highly illuminated Philosopher, Devine and Physitian, Arnoldus de Nova Villa who lived 400 years agoe, never seene in print before, but now by a Lover of the Spagyrick art made publick for the use of Learners, printed in the year 1611. Bodleian Library, MS Ashmole 1415.

A Compendium of Alchemical Processes Extracted from the Writings of Glauber, Basil Valentine, and Other Adepts. Kessinger, No Date.

Agoston, Tom. *Blunder! How the U.S. Gave Away Nazi Supersecrets to Russia.* New York: Dodd, Mead and Company, 1985.

Aquinas, Thomas. *Summa Contra Gentiles, Book Four: Salvation.* Notre Dame.

Bacon, Sir Francis, Lord Verullam. *On the Making of Gold,* from Century IV of *Sylva Sylvarum, or a Naturall Historie in ten Centuries,* London, 1627. This work was incorporated in Bacon's great unfinished work, the *Great Instauration* or *Instauratio Magna*. www.levity.com/alchemy/bacongld.html

Bernstein, Jeremy. *Hitler's Uranium Club: The Secret Recordings at Farm Hall.* Second Edition. New York: Copernicus, 2001.

Brooks, Geoffrey. *Hitler's Terror Weapons: From V1 to Vimana.* London: Leo Cooper, 2002.

Budge, Sir E.A. Wallis. *Amulets and Superstitions: The Original Texts with Translations and Descriptions of a Long Series of Egyptian, Sumerian, Assyrian, Hebrew, Christian, Gnostic and Muslim Amulets and Talismans and Magical Figures, with Chapters on the Evil Eye, The Origin of the Amulet, the Pentagon, the Swastika, the Cross (Pagan and Christian), the Properties of Stones, Rings, Divination, Numbers, The Kabbalah, Ancient Astrology, etc.* Oxford University Press, 1930.

Butler, Kenley, and Akaki Dvali. "Nuclear Trafficking Hoaxes: A Short History of Scams Involving Red Mercury and Osmium-187." www.nti.org/e_research/e3_42a.html

Cohen, Sam. "The Dove of War." *National Review* (December 17, 1995) 56–58, 76.

Cook, Nick. *The Hunt for Zero Point.* London: Century, 2001.

Cope, Freeman A. "Evidence from Activation Energies for Super-conductive Tunneling in Biological Systems at Physiological Temperatures," *Physiological Chemistry and Physics 3* (1971), pp. 403–410.

Cramp, Leonard. *UFOs and Anti-Gravity: Piece for a Jigsaw.* Kempton, Illinois: Adventures Unlimited Press, 1996. Reprint of the 1966 edition.

David Adam, "What is red mercury?" *The Guardian,* September 30, 2004. www.guardian.co.uk/science/2004/sep/30/thisweekssciencequestions1

Dröscher, Walter, and Jochem Häuser. "Current Research in Gravito-Electromagnetic Space Propulsion." *Institut für Grenzgebiete der Wissenschft.* Innsbruck, Austria.

Dröscher, Walter, and Jochem Häuser. "Guidelines for a Space Propulsion Device Based on Heim's Quantum Theory." *40th AIAA/ASME/SAE/ASE Joint Propulsion Conference and Exhibit* AIAA 2004-3700 (11–12 July, 2004).

Dröscher, Walter, and Jochem Häuser. "Physical Principles of Advanced Space Propulsion Based on Heim's Field Theory." *38th AIAA/ASME/SAE/ASE Joint Propulsion Conference and Exhibit*, AIAA 2002-4094 (7–10 July, 2002).

Farrell, Joseph P. *Giza Death Star Destroyed*. Kempton, Illinois: Adventures Unlimited Press, 2005.

Farrell, Joseph P. *Reich of the Black Sun: Nazi Secret Weapons and the Cold War Allied Legend*. Kempton, Illinois: Adventures Unlimited Press, 2004.

Farrell, Joseph P. *Secrets of the Unified Field: The Philadelphia Experiment, the Nazi Bell, and the Discarded Theory*. Kempton, Illinois: Adventures Unlimited Press, 2008.

Farrell, Joseph P. *The Cosmic War: Interplanetary Warfare, Modern Physics, and Ancient Texts*. Kempton, Illinois: Adventures Unlimited Press, 2007.

Farrell, Joseph P. *The Giza Death Star Deployed: The Physics and Engineering of the Great Pyramid*. Kempton, Illinois: Adventures Unlimited Press, 2003.

Farrell, Joseph P. *The SS Brotherhood of the Bell*. Kempton, Illinois: Adventures Unlimited Press, 2006.

Flammel, Nicholas. *Hieroglyphical Figures: Concerning both the Theory and Practice of the Philosophers Stone*. Kessinger, reprint of the 1624 Walsey edition.

Flowers, Stephen E., and Michael Moynihan. *The Secret King: The Myth and Reality of Nazi Occultism*. Los Angeles: Feral House, 2007.

Forman, Simon. Alchemical poem *Of the Division of Chaos*, Bodeleian Library, Oxford, MS. Ashmole 240.

Gardner, Sir Laurence. *Lost Secrets of the Sacred Ark*.

Gerlach, Prof. Dr. Walther. "Die Verwandlung von Quecksilber in Gold" ("The Transmutation of Mercury into Gold"). *Frankfurter Zeitung*, Freitag, 18, Juli 1924 (Friday, July 18, 1924).

Graddon, Nigel. *Otto Rahn and the Quest for the Holy Grail: The Amazing Life of the Real Indiana Jones*. Kempton, Illinois: Adventures Unlimited Press, 2008.

Greiner, Walter, and Aurel Sandulescu. "New Radioactivities." *Scientific American,* March 1990, pp. 58–67.

Haisch, Bernard, Alfonso Rueda, and H.E. Puthoff. "Inertia as a Zero-Point-Field Lorentz Force." *Physical Review A,* Volume 49, Number 2, pp. 678–694.

Hall, Manly P. *The Secret Teachings of All Ages.* Reader's Edition. Penguin, 2003.

Hargrove, Stephen (?). "Laser Technology Follows in Lawrence's Footsteps." www.llnl.gov/str/Hargrove.html

Helmenstine, Anne Marie, Ph.D. "What is Red Mercury?" chemistry.about.com/cs/chemicalweapons/f/blredmercury/htm

Hoagland, Richard C. Personal communication with the author. Tues. November 20, 2007, 9:29 p.m.

Hoagland, Richard C. Personal communication with the author. Wed. November 21, 2007, 2:23 p.m.

Holmyard, E.J. *Alchemy.* Minneola, New York: Dover Publications, 1990.

Hudson, David. "The Chemistry of M-State Elements." www.asc-alchemy.com/hudson.html

Hydrick, Carter P. *Critical Mass: The Real Story of the Atomic Bomb and the Birth of the Nuclear Age.* Internet-published manuscript, 1998. www.3dshort.com/nazibomb2/CRITICALMASS.txt

Iwamura, Yasuhiro, Mitsuru Sakano, Takehiko Itoh. "Elemental Analysis of Pd Complexes: Effects of D2 Gas Permeation." *Japan Journal of Applied Physics,* Vol. 41 (2002) pp. 4642–4650, Part I, No 7A, July 2002.

Kozyrev, N.A. "Causal of Nonsymmetric Mechanics in a Linear Approximation." Pulkovo, 1958.

Kozyrev, N.A. "Astronomical Observations Using the Physical Properties of Time." *Vzpykhivayaushchiye Zvezdy,* Yerevan (in Russian). pp. 209–227.

Kozyrev, N.A. "Causal Mechanics and the Possibility of Experimental Studies of the Properties of Time." *History and Methodology of Natural Science,* 2nd issue, Physics, Moscow. pp. 91–113.

Kozyrev, N.A., and V.V. Nasonov. "On Some Properties of Time Discovered by Astronomical Observations" (in Russian). *Problemy Issloedovaniya Vselennoi* 1980, #9.

Kozyrev, Nikolai A. "Possibility of the Experimental Study of the Properties of Time." www.abyme.net

Levich, A.P. "A Substantial Interpretation of N.A. Kozyrev's Conception of Time." www.chronos.msu.ru/EREPORTS/levich_substan_inter/levich_substan_inter.htm

Lim, C.S., R.H. Spear, W.J. Wermeer, M.P. Fewell. "Possible discontinuity in octupole behavior in the Pt-Hg region." *Physical Review C,* Volume 39, Number 1, March 1989, 1142–1144.

Lucas, Renée. "Nuclear Shapes." *Europhysics News,* (2001) Vol. 31 No. 7, p. 5. www.europhysicsnews.org/articles/epn/abs/2001/01/epn01101/epn01101.html

Macchiavelli, A.O., J. Burde, , R.M. Diamond, C.W. Baeusang, M.A. Deleplanque, R.J. McDonald, F.S. Stephens, and J.E. Draper. "Superdeformities in Pd." *The American Physical Society,* August 1988, pp. 1088–1091.

Manhattan Project, The (1986), starring John Lithgow, Christopher Collet, and Cynthia Nixon.

Marinov, A., I. Rodushkin, D. Kolb, A. Pape, Y. Kashiv, R. Brandt, R.V. Gentry, and H.W. Miller. "Evidence for a long-lived superheavy nucleus with atomic mass number A=292 and atomic number Z@122 in natural Th," unpublished internet paper.

Mayer, Thomas, and Edgar Mehner. *Hitler und die "Bombe."* Schleusingen, Germany: Amun Verlag.

McLean, Adam, transcriber. *Place in Space, the Residence of Motion: The Secret Mystery of Nature's progress, being an Elucidation of the Blessed Trinity, Father — Son — and Holy Ghost. Space — Place — and Motion.* Sloane MS 3797, folios 3–5, at www.levity.com/alchemy/place_in_space.html

McLean, Adam. *The Glory of Light,* MS Ashmole 1415.

Michrowski, Andrew. "Time and Its Physical Relationships." The Planetary Association for Clean Energy, Inc. (No date given)

Nachalov, Yu. V., and E.A. Parkhomov. "Experimental Detection of the Torsion Field." www.amasci.com/freenrg/tors/doc15.html

Nachalov, Yuri V. "Theoretical Basis of Experimental Phenomena." www.amasci.com/freenrg/tors/tors3.html

Nagaoka, H., "Transmutation of Hg into Au." *Naturwissenschaften,* 13:682–684 (1925); *Naturwissenschaften,* 14: 85 (1926); *Nature* 114 (August 9, 1924), 197; *Nature* 117 (#2952, May 29, 1926), pp. 758–760.

Nasonov, V.V. "Physical Time and the Life of Nature: A Talk at the Seminar on the Problems of Time in Natural Science." Moscow: University of Moscow, 1985.

Nelson, Robert A. "Adept Alchemy." www.levity.com/alchemy/nelson2_1.html

Paracelsus, "Concerning the Projection to be Made by the Mystery and arcanum of Antimony." *The Aurora of the Philosophers,* www.levity.com/alchemy/paracel3.html

Paracelsus, "Concerning the Projection to be Made by the Mystery and Arcanum of Antimony." *The Aurora of the Philosophers,* www.levity.com/alchemy/paracel3.html

Paracelsus, "The Aurora of the Philosophers." *Paracelsus and His Aurora & Treasure of the Philosophers, As also The Water-Stone of the Wise Men: Describing the matter of, and manner how to attain the universal Tincture. Faithfully Englished. And Published by J.H. Owen.* (London: Giles Galvert, 1659). Text may be found at www.levity.com/alchemy/paracel3.html

Paracelsus, *Concerning the Tincture of the Philosophers,* compiled and transcribed by Dusan Djordjevic Mileusnic from *Paracelsus, his Archidoxis: Comprised in Ten Books, Disclosing the Genuine Way of making Quintessences, Arcanums, Magisteries, Elixirs, &c., Together with his Books Of Renovation & Restauration, of the Tincture of the Philosophers; Of the Manual of the Philosophical Medicinal Stone; Of the Virtues of the Members; Of the Three Principles; and Finally his Seven Books; of the Degrees and Compositions of Receipts, and Natural Things, Faithfully and plainly Englished, and Published by J.H. Oxon,* London, 1660.

Paracelsus, *De Elemente Aquae,* Lib IV, Tract IC, c. 15.

Paracelsus, *The Apocalypse of Hermes*, 14.

Paracelsus, *The Aurora of the Philosophers, from Paracelsus, his Aurora & Treasure of the Philosophers, As also The Water-Stone of the Wise Men; Describing the matter of, and manner hot to attain the universal Tincture. Faithfully Englished, and Published by J.H. Oxon*. London: Giles Calvert, 1659. Transcribed by Dusan Djordevic Mileusnic.

Poovey, Kirt R.. "The Red Mercury Nightmare?" www.prisonplanet.com/analysis_poovey_122602_redmerc.html

Randeria, Mohit, Ji-Min Duan, Lih-Yor Shieh. "Bound States, Cooper Pairing, and Bose Condensation in Two Dimensions." *Physical Review Letters*, Volume 62, Number 9, 27 February 1989, 981–984.

Ruland, *On the Materia Prima*, from *Lexicon alchemiae sive dictionarium alchemisticum, cum obscuriorum verborum, et rerum Hermeticarum, tum Thephrast Paracelsicarum phrasium, planam explicationem coninens* (Frankfurt, 1612), transcribed by John Glenn. www.levity.com/alchemy/ruland_e/html

Schroer, Bert. "Physicists in Time of War," December 2005.

Schwaller De Lubicz, R.A. *Sacred Science: The King of Pharaonic Theocracy.* Rochester, Vermont: Inner Traditions International, 1988.

Shikhobalov, Lavrenty S. "N.A. Kozyrev's Ideas Today."

Shimizu, Y.R., E. Vigezzi, E., and R.A. Broglia. "Inertias of super-deformed bands," *Physical Review C* (Volume 41, Number 4, April 1990, 1861–1854).

Singley, Eric. "What is Red Mercury?" groups.google.com/group/sci.chem/msg/69a33ee6f25c5073?q=group:sci.chem+ins

Stark, Anne. "Researchers Move Closer to Switching Nuclear Isomer Decay On and Off." Department of Energy, Lawrence Livermore National Laboratory, April 6, 2007. www.eurekalert.org/pub_releases/2007-04/dlnl-rmc040507.php

Stevens, Henry. *Hitler's Flying Saucers*. Kempton, Illinois: Adventures Unlimited Press.

Summers, Chris. "What is red mercury?" news.bbc.co.uk/1/hi/uk/5176382.stm

Von Sebottendorff, Rudolf Baron. *Bevor Hitler Kam*. Munich: Deustula-Verlag Grassinger & Co. 1933, 1 Auflage.

Waite, A.E. "On the Philosophers' Stone," from *Collectanea Chemica* (London, 1893). www.levity.com/alchemy/collchem.html

Wegener, Franz. *Der Alchemist Franz Tausend: Alchemie und Nationalsozialismus*. KFVR: 2006.

Weinberger, Sharon. *Imaginary Weapons: A Journey through the Pentagon's Scientific Underworld*.

Wideröe, Rolf. *The Infancy of Particle Accelerators*, ed. Pedro Waloschek. www.waloschek.de/pedro/pedro-texte/wid-e-2002.pdf

Wilcock, David, and Sepp Hasslberger. "Aether, Time, and Torsion." blog.hasslberger.com/mt/mtview.cgi/1/entry/66

Wilcock, David. "The Aether Science of Dr. N.A. Kozyrev." *Nexus Magazine*, Volume 14, Number 3, May-June 2007. pp. 45–47.

Wilcock, David. "The Breakthroughs of Dr. Nikolai A. Kozyrev." *The Divine Cosmos*. www.divinecosmos.com

Wiligut, Karl Maria, a.k.a. Jarl Widar. "Zodiacal Signs and Constellations." *Hagal* 12 (1935), Heft 4. pp. 56–58.

Wiligut, Karl Maria, a.k.a. Jarl Widar. "Number." *Hagal* 11 (1934), Heft 8.

Wiligut, Karl Maria, a.k.a. Jarl Widar. "The Creative Spiral of the 'World-Egg'!" *Hagal* 11 (1934), Heft 9

Wiligut, Karl Maria. "Ancient Family Crest of the House of Wiligut." *Hag All All Hag* 10 (1933), Heft 2/3, pp. 290–293.

Wiligut, Karl Maria. "Description of the Evolution of Humanity from the Secret Tradition of our Asa-Uana-Clan of Uiligotis," SS Document, 17 Hune 136, Marked read by H.H. (Heinrich Himmer), Archival File for Wiligut/Weisthor, *Bundesarchiv Potsdam* NS 19/3671.

Wiligut, Karl Maria. "The Nine Commandments of Gôt."

Witkowski, Igor. "The Third Reich — A Key to Secret Technology," article in preparation for the German edition of *Nexus* magazine, personal communication to the author.

Witkowski, Igor. Email to the author, 15 April 2008.

Witkowski, Igor. Supplements from personal communications to the author to be added to the pending German edition of *The Truth About the Wunderwaffe*.

Witkowski, Igor. *The Truth About the Wunderwaffe.* Farnborough, Hampshire, England: European History Press. 2003.

Witkowski, Igor. *The Truth About the Wunderwaffe.* Farnborough, Hampshire, England: European History Press. 2003.

Yerkes, William. "Red Mercury — Thoughts and Speculation." www.chemistry.about.com/od/chemistryarticles/a/aa100404a.htm

Zunneck, Karl Heinz. *Geheimtechnologien, Wunderwaffe und die irdischen Facetten des UFO-Phänomens.* Schleusingen, Germany: Amun Verlag. 2002.

DARK MISSION:
The Secret History of NASA
by Richard C. Hoagland
and Mike Bara

6 x 9, 600 pages, w/ 16 color pages
ISBN 978-1-923595-26-0, $24.95

A Feral House Original Paperback

www.FeralHouse.com

For most Americans, the name NASA suggests a squeaky-clean image of technological infallibility. Yet the truth is that NASA was born in a lie, and has concealed the truths of its occult origins and its sensational discoveries on the Moon and Mars. *Dark Mission* documents these seemingly wild assertions.

Few people are aware that NASA was formed as a national defense agency adjunct empowered to keep information classified and secret from the public at large. Even fewer people are aware of the hard evidence that secret brotherhoods quietly dominate NASA, with policies far more aligned with ancient religious and occult mystery schools than the façade of rational science the government agency has successfully promoted to the world for almost fifty years.

Why is the US government intent on returning to the Moon as quickly as possible? What are the reasons for the current "space race" with China, Russia, even India? Remarkable images reproduced within this book provided to the authors by disaffected NASA employees give clues why, including spectacular information about lunar and Martian discoveries.

Former NASA consultant and CBS News advisor Richard C. Hoagland and Boeing engineer Mike Bara offer extraordinary information regarding the secret history of the National Aeronautics and Space Administration and the astonishing discoveries it has suppressed for decades. Co-author Mike Bara is an engineer who has worked for Boeing and other aeronautic firms.

The Freemasonic flag seen on the cover was brought to the Moon by 32° astronaut Buzz Aldrin, and later ceremoniously presented to Scottish Rite headquarters in Washington D.C.